Design and Science

Design and Science

Edited by

Leslie Atzmon

BLOOMSBURY VISUAL ARTS
LONDON • NEW YORK • OXFORD • NEW DELHI • SYDNEY

BLOOMSBURY VISUAL ARTS
Bloomsbury Publishing Plc
50 Bedford Square, London, WC1B 3DP, UK
1385 Broadway, New York, NY 10018, USA
29 Earlsfort Terrace, Dublin 2, Ireland

BLOOMSBURY, BLOOMSBURY VISUAL ARTS, and the Diana logo are trademarks of Bloomsbury Publishing Plc

First published in Great Britain 2023

Cover design: Joslyn Snell
Cover image © Joslyn Snell

A catalog record for this book is available from the British Library.

Library of Congress Cataloging-in-Publication Data

ISBN: HB: 978-1-3500-6192-7
ePDF: 978-1-3500-6194-1
eBook: 978-1-3500-6193-4

Typeset by RefineCatch Limited, Bungay, Suffolk

To find out more about our authors and books visit www.bloomsbury.com and sign up for our newsletters.

Contents

Contributors

Shivanthi Anandan is a scientist, inventor, and educator. Her early training in botany was enhanced by doctoral training in plant molecular biology and post-doctoral training in cyanobacterial molecular genetics. The overarching focus of her research is the response of plants and cyanobacteria to light: a critical environmental parameter. Her research program and scholarly publications have focused on deciphering the biogenesis and composition of the plant photosynthetic light harvesting machinery, cyanobacterial transcriptional gene expression in response to light intensity, and the role of a novel molecule in filament integrity in some cyanobacteria.

Currently, she focuses on researching and innovating modular, hydroponic units for growing plants in the home. Shivanthi, and collaborator Professor Diana Nicholas, have established the research group URBN STEAMlab to innovate, design, and test strategies for healthy, sustainable living. Their in-home hydroponic system, called Garden Fresh Home, utilizes cyanobacteria to help the growth of plants. They have two patent applications pending for their hydroponics system. As one of Drexel's highest ranked teachers, Shivanthi received the Christian R. and Mary F. Lindback Award for teaching excellence. She currently serves as Drexel University's Vice Provost for Undergraduate Education and collaboratively establishes best practices for an academically excellent and diverse student experience.

Leslie Atzmon is a designer, design historian, and design critic who teaches at Eastern Michigan University in the US. She has published in the journals *Eye*, *Design and Culture*, *Communication Design*, and *Design Issues*. Atzmon edited *Visual Rhetoric and the Eloquence of Design* (Parlor Press 2011) and co-edited *Encountering Things: Design and Theories of Things* (Bloomsbury 2017) with industrial designer Prasad Boradkar and *The Graphic Design Reader* (Bloomsbury 2019) with Teal Triggs of the Royal College of Art. In 2016, she was a Fulbright Fellow at Central Saint Martins in London investigating the topic of Darwin and design thinking. Atzmon and colleague Ryan Molloy were awarded a Sappi Ideas that Matter Grant in 2018, which supports design that changes lives for the better. Collaborating with students, they rebranded Ypsilanti's non-profit Riverside Arts Center as a community arts hub and designed a creative-project "Toolkit" for children. In 2019, Atzmon curated an exhibition, also entitled *Design and Science*, which ran at the University Gallery at Eastern Michigan University from September 11 to October 17, 2019, and then at Esther Klein Gallery at the Science Center in Philadelphia from February 13 to March 28, 2020.

Nurit Bar-Shai is an interdisciplinary artist and researcher, working at the intersection of art, science, and technology. Her research and artistic practice look into biological systems, including microorganism networks and communication systems, social and collective collaborations, emergence *in vitro* ecology, and biomaterial fabrications. She conducts experiments through creative collaborative inquiries and addresses the ethics and the emerging practices of Do-It-Yourself biology, Citizen-Science, and environmental practices. Bar-Shai cofounded Genspace NYC community biotech lab, where she founded and led the Arts and Culture program, bridging arts, design, and biological sciences through practice, educational, and outreach programs. Her work has been exhibited at ZKM, Science Gallery, CID Grand-Hornu, and OK-Center, and has been featured in the MoMA book *BioDesign: Nature, Science, Creativity*, the Princeton book *Exploring the Invisible: Art, Science, and the Spiritual*, *Design with Life: Biotech Architecture and Resilient Cities*, the Leonardo ebook: *Meta-Life: Biotechnologies, Synthetic Biology, ALife and the Arts*, and in media outlets including *CLOT*, *ARTNews*, and BBC World News. Bar-Shai held residencies at the biomedical engineering company Epibone. Her artwork was commissioned by Turbulence.org, awarded a Prix Ars Electronica, and included in the collection of the Rose Goldsen Archive of New Media Art at Cornell University.

Prasad Boradkar serves as User Research and Sustainability Lead on a project team at Google ATAP in Mountain View, CA. He is also Professor Emeritus in Industrial Design at Arizona State University (ASU) in Tempe. At ASU, he served as co-director of InnovationSpace and co-director of the Biomimicry Center. Boradkar is the author of *Designing Things: A Critical Introduction to the Culture of Objects* (Berg 2010), and co-editor, along with Leslie Atzmon, of *Encountering Things: Design and Theories of Things* (Bloomsbury 2017). His research focuses on anthropology, material culture, and design theory.

Deepa Butoliya is an assistant professor at Stamps School of Art and Design at the University of Michigan. She is a researcher and educator in industrial design and biodesign, and currently advises graduate students. Butoliya's interest in balancing social, technical, and artistic perspectives led her to pursue architecture and then design. During that period, Butoliya was drawn to design practice at the grassroots level, and she worked with Development Lab at MIT, Boston to create design solutions for international development, which gave her experience in design for social innovation along with working as an industrial designer.

Over the past ten years, her research has focused on critical perspectives and practices in design that are located at the intersection of models of knowledge and critical thinking emerging from multiple and global perspectives. She researches critique in design from pluriversal perspectives. Her research is rooted in exploring critical design practices from the global south and involves exploring the criticality embedded in the ubiquitous and very human practice of Jugaad—which is a Hindi term for make-do under severe constraints. As a part of her research, she has co-curated an international exhibition and organized a symposium to explore the current practices in speculative and critical design by bringing in global perspectives in the knowledge, research, and practice of this genre of design. Butoliya has presented her research and conducted workshops at several conferences, such as IDSA, Making Futures, EPIC, CHI, IASDR, and Primer.

Emily Candela is a historian of design and science who explores interdisciplinary relationships between design, science, and technology, and the stories told about them. Her research focuses on how knowledge is generated and visualized at the intersections of design and science practices. Working across writing, sound, and curating, her work also contributes to critically rethinking the ways in which the histories of design and its interdisciplinary relationships are studied and communicated.

She received her PhD in the History of Design and History of Science from the Royal College of Art and the Science Museum, undertaken through an AHRC Collaborative Doctoral Award. Her research has been published in *Historical Studies in the Natural Sciences*, the *Journal of Visual Culture*, *Current Opinion in Chemical Biology*, and *The Journal of Education Through Art*. She is currently Senior Tutor in the School of Communication at the Royal College of Art, where she leads the Communication Design Pathway of the College's Masters of Research (MRes) program. She is also a Visiting Lecturer in the Royal College of Art/Victoria & Albert Museum History of Design MA program. She produced the podcast *Atomic Radio* (Science Museum/Resonance FM), was awarded the 2021 Design Writing Prize by the Design History Society, and is formerly Curator of Twentieth-Century and Contemporary Furniture and Twentieth-Century Product Design at the Victoria & Albert Museum.

Michael Chandler completed his undergraduate studies at the University of Michigan and his medical studies at Wayne State University School of Medicine in Detroit. He went on to train in internal medicine and allergy-immunology at Northwestern University Medical School in Chicago. He is currently affiliated with the Weill Cornell Medical School and is a member of the attending staff at New York Presbyterian Hospital in New York City. Dr. Chandler has maintained a lifelong interest in history and medical history. He is also the co-founder of D Scope Systems, a digital medical video archive and analytic system for endoscopic procedures.

K. Lee Chichester's research focuses on relationships between art and science since the early modern period under the aspect of embodiment and designerly practices. She wrote her MA thesis on the origins of empirical science in early modern artisanal workshops, focusing on the rediscovery of ancient geometry. Her current work explores the relevance of the Arts and Crafts Movement and British Modernism to the development of organicist biology in the early twentieth century.

Chichester studied Art History and Biology in Berlin and New York, completing her PhD in 2022 at Humboldt-Universität zu Berlin. After having worked as research associate at the Cluster of Excellence "Image Knowledge Gestaltung" in Berlin from 2014 to 2017, she is currently Postdoctoral Researcher at Ruhr-Universität Bochum. She has curated various exhibitions, among them *August Gaul: Moderne Tiere* at the Kunstmuseum Bern in 2021.

Matt Cornell is an Asia-Pacific based choreographer who grew up in Darwin, Australia on Larrakia land. His work interrogates how we embody systems—social, cultural, political, or technological—and in turn how these systems embody us by forming communities and informing identities across and between forms including dancing, performance, sound composition, writing, podcasting, and curation, and which pass through contexts including theatres, galleries, public spaces, and online. His work entails creating sacred spaces, pivotal events, and transcendent experiences among which we can come together to share something that might give rise to deeper ways to know ourselves and each other and the stories at our foundation—so that we may get better at living together.

He is also founder of *Wombat Radio (podcast est. 2013)*, a curator for *How Did It Come To This?* zine in partnership with Darwin Community Arts, an occasional guest speaker

(TEDx) and panel chair (MetroArts), and the 2021 Digital Artist-Curator at Critical Path choreographic research center. www.MattCornell.com.

Daniel Eckert currently works as a CMC manager at BioNTech SE, a company he joined in early 2022. He coordinates CMC operational activities for drug substance manufacturing across different units, e.g., process and analytical method development, assessment of critical quality attributes, and critical-process parameters manufacturing.

Eckert holds a BSc in Biochemistry (Hochschule Mannheim University of Applied Sciences, Mannheim, Germany) and a MSc in Molecular Biotechnology (FH Campus Wien University of Applied Sciences, Vienna, Austria). During his studies at the Children's Hospital at Westmead in Sydney, Australia and at the Medical University of Vienna, Austria, his research focused on the electrophysiology of the heart, investigating the (mal)function of ion channels in dystrophic heart muscle. After his academic research, Eckert worked in the field of gene editing. In 2016 he joined Horizon Genomics in Vienna where he used the cutting-edge CRISPR/Cas9 technology to produce knockout cell lines. As part of the company's R&D team he helped to develop Induced Pluripotent Stem Cells carrying tumorigenic point mutations or fluorescent tags at specific target proteins of these cells.

Jan Eckert is Head of the Living Lab at the Swiss Center for Design and Health – Switzerland's national Competence Centre at the interface among people, healthcare, design and architecture. Previously Jan worked as Head of the Design Unit at the University of Gothenburg, as Full Professor and Head of the Master's Programmes in Design, Service Design and Digital Ideation at Lucerne University of Applied Sciences and Arts, as a Senior Researcher at Lucerne's Competence Center for Typology & Planning in Architecture, as Senior Strategy Consultant at Mint Architecture in Zürich, Switzerland, and as an Interior Architect both in Switzerland and Germany.

He holds a PhD in Design Sciences obtained from IUAV University of Venice and an international Master's Degree in Interior Architectural Design from the University of Applied Sciences in Stuttgart, Germany, the Edinburgh College of Art, Scotland, and the University of Applied Sciences and Arts in Lugano, Switzerland. Jan studied Interior Architecture at the University of Applied Sciences in Stuttgart, Germany and at the École Nationale Supérieure des Arts Décoratifs in Paris, France.

Victoria Geaney is an Associate Lecturer for Master of Arts in Biodesign at Central Saint Martins and Visiting Lecturer at Royal College of Art on the Master of Arts in Fashion. She works as a Research Associate for Future Fashion Factory at the Royal College of Art and holds a PhD from the School of Design, Royal College of Art. Through her fashion-led research, Victoria interrogates and explores collaborative, interdisciplinary fashion and biology approaches. Her research is driven by process and her work operates at the intersections of microbiology and fashion design research.

Geaney is part of the London Doctoral Design Centre (Arts and Humanities Research Council) and has initiated collaborations with synthetic biologists and microbiologists at Cambridge University, Surrey University, and Imperial College London. She is regularly invited to speak about her research, including at: Focus Textil Fashion Summit 2020; Victoria and Albert Museum, London; the Fashion Institute of Technology, New York; Amsterdam's Waag Institute; L'École Nationale Supérieure des Arts Décoratifs , Paris; Munich Fabric Start;

and the University of Western Australia. Her practice has featured in a number of national and international exhibitions, including: Vienna Design Week, the 3rd Istanbul Design Biennial, SHOWstudio Fashion Film Award, and as part of Harvard's entry in the National Science Foundation's "Big Ideas" Competition, as well as in publications including *Wired*, *Nylon*, *U+Mag*, *Higgs* magazine, and *Design Exchange*. Geaney's *Harvested Sunlight* project with Dr Simon Park is included in Rachel Armstrong's *Experimental Architecture: Designing the Unknown* book (2020).

Catherine Griffiths is a media artist and researcher exploring critical code and algorithmic aesthetics in the context of machine learning ethics. By creating simulations, short films, and software applications, her hybrid theoretical-creative research attempts to make palpable invisible computational forces that shape power and social dynamics. Drawing on the legacy of generative art, the rise in artificial intelligence, and critical theory, she seeks to contribute to an emerging arts knowledge.

She received her PhD in Interdisciplinary Media Arts from USC's School of Cinematic Arts, her MArch in Architectural Design from The Bartlett, University College London, and her BA in Fine Art from the University of the Arts London. Her research has been exhibited in the Centre Pompidou, Paris, the Arnolfini Centre for Contemporary Arts, Bristol, and the Tokyo Game Show at Geidai University, Tokyo. Her research has been published in the *Journal of Digital Culture and Society* and the *Journal of Science and Technology of the Arts*. She is currently an assistant professor at the University of Michigan with a joint appointment between Taubman College of Architecture and Urban Planning and the Digital Studies Institute.

Lina Hakim is a researcher, educator, and artist exploring theoretical and practical models of object-led inquiry, with a particular interest in overlaps between the material cultures of design, technology, art, science, craft, pedagogy, and play. Her current pursuits include an archival and editorial project aiming to reinstate André Breton as a key thinker on object-led research, and a proposal rethinking the history and theory curriculum for undergraduate design studies in Lebanon. She most recently authored a book on the graphics-based work of renowned Iraqi artist Dia Al-Azzawi (Khatt Books, 2017).

Hakim holds a PhD and MRes in Humanities and Cultural Studies (London Consortium, Birkbeck, University of London), an MA in Book Arts (Camberwell College of Arts, London), and a BA in Graphic Design (American University of Beirut). Her doctoral project, "Scientific Playthings: Artefacts, Affordance, History," considered three nineteenth-century scientific instruments that became toys to explore the kinds of thinking that things afford at the levels of encounter, production, use, and re-appropriation. Hakim recently held an Andrew W. Mellon Postdoctoral Research Fellowship at the Victoria & Albert Museum, where she co-coordinated the V&A Research Institute (VARI) Pilot Project. She is currently Senior Lecturer in Visual and Material Culture at Kingston University, London.

Nicole Koltick currently serves as an Associate Professor and Associate Director for the MDes in Design Research program in the Department of Architecture, Design and Urbanism at Drexel University. She founded the Design Futures Lab in 2012 to explore the convergence of *Machines, Materials,* and *Narratives.* Her lab develops design research projects from speculative provocations into highly resolved prototypes of objects, experiences, and environments for the near and far future. Her work engages a vast array of complex, multi-disciplinary problems that fall outside the scope of traditional design practice. Recent lab projects include scent communication devices, synthetic biological interior surfaces, hybrid bio-computational human interfaces, and a multi-year investigation into speculative ambient robotic systems.

Koltick's work embodies her commitment to anti-disciplinarity, operating at highly variable scales and deploying widely divergent methodologies. The products of her research include essays, tangible design objects, experiences, environments, and films. She is a MacDowell Fellow, has delivered a TEDx talk on the work of the Design Futures Lab, and her work has been covered on leading websites including *Fast Company*, *Dezeen*, the Creators Project, and Architizer. Professor Koltick holds a BFA in Art from Carnegie Mellon University, and a Master of Architecture from UCLA. She is currently finishing a PhD in Philosophy, Art, and Critical Thought with a focus in Digital Design at the European Graduate School in Saas-Fee, Switzerland.

Clarissa Ai Ling Lee is an art-science scholar and practitioner who is currently a fellow at the Käte Hamburger Kolleg Aachen's "Cultures of Research" programme at RWTH Aachen University, and formerly a senior lecturer at the Faculty of Creative Arts at Universiti Malaya. She trained in critical philosophy and critical theory, and is now using that training to develop a transdisciplinary approach to a philosophy of artscience and creative research. Besides writing academic articles and working on a monograph, she has a strong interest in outreach work that has led her to produce a series of pedagogical videos on themes ranging from sustainability to speculative design to infodemic. She is the author of *Artscience: A Curious Education*, published under the imprint of Gerakbudaya Essentials in Malaysia. At present, she is developing an education-centric non-profit called EPICURUS—which aims to promote play and experiential learning at the core of STEM+A education—with a start-up grant from the UNESCO Commission in Malaysia. Her portfolio of work and engagement can be accessed at https://modularcriticism.blogspot.com.

Sharon Lin is a poet and essayist. Her work appears in *The New York Review of Books, WIRED, Wildness, Sine Theta*, and elsewhere, and is anthologized in *Best New Poets 2021 and Voices of the East Coast* (Penmanship Books). She lives in New York City.

Christine Marizzi graduated from the University of Vienna, Austria with a PhD in Genetics. Her current position as Director of Community Science at BioBus, Inc. allows her to combine her extensive training in genetics and microbiology with her passion for science communication and outreach in a nonprofit environment. With more than fourteen years in national and international education, she dedicates her time, energy, and intellect to build frameworks that help citizen scientists realize their greatest potential in STEAM fields, and to provide traditionally under-represented minorities with multiple entry points to STEAM disciplines. She has published several peer-reviewed articles about citizen science infrastructure as well as art-based education.

Her deep interest in biological systems and data visualization has led to several collaborations with artists and architects in Vienna, Tel Aviv, London, and New York, designing custom visuals and hands-on activities for their events. Resulting artworks were critically acclaimed and exhibited widely, including at the Architecture Biennale Venice (Italy), Vienna Science Festival (Austria), Aspen Ideas Festival (USA), the American Society for Microbiology

(USA), St. Francis College New York (USA), the School of Visual Arts New York (USA), ACADIA (Mexico), and the United Nations General Assembly 74 in New York (USA).

Diana Nicholas holds a Bachelor of Architecture from Carnegie Mellon University and an MFA from the University of the Arts. She is a socially responsive designer and researcher. A registered architect, fine artist, and certified interior designer, her research work is focused in two interconnected areas: 1) Design and Design Processes: this includes inter-professional creative collaborations with health and STEM researchers that include mentorship of novice designers in advanced manufacturing, designing, ideating, and advocating for change, 2) Focusing on home as a base for urban families: this involves defining conceptions of home and developing health-oriented solutions for urban substandard housing and housing insecurity issues in the US.

In 2013 Nicholas established Integral Living Research (ILR), which supports research on housing and process, as well as inter-professional teaching that connects scholarship, teaching, and service. ILR mentors students in the Master of Science in Design Research program, of which Nicholas is the founding director. Funded ILR work includes Garden Fresh Home (Patent Pending) with microbiologist Professor Shivanthi Anandan, and Health and Design Research with epidemiologist and biostatistician Yvonne Michael. Nicholas's work has been published and exhibited in a variety of academic and professional journals including *ARCH IN-Form*, *Context*, *ARCC,* and *ACSA*.

Clint Penick is an Assistant Professor of Biology at Kennesaw State University, where he studies the evolution and ecology of social insects. A driving theme in Penick's work has been a connection between science and design, including his work on the bat brolly. Penick previously served as the lead biologist of the Biomimicry Center at Arizona State University, where he worked alongside industrial designers and engineers to translate concepts from biology into human applications. Current topics of research in his lab range from how ants use antibiotics to fight disease, to how the hexagonal-celled nests of wasps and bees may inspire the design of lighter-weight aerospace parts.

Miriam Simun works at the intersection of ecology, technology, and the body. Her practice spans multiple formats including video, performance, installation, and communal sensorial experiences. Simun is concerned with the collision of bodies (human and non) with rapidly evolving techno-ecosystems. If collision can be understood to be a form of disturbance (in the ecological sense), then in disturbance we move through damage to an opportunity for renewal. Simun works with the sensual conditions of this renewal. Trained as a sociologist, Simun spends time in communities of experts ranging from biomedical engineers to botanists; hunters to human pollinators; octopuses to breastfeeding mothers. Taking on the role of "artist-as-fieldworker," much of her process is rooted in research as lived experience, forefronting corporeal and sensorial ways of learning, listening, and knowing.

Simun's work has been supported by Creative Capital, Robert Rauschenberg Foundation, Joan Mitchell Foundation, Onassis Foundation, Gulbenkian Foundation, and the Foundation for Contemporary Arts. Her work has been recognized internationally in publications including the BBC, *The New York Times*, *The New Yorker*, CBC, MTV, *Forbes*, *Art21*, and *ARTNews*. Simun is a graduate of the MIT Media Lab, ITP at NYU Tisch School for the Arts, and the London School of Economics and Political Science.

Audrey Speyer completed her undergraduate work in textile design, and MA degree in Materials and Design at ESAA Duperré in Paris. She earned a second graduate degree, MA Material Futures, at Central Saint Martins in London in 2016. Speyer is interested in crossover projects between design and science through a common focus on the environment. For her MA Material Futures, Speyer researched the biotechnology of fungi that break down contaminants found in soil. For her degree project, Speyer designed PuriFungi's MycoPod—a bioremediation system using mushrooms—while working with a laboratory in the UK and with scientific researchers from Kew Gardens (UK, Richmond) and CNRS (France, Paris). Speyer exhibited PuriFungi in Milan, Venice, London, and Paris in spring and summer 2016.

Speyer deployed the mycoremediation technique on cigarette butts in order to treat this toxic waste and transform it into a clean biopolymer. She created myco-ashtrays made of mycoremediated cigarette butts, for which she won the European prize (DEMO Interreg) in 2019. Her work was published in *75 Designers for a Sustainable World* (Editions de La Martinière), and on the European Circular Economy Stakeholder Platform. In 2019, she founded the company PuriFungi and patented her technique of treating cigarette butts with mycoremediation and recycling them into a biopolymer. Since then, Speyer has implanted PuriFungi's collection and recycling systems, and has distributed her myco-ashtrays at various festivals, companies, and municipalities in Belgium, France, and Luxembourg.

Karin von Omptеda is Associate Professor of Graphic Design at OCAD University in Canada. Her background is in both science and design, having undertaken an MSc Biology (University of Toronto) and BDes Graphic Design (OCAD University). She holds a PhD in Visual Communication (Royal College of Art). Von Ompteda's doctoral research is focused on integrating scientific and design knowledge on typeface legibility, with a focus on low-vision readers. Working with data plays a central role in her research, practice, and teaching, and she has developed her data manifestation pedagogy internationally over the past decade through courses and workshops with students, professionals, and the public. Her work has been presented internationally through conferences (IEEE VISAP, RGD DesignThinkers), exhibitions (Brno Biennial, BIO Biennial), publications (Laurence King, RotoVision), and press (BBC's *The Forum*, China Daily). Projects have been funded through research councils in both Canada and the United Kingdom, and clients have included BBC Research and Development and the National Film Board of Canada.

Illustrations

Figures

Tables

Introduction

LESLIE ATZMON

I've been thinking about doing a project on design and science for a long time. After obtaining an undergraduate degree in biology, I spent several years testing polluted water. When I decided to leave my water-chemistry job to return to school to study graphic design, people were confused. What a switch! How, *exactly*, they asked me, does design relate to biology? Won't your biology background be wasted? Well . . . no, it wasn't. The relationships between design and science are not obvious; most people consider design and science to be distinct fields that function in very different ways. Having experienced both worlds has given me a unique perspective on design and science together. I could clearly see, for example, commonalities in designerly and scientific thinking and making processes. The conventional wisdom is that the goal in science is to set up a theory or hypothesis about how something works, and then try to demonstrate that it has merit, and that the goal in design is to produce a functioning object that serves specific preferred purposes.[1] But there is much more to scientific and design work than this. So, I decided to do a project in which I could articulate the common ground between design and science. The resulting collection emphasizes the creative features that scientific inquiry shares with design: the processes by which both sorts of practitioners devise and explore ideas and outcomes.

What do we gain from studying the overlapping areas between design and science? We discover that both scientists and designers frequently utilize dynamic creative processes that lead to innovative ideas and outcomes. Architect Kyna Leski writes that creative processes across disciplines are always about "knowing, making, or discovering something that does not yet exist" (2011: n.p.). We can find areas in which design and science converge, and fertile ground for interdisciplinary projects or collaboration between designers and scientists, in the *methods and means* that they use for discovering, knowing, and making. What are these methods and means? The creative process in design and science is messy; it is "nonlinear, experimental, and serpentine" (Gregoire and Kaufman 2015). To carry out innovative work that gives rise to new knowledge, both designers and scientists need first to suspend preconceived notions, and then consider inventive solutions. Designers and scientists both use visual metaphor to help them develop associations between disparate things or seemingly distinct phenomena during creative thinking and doing. They both employ visual imagination, and use iterative and reflective practices while thinking and doing, and they externalize their partially formed ideas in sketches or models. And designers and scientists also seek patterns to help them synthesize outcomes. In this Introduction I discuss how designers and scientists use unlearning and illumination, visual imagination, visual metaphor and associative thinking, iterative and reflective practices, models and sketches, and pattern synthesis in their various practices. I also consider how these methods and activities make space to carry out interdisciplinary projects and also collaborative work between designers and scientists. Finally, I introduce the essays in this volume in the context of the above ideas.

Messy: Unlearning and Illumination in Creative Work

Researchers have argued that the creative process is invariably messy, but what, exactly, does that mean? And what role does messiness play in catalyzing creative interdisciplinary work? Leski argues that creative thinking first requires "unlearning," in which designers or scientists must banish preconceived notions about the work to be done. Despite what the term may suggest, unlearning does not yield a *tabula rasa*. Instead, unlearning in the creative process "disturbs our sense of what we know." What we know doesn't disappear with unlearning—it becomes a bit fuzzy and unmoored. This unmoored state of unknowing, Leski explains, "compels us to go forward in search of knowing" something new (2011: n.p.).

Unlearning isn't a total cleansing of ideas; rather, it opens up space for their novel recombination. American physicist Murray Gell-Mann discusses how critical it was to Einstein's work that he be able to let go of preconceptions. He describes how Einstein had to first "break away from the accepted but erroneous idea of absolute space and time." Only then, Gell-Mann argues, could Einstein "take seriously as a general principle the set of symmetries of Maxwell's equations for electromagnetism—the symmetries that correspond to special relativity" (1994: 261). Scientist and philosopher Michael Polanyi calls this sort of thinking "illumination," a process during which scientists and designers first unlearn and then imagine various possibilities in order to "gain a foothold in another shore of reality" (2012: 123).

Illumination frequently entails a kind of thought experiment. In thought experiments, hypothetical or fictional situations are used to imagine how concepts or theories may play out in the real world (see essays by Chandler and Lee in this collection). Like many scientists, Charles Darwin used thought experiments, which he called "imaginary illustrations" (Love 2010), to play out scenarios around natural selection. In *On the Origin of Species* Darwin writes: "In order to make clear how, as I believe, natural selection acts, I must beg permission to give one or two imaginary illustrations" (1964 [1859]: 90). Natural selection, mutation, and evolution are not visible, and they have no set beginning and end. Darwin had to imagine in his mind's eye a new kind of "shore of reality": that is, various possibilities for how previously undescribed, invisible, and time-based processes might play out (Atzmon 2015).

Unknowing is indeterminate. Architect and design researcher Gabriela Goldschmidt argues that creative projects have "ill-structured" problems with both unformed and unfixed beginnings and outcomes, as in the Darwin example above (Goldschmidt 1997: 441). The agitation of unknowing is shared by designers and scientists, which makes a collaborative space for interdisciplinary projects that are both design and science. The clutter of unknowing diminishes preconceived ideas about the other discipline, and opens up ways for those involved to generate visual metaphors across disciplines that can lead to novel concepts.

Visual Metaphor and Associative Thinking

How does visual metaphor, which can reveal tacit likeness, lead to novel ideas? As he worked to understand the hidden processes of evolution, Darwin rendered thinking sketches in which he used visual metaphors of trees, coral, and seaweed to help him figure out possible mechanisms for species divergence, emergence, and extinction. Leonardo da Vinci developed the visual metaphor of a ripe seed or rich-brown nut to depict his not-quite-accurate womb diagrams, and physicist Richard Feynman created schematics in which he used various sorts of lines as metaphors to represent the quantum-mechanical world (Frankel 2003). These scientists used visual metaphor as a way to think through the hidden mechanisms of real-world processes just as designers use visual metaphor to imagine hidden connections that can lead to possible outcomes.

"Essentially," write psychologist Karin Lindgaard and economist Heico Wesselius, metaphor allows us to "use certain aspects of our experience to organize our understanding of phenomena that are less clear to us." Metaphorical thinking highlights or conceals "aspects of phenomena in ways that make both our understanding and our experience deeply interpretive" (2017: 85). Physician Michael Chandler presents just such an interpretive case from his practice. In his essay for this collection, "Visual Thinking and the Art of Medical Diagnosis," Chandler describes how metaphorical thinking, in which he compared the lungs to an upside-down tree and a pipe organ, led him to an association that saved a patient's life. He caught the patient's lung tumor by first utilizing the metaphors of an upside-down tree and a pipe organ, and then applying associative thinking to his observations.

Architect Hernan Casakin explains how Daniel Libeskind used visual metaphor—in this case, a spiral—and association in the Victoria and Albert Museum's Boilerhouse Extension to spark associative thinking for the design. Casakin explains that the

> "spiral progression of art and history," constitutes the basic metaphor for the design of the museum. The "spiral dynamics of art and history" can be observed in the outward appearance of the building, as well as in the internal circulation pattern of the different floors of the museum revolving around an uneven vertical axis.
>
> 2011: n.p

Libeskind's metaphorical and associative thinking led to a literal pattern—a spiral-based architectural space. Libeskind certainly began by modeling his metaphorical concept. Visual metaphors can suggest novel outcomes through sketches, models, prototypes—or scientific theories, which can also be thought of as models. "Every metaphor," writes philosopher Max Black, "is the tip of a submerged model" (M. Black as quoted in Miller 1996: 115).

Externalization: Sketches and Models

How do designers and scientists utilize visual metaphors in their work? Scientists James Watson and Francis Crick—who were scrambling to discover the yet undetermined molecular structure of DNA—were intrigued by Rosalind Franklin's X-ray crystallographic work on DNA and Caltech scientist Linus Pauling's research on the structure of proteins.[2] Franklin used X-ray crystallography, in which X-rays directed at a molecule are diffracted (scattered) by the "solid" parts of the molecule, and pass through where there is no solid material (see essay by Candela in this volume). This process reveals the 3D structure of molecules—in Franklin's case, of DNA. Using a set of molecular models to explore his ideas about protein structure, Pauling had discovered that they folded into a "single helix coiled like a spring" (Mukherjee 2017: 148). Watson describes how they adapted Pauling's method:

> The essential trick was to ask which atoms like to sit next to each other. In place of pencil and paper, the main working tools were a set of molecular models superficially resembling the toys of preschool children. We could thus see no reason why we should not solve DNA in the same way. All we had to do was construct a set of molecular models and begin to play.
>
> 1969: 38

Watson and Crick brought together Franklin's X-ray imagery and Pauling's protein modeling method to create prototypes for DNA. Tapping into these two visualizations at once allowed Watson and Crick to explore possible DNA molecules that had two, three, or four helical strands. Their "play" with trial-and-error modeling led them to the structure of DNA, a construction that also suggested how DNA functions.[3]

Darwin similarly played with "tree-of-life" diagrams to help him determine the nature of evolutionary systems. Beginning with a tree metaphor, Darwin used sketching, information visualization, and graphic representation as mechanisms for both externalizing his thoughts, and for communicating his ideas to the public. Darwin, in fact, felt that such sketches frequently were the best way to express his complex ideas (Atzmon 2019: 2).

When conceptualizing possible outcomes, designers also typically make associations between disparate things or seemingly distinct phenomena—what Einstein called "combinatory play" (Stevens 2014: 100). Graphic designer Lori Young designed a neo-rococo ornamental font called Magneto Motivity (2005) (Figs. 0.1 and 0.2) by literally combining Helvetica letterforms with copies of the same letterforms: for example, adding or subtracting medium weight uppercase Helvetica Hs from one uppercase medium weight Helvetica H. Helvetica, which was designed in the 1950s,

Fig. 0.1 Lori Young, visual essay featuring Magneto Motivity font, *Visual Communication* 4 (2), 151–7, 2005.

Fig. 0.2 Lori Young, visual essay featuring Magneto Motivity font, *Visual Communication* 4 (2), 151–7, 2005.

was based on Modernist principles that purposely eschewed ornament. She created ornamentation only by adding or subtracting the same Helvetica character from itself. Using combinatory play, and iteration, reflection, and reiteration, Young produced a richly decorative font that comments on the cultural meanings inherent in ornament.

Both designers and scientists externalize unformed ideas in order to know, make, and discover. Microbiologist Shivanthi Anandan and architect Diana Nicholas describe this process in their essay for this volume, "URBN STEAMlab and Biophilic Environments: Science, Art, and Design." Their URBN STEAMlab team, which includes students from both design and science, collaborated on the design of a hydroponic plant growth system for indoor living spaces. Modeling was a crucial part of their process. They first prototyped their system as a free-standing unit that holds a 3D printed vase and live microbial biomix on which food plants can grow. Then they carried out further experiments, producing (and sometimes rejecting) many small vase-shaped prototypes to hold the biomix. This iterative process revealed that how well the biomix substrate grows is dependent on the form and function of the vase structure.

The team used the British Design Council's Double Diamond Process—which encourages discovering, defining, and developing—for producing prototypes and solutions. This open process encouraged the design team members to assimilate scientific points of view and the scientists to integrate designerly approaches. Creating models together, according to Anandan and Nicholas, spurred the team members to "embrace reflection and iteration," to co-design with each other, and to set aside the boundaries between design and science.

Creation, Reflection, and Reiteration

Creative thinking in design and science may be characterized by productive disorder, disarray, and indeterminacy, but how do designers and scientists ultimately extract order from this chaos? Both designers and scientists create, reflect, and reiterate, in a process that leads to new ideas and outcomes. Using ideas about "actual or virtual worlds," philosopher Donald Schön argues, designers construct and re-construct the "objects and relationships . . . determining 'what is there' for the purposes of design" (1992: 4).

Schön, in fact, is best known for his ideas on reflective practice in design, in which designers reflect on their actions and then respond to these reflections in processes of continuous learning. Bioinformatics scholar Luke Mathieson argues that science employs a version of Schön's reflective practice in design, what he calls "critical reflective practice." The scientific method, he explains, involves a "repeating cycle of observation, hypothesis, prediction, and testing, or in the terminology used to describe critical reflective practice, experience, reflection, formation of abstract models, and testing" (2016: 8). In both design and scientific reflective practices, practitioners formulate claims or models and test them "to discern their applicability;

they are compared against observation and either empirically falsified or partially validated" (2016: 8).

This circular system, in which outcomes are used as feedback, is also true of cybernetics, which was first presented by mathematician Norbert Wiener in the 1940s. Design theorists Hugh Dubberly and Paul Pagaro argue that in both design and science, cybernetics "follows the circular process of observe→reflect→make that is common to the recursive and accumulative process of learning in service of effective action" (2015: 74). This circular process encourages reiteration.

"Reiteration," writes Leski, "is common to a storm and creativity . . . In the creative process as in a storm, the cycle includes editing and [re]production" (2011, n.p.). And it includes failure. Experimental fashion designer Victoria Geaney details the reiterative processes used in her collaboration with two biologists in her essay for this collection, "Vital Matters: Growing Living Materials." Geaney describes how all three first brainstormed about and then experimented with incorporating bacterial cellulose (cellulose produced by bacteria) into sculptural garments.

> I showed them technical drawings and designs that used bacterial cellulose in a highly sculptural way, and then worked with the scientists to blend the bacterial cellulose into a substance that I sculpted into small pyramid shapes . . . team and I then tested our brainstorming ideas, including growing the cellulose and combining it with other materials and substances, such as newspaper, flower petals, and fabric samples. Although these attempts were unsuccessful, they showed the capabilities and boundaries of working first-hand with the cellulose.

Like unlearning and illumination, creation, reflection, and re-creation generated a shared space in their collaboration for ideas considered by both the scientists and Geaney to coalesce.

The Anandan and Nicholas URBN STEAMlab team similarly produced a whole host of experimental prototypes—some of these were partially or fully successful, and some were not. All of the prototypes, however, moved the process ahead. The team learned from unsuccessful prototypes; they adapted successful prototypes, in an iterative process that led to new concepts for their project, and to yet other prototypes. "Iterative processes," physician Michael Chandler argues, add "depth to medical practice and prevent[s] physicians from locking in on diagnoses too early, which can cause problematic outcomes." In science, as in design, iterative processes increase the opportunity for successful outcomes.

Synthesizing and Patterning

The processes of finding and then employing systematic patterns that come out of thinking and modeling are shared by designers and scientists. "A designer," Schön writes, "sees, moves, and sees again. Working in some visual medium . . . the designer sees what is 'there' in some representation," then creates externalized models in relation to this representation, which then inform further designing. "In all this 'seeing"' Schön argues, the "designer not only visually registers information, but also constructs its meaning; he/she identifies patterns, and gives them meanings beyond themselves" (1992: 5). Designer Jon Kolko calls these processes "sensemaking." Kolko argues that experimentation/iteration and "sensemaking" together are synthesis processes in which designers integrate their thinking about, and experimentation for, a project (2010: 18).

Blurring the Boundary between Natural and Human-made Phenomena

Blurring the boundary between the natural and the human-made world is yet another critical factor for successful interdisciplinary work and collaboration. In their project *Hyphae (3D)*, designers Jessica Rosenkrantz and Jesse Louis-Rosenberg (they call themselves Nervous System) use an algorithm pattern to synthesize the dynamic process by which leaves develop their vein networks. They use digital rendering and fabrication to turn these algorithmic outcomes into branching sculptures (Atzmon 2019: 6) (Fig. 0.3). Bioengineer Jordan Miller, who researches ways to create human biomaterials for medical purposes, uses 3D printing to fabricate engineered living tissues that have vascular architectures (blood circulation structures).[4] Miller discovered Nervous System's *Hyphae* work online and contacted Rosenkrantz and Louis-Rosenberg. Galvanized by their algorithmic rendering of leaf venation, Miller suggested that they "work together . . . to design synthetic living tissues and organ replacements for human patients" (Atzmon 2019: 6).

The collaboration between Miller and Nervous System led to the speculative design project *3D-Printed Organ Research*, which utilizes the *Hyphae* algorithm to generate "entangled vessel networks, air sacs, and blood vessels" and then to "3D-print artificially engineered lungs" (Atzmon 2019: 6) (Fig. 0.4). To produce this artificial lung tissue, Miller's lab and Nervous System worked together to adapt the *Hyphae* algorithm to a form that could be used to embed 3D-printed live cells into "soft gels containing very small, intricate blood vessels down to 300 microns in diameter" (Atzmon 2019: 40).

The messy, iterative, and reflective processes that both scientists and designers use can blur the boundaries between the human-made and the natural worlds. Working this way made it possible for the Miller lab and Nervous System collaborators to produce an ingenious design and science outcome: living 3D-printed lung tissue.

Designer Audrey Speyer similarly works in the gray area between the natural and human-made realms in her soil mycoremediation project PuriFungi. In her essay for this volume, "PuriFungi, A Natural First Aid Kit for the Earth," Speyer explains that she "embraced the notion of co-working with living nature," in order to help her see how fungi and her biodegradable fungi incubator could be deployed together as partners in soil remediation. It became clear to Speyer that she needed the "natural traits of resilience, generation, circularity, and symbiosis in nature" to underpin the design of

Fig. 0.3 Nervous System, *Hyphae* (*3D 2*), 8 in x 8 in x 8 in, nylon 3D printed by Selective Laser Sintering, 2010.

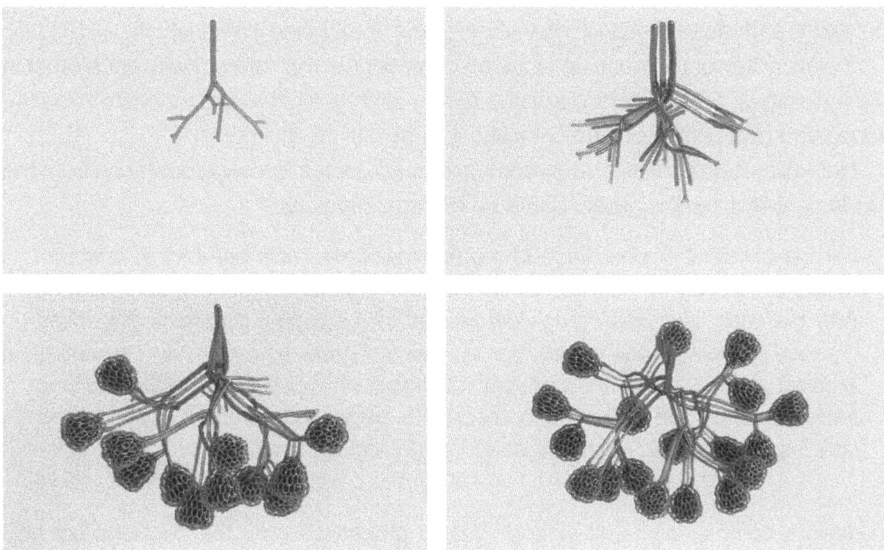

Fig. 0.4 Nervous System and Jordan Miller, 3D Printed Organ Research video screen grabs, human cells 3D printed by SLATE, 2019.

the total PuriFungi system. Using PuriFungi involves implanting a group of incubators, in which fungal media is cultured, in a polluted area to create a PuriFungi incubator network system. The fungi mycelia that emerge from the incubators also grow into a fungal network in the environment, and it is these mycelia that pull pollutants out of the soil. In essence, in the PuriFungi system the biodegradable incubator and the fungi function together seamlessly, and by doing so PuriFungi muddies the boundaries between human-made and natural design structures and functions. Approaching both human-made and natural design together suggests ways that biological and human-designed structures and functions can merge into such hybrid systems. In PuriFungi, mutual processes are key to soil remediation.

Emergence and Complex Systems

What is going on when designers, or designers and scientists, work in ways that blur the boundaries between natural and human-made phenomena so that they may function as shared mutual processes? How does the unlearning and Illumination that can provoke visual metaphorical and associative thinking, and the externalization of ideas in sketches and models, work? In what ways does creation, reflection, and reiteration give rise to synthetic and patterned thinking? All of these undertakings engage emergent processes that function as parts of complex systems.[5]

In emergence, "All forms of nature and all forms of civilization have 'architecture,' an arrangement of material in space and over time that determines their shape, size, behavior and duration, and how they come into being" (Weinstock 2010: 11). Complex systems are made up of varied components that interact with each other in dynamic ways. These interactions give rise to collective behaviors that determine the variegated relationships of the complex system with its environment.

Natural systems entail innately complex issues, as the following quotation from the fundamentally interdisciplinary Santa Fe Institute suggests:

> For the physicist or chemist studying emergent electronic behavior in quantum matter or turbulence in fluids, the gateways might include growing and studying new materials and developing new probes to measure fluctuations that might disclose universal scaling behavior or new coherent and possibly competing ordered states . . . For the biologist, biological physicist, or ecologist studying living systems, the collective components begin with proteins, neurons, or species and go on to cells, brains, and ecological dysfunction
>
> 2014

In design, complex systems extend "design processes from the development and fabrication of a singular static artefact or building to families of variant forms that can respond to varying conditions" (Hensel, Menges, and Weinstock 2010: n.p.). To be emergent, strategies for design must "include iterations or physical modeling . . . [and] the self-organizing material effects of form finding" (2004: 7). Emergence requires understanding "singular fixed" design projects as part of a nexus with other components in an "energy and materials system" that evolves, exists, and ends (2004: 7). These design systems are part of the natural-plus human-made environmental network.

The complex systems that are integral to emergent design and emergent natural systems create a productive shared space for interdisciplinary thinking and doing. Emergence explains how natural systems both evolve and maintain themselves, but it also suggests models and processes for design that make obsolete arbitrary distinctions between complex natural and human-developed processes and forms. Indeed, as architect Michael Weinstock observes, "Energy, information, and material flow through all the forms of life, through humans and all their works" (2010: 9).

In their discussion of emergence, systems engineers Patrick T. Hester and Kevin MacG. Adams distinguish between systematic and systemic thinking or analysis (2013). The authors define systematic thinking as a step-by-step data-gathering and problem-solving process to solve a "singular problem" (2013: 313). In contrast, they argue, emergent complex problems require systemic thinking for a "larger, more abstract unit of analysis, that of a mess . . . a system of problems" whose "analysis is not merely summative" (2013: 313).

Although Hester and Adams argue that systematic approaches are obsolete, the essays in this collection demonstrate that both sorts of approaches are crucial for understanding complex emergent processes in design and science. Physician Michael Chandler's systematic diagnosis of a patient's lung tumor, or Prasad Boradkar and Clint Penick's data gathering and analysis for the design of their bat-wing umbrella in their essay for this volume, "Design Inspired by Nature: The Bat Brolly," are part and parcel of the complex multifarious systems that make it possible for the human body or designed objects to function. They are no less crucial than the multiple unrelated systems that are brought together in Victoria Geaney's living bacteria garments.

The Essays

I have divided the essays in this collection into four themed sections: Visual Metaphor, Conceptualization, and Modeling Ideas in Design and Science; Biomimicry and Biodesign; Makers and Users in Design and Science; and Data Manifestation in Design and Science. In the effort to assign these essays to single sections, I discovered that their interrelated content made my decisions discretionary, if not arbitrary. Jan and Daniel Eckert's essay "Agile Artifacts—Designing the Incomplete," which is in the Visual Metaphor, Conceptualization, and Modeling Ideas in Design and Science section, reveals how "final" artifacts can act like models in design and science. But the essay also spotlights just how crucial users are in this process. This essay could also have been placed in the Makers and Users in Design and Science section. In fact, most of the essays could have been placed in more than one section. Conceptual overlap among the work in these themed sections is not really surprising. The components of complex systems are varied, and the processes that produce emergent design and science are permeable. With this caveat in mind, in the next sections of this Introduction I gloss over the themes and consider how the essays do them justice.

Part 1. Visual Metaphor, Conceptualization, and Modeling Ideas in Design and Science

"Models are metaphors" writes financial engineer Emanuel Derman, they are "relative descriptions of the object" that aid in conceptualization by comparing the situation at hand to "something similar [that is] already better understood" (2010). The essays in this part explore how designers and scientists use visual metaphor, visual imagination and conceptual thinking, and modeling to sharpen propositions and derive solutions. The authors deal with how relationships among designers, scientists, artifacts, and environments can lead to emergent concepts, models, and outcomes.

In her essay "Matters of Mathematics: Designerly Practices in Geometry," K. Lee Chichester demonstrates that externalizing ideas through physical modeling can lead to solutions and outcomes that may be missed by using purely theoretical methods. By "getting their hands dirty" working directly with materials, Chichester argues, early modern artisans became better than mathematicians at "envisioning complex shapes and understanding the laws of nature." Around 1900, though, "physical models and visualizations" started to be discredited—a change, that Chichester notes, "accompanied the rise of non-mechanical explanations in physics, specifically in electromagnetics and quantum theory." Non-mechanical explanations, she contends, lead to "designs that imitate science from a distance, without being truly involved in the research process." Spurred by a desire for a new unity between design and science, mid-twentieth-century historians of science and art examined artisans' contributions to early modern collaborations with scientists, spotlighting the "recognition of design-specific ways of knowing." Intellectual precision, pictorial imagination, and new scientific ideas go hand in hand, Chichester points out, because the adaptable nature of physical representation and externalization of ideas in models sharpens scientific thinking.

Purposely flexible outcomes or artifacts can likewise function as adaptable models. "Agile Artifacts—Designing the Incomplete," the second essay in this section, explores how incompleteness in artifacts can create a space for collaboration and ongoing design innovation. Both of the authors—design scholar Jan Eckert and his brother molecular biologist Daniel Eckert—work with artifacts that are characterized by incompleteness that opens spaces for "interaction, interpretation, and mutation." The authors compare Jan Eckert's theory of *Handlungsspielraum*—which investigates intentional or unintended areas of incompleteness in designed artifacts—with Daniel Eckert's genetic editing experience using knockout cell lines, which produce cells that deliberately carry a defect (e.g., a gap) in their genetic code. Knockout cells are cells in which a gene is made inoperative, that is, the gene is "knocked out." Both kinds of gaps—intentionally incomplete designed artifacts and knockout cells—are targeted to users who can utilize them to adapt the artifact to their needs. The authors call this kind of designed or genetically engineered user-focused, adaptable artifact an "agile artifact." They explain that agile artifacts embody an "ongoing process that foregrounds continuous updating of the artifact in its use contexts and with its users." Agile artifacts, they conclude, "include the designer, the users, and the artifact in a circle of continuous creation where all agents act autonomously and embrace the imperfect as part of an ongoing journey."

Jan and Daniel Eckert discovered compelling affinities between design artifacts with intentional user-focused gaps and knockout cell lines with intentional user-focused gaps. In the next essay, Clarissa Ai Ling Lee considers how emergent diagrams—which function as thinking models rather than as merely data-rendering vehicles—"translate information from different fields into a shared visual language, revealing commonalities among various kinds of knowledge, and bridging disparate information grammars." In her essay "Diagramming ArtScience: Designing at Knowledge Intersections," Lee argues that diagramming shapes emergent knowledge in both design and science by "making the invisible visible, generating new perspectives, and suggesting new ways of asking questions." She defines the term "diagram" broadly to include computational imaging, ancient astrological charts, and anatomical diagrams in early modern flap books. Lee concludes with a series of participatory design-fiction thought experiments that immerse participants into the invisible realm of sub-atomic particles. Turn-of-the-twentieth-century zoologist and mathematician D'Arcy Wentworth Thompson, who Lee discusses in her essay, regarded the material forms of living things as a diagram of the forces that have acted on them (Weinstock 2010: 20). A trip through Lee's "exhibitions" reveals how "biological and quantum physical phenomena" are expressed through participatory and emergent diagrams that "create a shared space" for ideas in both fields.

Whereas Lee portrays fictional quantum-physics environments in which humans can participate, in "Phenomenal Machines," architect Nicole Koltick presents a dynamic physical cybernetic environment with only non-human participants.[6] Her speculative design project *Phenomenal Machines* includes only robots, water, and crystals. The robots determine the "form and color of sodium chloride salt crystals," Koltick explains, and "crystal formation and growth are affected by the environment," especially by water. The interdisciplinary team that designed *Phenomenal Machines*—which included a philosopher/engineer, three graduate students, and Koltick—developed this speculative environment. While conceptualizing and modeling the project, team members asked themselves "what does taking in information, translating it, and then putting it into action look like for a non-biological entity? What does robot cognition look like?" *Phenomenal Machines* obliges viewers to consider what robots experience and to observe the outcomes of robot and crystal interactions in an environment without human input.

Part 2. Biomimicry and Biodesign

"The intentionality behind science and design," urges designer and biologist Daniel Christian Wahl, "needs to shift from aiming to increase prediction, control, and manipulation of nature as a resource, to a transdisciplinary cooperation in the process of learning how to participate appropriately and sustainably in Nature" (2006: 289). Both biomimicry and biodesign push back against purely anthropocentric design by leveling the playing field among living things and non-living things. Biomimicry is a "practice that learns from and mimics the strategies found in nature to solve human design challenges" (Biomimicry Institute). Engineers K.D. Jones and M.F. Platzer use biomimicry in their research, which demonstrates the efficiency of flapping-winged flight that is based on birds, insects, and fish. They argue that flapping-winged flight may replace propellers or jet engines in micro air vehicles (2006). Critic William Myers astutely observes, though, that biodesign doesn't just emulate nature as biomimicry does; instead, biodesign integrates design with natural systems (2012: 10). The implication here is that biodesign goes "beyond" biomimicry to incorporate design and natural processes. I would argue, though, that biodesign is not beyond biomimicry, but beside biomimicry: each of these approaches has value. Biomimicry and biodesign are symbiotic and complementary. They each draw valuable connections between design and science, and both challenge an overly anthropocentric model of design.

As Nicole Koltick does in *Phenomenal Machines*, industrial designer Prasad Boradkar and biologist Clint Penick call into question human-centeredness in design in their biomimetic bat-wing umbrella. In their essay "Design Inspired by Nature: The Bat Brolly," Boradkar and his graduate student Penick consider how biomimetic artifacts, such as their umbrella, emulate "nature's strategies" in order to "create sustainable solutions to human challenges." A principal founder of the field of biomimicry, biologist Janine Benyus explains that in a "biomimetic world, we would manufacture the way animals and plants do." Nature, she points out, offers the models for

> Solar cells copied from leaves, steely fibers woven spider-style, shatterproof ceramics drawn from mother-of-pearl, cancer cures compliments of chimpanzees, perennial grains inspired by tallgrass, computers that signal like cells, and a closed-loop economy that takes its lessons from redwoods, coral reefs, and oak-hickory forests.
>
> [1997] 2002: 2–3

Boradkar and Penick considered various organisms as models, such as broad-leaved plants, but the authors found that the structure and function of bat wings worked best for their umbrella. The duo used the four-part biomimicry design framework called Life's Principles, which utilizes scoping, discovering, creating, and evaluating processes that are gleaned from nature's strategies. These steps are commonly used in an interchangeable order, depending on the project.

Boradkar and Penick's bat-wing umbrella design is still in the midst of a prototyping, and as is common in prototyping, the umbrella design will require further iteration, modeling, and testing. And, as is typical in in-process design and scientific research, the project demands many rounds of trial and error. The authors have found that the available materials that they have tried to date "present difficulties for accurately mimicking . . . the bat's membranous wings and articulating joints"; as a result, they continue to experiment with different materials. Boradkar and Penick conclude by reiterating that although bats and other organisms can aid designers in solving difficult problems, the "larger goal" of biomimicry is "to develop and explore further the idea of life-centered design."

In biomimicry, designers use processes and structures from nature to inspire the form and function of designed artifacts and systems. Designed artifacts can also be configured to be a functioning constituent of natural systems in biodesign, such as those formed by fungi (mushrooms) in the wild. Like Boradkar and Penick, Audrey Speyer first identified "natural phenomena"—in her case, fungi—and then analyzed how they function in order to figure out how to deploy them. Speyer then designed a system called PuriFungi that utilizes fungi to break down industrial pollutants.[7] "Mushrooms [fungi], the primary natural recycler," Speyer explains, "have a . . . system that absorbs, digests, and stocks both organic and inorganic toxic substances, such as industrial waste. The waste is absorbed through the mycelium, which is the [filamentous] root system of mushroom cultures." This process is called mycoremediation, and PuriFungi uses a system of networked pods that work as a system to treat soil, after which plants and animals eventually return to what was a formerly polluted site.

Lo Lamento by designer Victoria Geaney merges design and biology as Speyer does. Geaney creates a "living" garment—in collaboration with two biologists—that gets its soft blue glow from living bioluminescent bacteria *Photobacterium kishitanni*. Not as conspicuously practical as Speyer's mycoremediation, *Lo Lamento* falls into the category of speculative design, which blurs "distinctions between the 'real' real and the 'unreal' real," giving "form to the multiverse of worlds our world could be" (Dunne and Raby 2013: 159).

In her essay, Geaney describes in detail how she collaborated while a PhD student at the Royal College of Art with the two synthetic biologists to design *Lo Lamento.* Synthetic biology involves the design and production of new biological components and systems, or redesigning existing biological systems "for useful purposes." Although *Lo Lamento* doesn't use genetic engineering per se, design-driven genetic engineering is most common in synthetic biology (Roberts et al. 2013). The process and outcomes that Geaney describes in her collaboration with two synthetic biologists, though, do offer insights into interdisciplinary collaboration in biodesign. This project also suggests the ecological, social, philosophical, and political implications of utilizing living materials in design, "while also making beautiful microbial worlds visible."

Although *Lo Lamento* reveals unseen microbial worlds, Miriam Simun and Sharon Lin's essay "Follow Your Nose," shows how critical and underappreciated the sense of smell is to how we interface with the natural world. Simun's project, the food truck *GHOSTFOOD*, offers participants foods with "soon-to-be-unavailable" flavors that come from plants and animals that will be lost because of climate change. The menu pairs scents of at-risk natural resources with unrelated foodstuffs: "*chocolate* served as chocolate milk," which is actually "chocolate scent paired with sweetened cow milk; *cod* served battered and fried," which is actually "cod scent paired with vegan fish substitute; and *peanut butter* served with jelly and bread," which is actually "peanut butter scent paired with bread, jelly, and soy butter." Simun explains that this "scent-food pairing" menu includes foods that are threatened with extinction from climate change, and are "particularly common in American childhoods." Familiar childhood foods are likely to conjure strong memories. "At the same time," she confirms, "they suggest issues around availability and future memories of these foods for children to-come."

The second project presented in this essay, *Agalinis Dreams*, liberates an "un-smelled" fragrance. Here Simun considers *Agalinis acuta*, a flower so small that its scent has never been perceived by humans. Using living-flower headspace technology, Simun captured the scent of the *Agalinis* and recreated it for human perception using her specially designed adoro nosepiece. In both projects, the designed nosepieces offer users insights into the underappreciated function of the sense of smell as well as into difficult issues around climate change and species extinction.

Part 3. Makers and Users in Design and Science

The essays in this section spotlight how design and scientific makers and users employ tacit, embodied thinking to precipitate novel ideas and things. Scientist and philosopher Michael Polanyi explains that tacit knowledge comes out of "exploratory acts" that employ "conceptual and sensory information and images" to "make sense of something" (Smith 2003). The theory of embodied knowledge maintains that the brain and mind are not abstracted from bodily processes of "feeling, movement, and perception"; rather, embodied experience is significant to the mind and brain—and to creating ideas and knowledge (Lindgaard and Wesselius 2017: 86). The designers, scientists, citizen scientists, and users that are discussed in these essays imagine new scenarios. They play and carry out "exploratory acts" (Smith 2003) that come together in new models, theories, speculative artifacts, or creative design hacks.

In the 1970s, design theorists Horst Rittel and Melvin Webber, who studied problem solving in various disciplines, coined the term "wicked problems." Wicked problems can be understood as complex systems. They are a "class of social system problems which are ill-formulated, where the information is confusing, where many [shareholders] have conflicting values, and where the ramifications in the whole system are thoroughly confusing" (Forlizzi 2011: 49). Rittel also argued that wicked problems could best be addressed by a participatory design process in which designers work together with various stakeholders, specifically users.

The first essay in this section presents the creative explorations of the interdisciplinary team URBN STEAMlab at Drexel University in Philadelphia. The team of students and faculty members was user-focused from the start. Their goal was to tackle the "wicked problem" of obesity in some urban environments by making fresh food available to underserved people in Philadelphia. In "URBN STEAMlab and Biophilic Environments: Science, Art, and Design," authors and team leaders architect Diana Nicholas and microbiologist Shivanthi Anandan discuss the processes that led to their indoor hydroponic plant growth system—which consists of a "smart vase," with a "proprietary biomix of plant microorganisms" on which food plants can grow.

Nicholas and Anandan describe how the designers and scientists on the team worked together on "creative trajectories," individual skills that frequently tap into tacit knowledge and embodied making. The authors elaborate how team members with design and science backgrounds shared ideas. They describe how those involved used design "lenses" to help develop shared group priorities and to develop and refine prototypes.

This interdisciplinary team purposely approached the project in a user-focused and human-centered manner, taking into consideration the specific circumstances of the people who would be using this system.

> In urban Philadelphia, many rowhomes in underserved neighborhoods lack access to working plumbing or appliances . . . One research student, therefore, spent a summer examining mobile ideas . . . One mobile project that came out of this research is a kitchen unit that can be wheeled out onto the porch of a home whose kitchen is not functional.

The team retained a strong user-focus throughout the collaboration, which led to innovative outcomes that incorporated natural, artificial, and human considerations in seamless and systemic ways.

In "Designing a Scientific Instrument: Lessons from the Crookes Radiometer," Lina Hakim similarly considers both the creation and user reception of the evocative nineteenth-century scientific instrument, the Crookes Radiometer. She begins by describing what it is. The Radiometer, or "light-mill," consists of four rotating vanes—also called the "fly"—within an evacuated glass bulb. "When the instrument is exposed to light, Hakim explains, the "fly rotates"; the "spin intensifies with the length of exposure to light, and it slows down then stops when the light source is taken away."

Hakim next investigates how this instrument expresses meanings through "how it was made . . . why it was made . . . and what it does." Crookes's accounts of his research and embodied making processes, Hakim argues, offer "rare documentation" of Crookes's tacit thinking during "complex play with materials, phenomena, and theories" that led to the working device. Other scientists were users: they tested the Radiometer and debated how it worked, often trying to disprove Crookes's own explanations. Hakim also describes how the device, which seemed to magically convert light into kinetic energy, enthralled everyday users at public demonstrations, and even other scientists. These scientific disputes, and the awestruck public responses to the device, Hakim argues, highlight the critical role makers and users play in "both science and design as practices that make sense of the world."

"Bringing non-scientific making techniques together with biological processes" explain Christine Marizzi and Nurit Bar-Shai, "can spur novel ideas and outcomes." In "From the Laboratory to the Studio: Microorganisms in Art and Design," microbiologist Marizzi and bioartist Bar-Shai introduce visual projects featuring bacteria that are biotechnologically modified to produce color. Using a series of innovative projects as case studies, they describe exploratory making processes that yield inventive "living" art and design.

Biotechnology commonly makes use of genetic modification, and it can be controversial, so the authors also address ethical issues around genetically modified organisms. Community science labs that encourage public participation in scientific research have popped up in response to these ethical concerns. One crucial goal of community science labs is to make the public aware of biotechnological issues that affect them as users or consumers. But these labs also strive to share knowledge through participatory, hands-on, public-outreach programs—such as Marizzi and Bar-Shai's project *The NYC Biome MAP*. For this project, participants "selected an area from a grid map of NYC" and matched it with a "Petri dish prepared with sterile laser-cut stencils that matched the topography of the assigned grid map area." The NYC Biome MAP team then assembled these images into an urban biome map of Manhattan. "New Yorkers shape NYC's urban biome," Marizzi and Bar-Shai point out, and "this unseen microbial world has a significant impact on them." The everyday New Yorkers who participated in *The NYC Biome MAP* project are literally embodied makers and users.

In "Making Justice with Biodesign: A Pedagogical Approach," Deepa Butoliya likewise considers how makers and users can collaborate to produce socially conscious biodesign. For her biodesign class at the University of Michigan, Butoliya's user-focused pedagogical approach is informed by her research. She calls her research critical *jugaad*: a Hindi word for "DIY [design-hack] processes" used by "people in marginalized communities" that "build resilience in the face of adversity." Butoliya uses *jugaad* as a critical springboard for analyzing Western speculative and critical biodesign from a non-Western point of view. She shares this method of reconsidering biodesign with her students.

Butoliya's biodesign class is divided into collaborative teams. Before the student teams begin designing, she introduces them to issues and ideas in biodesign and biotechnology, and especially to the idea of social-justice-oriented biodesign. At the same time, she encourages creative making processes that activate embodied making and tacit thinking. Her student teams design generative tools—in the biodesign class she describes, for example, they created and then played with relevantly themed physical card games.

I would argue that these kinds of generative tools encourage play, a process that solicits tacit knowledge and demands embodied thinking. As discussed above,

Watson and Crick played with models to discover the structure of DNA, and designers and scientists both use "combinatory play" in their projects. Indeed, anthropologist Daniel Lende explains that play drives "extensions from 'noncognition' [tacit, embodied cognition] to cognition" (illumination, as discussed earlier in this introduction) (Lende 2008). The generative tools that Butoliya's biodesign students used helped them to come up with four thought-provoking final projects. Two of the projects, *Chitonet* and *Gnosis*—one of which dealt with invasive species, the other with ways to diagnose and destigmatize STIs—engaged users in the process and had a critical impact on users. This focus on users, and the "learning, critiquing, thinking, and making processes" the students used, Butoliya contends, yielded "self-reflective, critical, adventurous, and thoughtful" projects with a "strong ethos of social justice."

Part 4. Data Manifestation in Design and Science

"Data analysis," computer scientist Colin Ware explains, "is about finding patterns that were previously unknown or that depart from the norm" (2021: 183). The essays in this section explore ways that analyzing and representing data and information can lead to inventive ideas or insightful conclusions. Ware also insightfully points out that scrutinizing and portraying data requires visual query (2021: 183). I would argue, though, that analyzing and representing data effectively often also requires real-world or virtual multi-sensory queries.

In her essay for this volume, Karin von Ompteda defines data manifestations as designed objects that "communicate data through a range of sensory experiences." In this section, I expand the term data manifestation to encompass data that emerge from multi-sensory analysis or are represented using multi-sensory means. In some of the essays in this section, multi-sensory analysis or representation of data is physical, and in other cases it is virtual—for example, in imaginary sensory experiences that are part of thought experiments, or in digital animations of soil erosion or flowing water.

The data manifestations discussed in this section are also situated entities: they are shaped by the contexts from which the data emerge. The meanings gleaned from data manifestations are also colored by the contexts in which they are apprehended. Brown et al. describe how a schematic representation of a "complex machine in a manual is distinctly different from how the machine actually looks" and works, but in an "intriguing way you need the machine to understand the manual, as much as the manual to understand the machine" (1989: 36). Although the manual representation is visual, and perhaps also textual, the user needs to see, hear, and otherwise experience the machine in action to fully comprehend its representation. Data manifestations are also narrative. The "embedding circumstances" for the knowledge that is derived from data representations, Brown et al. contend, "provide essential parts of its structure and meaning" (1989: 36).

Several of the essays in this section demonstrate that creating and accessing data manifestations is multi-sensory, situated, and narrative. The authors argue that data are frequently best communicated through multiple senses, that empiric medical data are most effective when framed in the context of imaginative medical observation and reasoning, that meanings gleaned from data manifestations can vary among social and intellectual networks, and that cinematic data manifestations can recontextualize abstract scientific data.

In “Data Manifestation: Climate Change Data in the Home and on the Body,” Karin von Ompteda maintains that abstracted, two-dimensional graphic renditions are not very effective for communicating climate change data to the public. This is due, in part, to the fact that climate change is a narrative process that “occurs on a global scale, changes at a relatively slow pace, and isn’t immediately tangible.” Von Ompteda argues for alternatives to two-dimensional screen- or paper-based data representations. She calls instead for data manifestations that utilize the ways that “humans experience and interpret the world around them,” in particular, by employing physical objects. She presents two data object case studies completed by her students: a wool blanket and a t-shirt that are designed to represent and transmit specific climate change data. In *The Comfort of Ignorance*, for example, a “bar chart that is woven into the blanket illustrates global surface temperature from 1965 to 2015, relative to average temperatures from 1951 to 1980.” This data is extremely disconcerting, and a graphic bar chart would not do justice to its seriousness. The blanket, though, is meant to be used. But it is very warm and the wool is prickly—as von Ompteda explains: “it literally engulfs [users] in the discomfort of this data.” Like *The Comfort of Ignorance* blanket, the second project, *Sea Level Rise T-Shirt*, projects climate change data narratively and contextually through personal experience, and in particular, through physical interaction with and emotional reaction to the data objects.

Emily Candela likewise argues that how data manifestations are “used” determines how they are apprehended. Her essay, “The Scientist’s Social Network: Reimagining Crystallographic Diagrams Ahead of the 1951 Festival Pattern Group Collaboration,” looks at the cross-disciplinary exchange between scientists and designers instigated by mid-twentieth-century X-ray crystallographer Helen Megaw. In crystallography, “X-rays are first directed at molecular structures; they are then diffracted (scattered) by the ‘solid’ parts of the molecule, and pass through where there is no solid material,” rendering an image of the solid parts of the molecule (Atzmon 2019: 3). 3D molecular structures are invisible to the naked eye; they appear as a complex series of dots and dashes in crystallography. These data require expert interpretation. Mid-twentieth-century crystallographic data manifestations commonly needed complementary tactile or physical data, such as “hand-drawn diagrams and three-dimensional modeling,” to fully contextualize molecular structure.

But molecular structure wasn’t the only content that emerged from Megaw’s crystallographic diagrams. Megaw shared her crystallography diagrams with designers who were part of the 1951 British Festival Pattern Group. According to Candela, Megaw’s diagrams functioned as a network-based narrative “trading zone”: that is, as a “catalyst and meeting place for [cross-disciplinary and] cross-cultural exchange." Each network of people that was involved in the pattern-design project interpreted the narrative meaning of Megaw's crystallographic diagrams. And they did so through their own political, social, aesthetic, and theoretical frames of reference—for example, as representative of "unbridled but 'parallel' progress in science and the arts," or as geometric forms that symbolized both aesthetic and social utopian order and universality.

Like X-ray crystallography, medical diagnosis often infers what is at least partly unseeable. Frames of reference are also critical to medical diagnosis. Physician Michael Chandler, who laments the erosion of imaginative thinking in medical reasoning, argues that "reliance on data points over reasoning" can lead to "simplistic, and possibly incorrect, diagnoses." In "Visual Thinking and the Art of Medical Diagnosis," Chandler describes the most effective medical reasoning as a "probability assessment [a medical Bayesian analysis] blended with a . . . thought experiment." In thought experiments, which are performed in the imagination, we "set up some situation, we observe what happens, then we try to draw appropriate conclusions" (Brown and Stuart 2020). Chandler's essay affirms that medical diagnosis—which employs visual imagination in the undertaking of "weaving" data from physical examination and medical tests and into a complete picture—is a multi-sensory process that constructs a contextual diagnostic narrative.

Multi-sensory techniques and narrative are crucial to cinematic data manifestation, a design process that utilizes "photography, motion, framing, depth of field, and computer-generated effects" to render scientific data in motion. In "Cinematic Data Visualization," Catherine Griffiths shows how "multi-sensory modalities" that also take "context into consideration" enhance scientific data manifestations. She details two of her own cinematic data manifestation projects—*Alluvium* and *LA River Nutrient Visualization*—in which she utilizes cinematic techniques to construct situated, multi-sensory, screen-based data narratives. *Alluvium* presents geological data showing the "impact of diverted flood waters on a sediment channel in Death Valley, California." Griffiths uses multi-layered imagery to render the extreme effects of this diversion: concomitant arid topography and flash flooding. Griffiths also uses layered cinematographic techniques in *LA River Nutrient Visualization*, which represents changing pollution levels over time in Los Angeles River water in its urban context. Both of these virtual projects simulate real-world experiences, engaging the senses if only in the imagination. Contextual information of the sort used in these two projects is critical to thorough communication, Griffiths concludes, but engaging the senses, rendering narrative, and depicting context in time-based media can also alter how we perceive information.

Conclusion

The essays in this collection call into question strict delineations between design and science. Of course, there are ways that design and science are quite distinct: the needs of users, for example, almost always play a role in design outcomes, and user needs are not typically part of conventional basic scientific research. Designers traditionally begin with an issue that needs resolution, they consider possible solutions to that issue, and then they design and produce suitable outcomes. Scientists, on the other hand, begin with a research question, formulate a hypothesis, test that hypothesis by conducting experiments, and then analyze the data from these experiments.

I would argue, though, that there is nothing intrinsically whole or distinct about these purportedly discrete disciplines. Rather, the boundaries between them are often fuzzy, and are often more a matter of hermeneutic convenience. *Design and Science* spotlights how certain shared ways of thinking and doing can function as productive spaces for interdisciplinary and collaborative work in design and science. These shared spaces provide a fertile ground for work that incorporates design and science, creative thinking and making, and novel ideas and outcomes that are not likely to happen in either discipline alone. As the essays in this collection attest, tapping into the spaces where design and science permeate each other can yield exciting new hybrid configurations of knowledge.

Notes

1. Political scientist Herbert Simon, author of the book *The Sciences of the Artificial*, differentiated design and science based on this sort of conventional wisdom. He contended that science seeks to understand "things: how they are and how they work" and that design formulates "courses of action aimed at changing existing situations into preferred ones" (Simon 1982: 129). I would argue, rather, that science and design each operate in both of these ways. Simon's ideas had a tremendous impact on design theory and pedagogy in the 1960s and 1970s and beyond.
2. They were more than intrigued by Franklin's work—they actually stole her research. See Howard Markel, (2021), *The Secret of Life: Rosalind Franklin, James Watson, Francis Crick, and the Discovery of DNA's Double Helix*, New York: Norton.
3. Hannah Star Rogers argues that the relationship between form and content plays out in a material-rhetorical complex in both science and art. This process, she explains, is commonly called experimentation. "Experiments," according to Rogers, "propose worlds by organizing materials that stand in for things not present . . . in short, ideas are represented through materials" (2022: xxviii).
4. See also Bártolo, P. J. S. (2006) "State of the Art Solid Freeform Fabrication for Soft and Hard Tissue Engineering." In C.A. Brebbia, ed. *Design and Nature III: Comparing Design in Nature with Science and Engineering* 87, (Southampton: WIT Press), (233–44).
5. In "Design Science and Design Education," designer Ken Friedman proposes educating designers to practice a science of design, which he defines as a complex-systems "comprehensive design process" that is a "rich . . . integration of the scientific and the sensual, the intellectual and the intuitive" (1997: 9). Friedman explains how this process engages complex systems and design: "The ambiguity that exists in complex systems enhances the robust quality that we identify in evolutionary terms as fitness." Evolved

systems, he argues, are better "able to reshape the environment in such a way that the environment becomes better suited to their needs (Lewin 1993: 104). Design education based on critical inquiry engages the complexity inherent in the empirical world to achieve its goals" (1997: 29).

6. In his 1948 book on cybernetics, *Cybernetics: Or Control and Communication in the Animal and the Machine*, Norbert Wiener "laid the theoretical foundations" for the "study of controlling the flow of information in systems with feedback loops, be they biological, mechanical, cognitive, or social" (1948).
7. PuriFungi was initiated as a graduate thesis project in the Material Futures Programme at Central Saint Martins in London.

References

Atzmon, L. (2015), "Intelligible Design: the Origin and Visualization of Species," *Communication Design* 3 (2): 142–56, http://dx.doi.org/10.1080/20557132.2016.1199366 (accessed on September 16, 2021).

Atzmon, L. (2019), *Design and Science*, exhibition catalogue, Ypsilanti: MI, Eastern Michigan University Galleries.

Bártolo, P. J. S. (2006), "State of the Art of Solid Freeform Fabrication for Soft and Hard Tissue Engineering," in C.A. Brebbia, ed. *Design and Nature III: Comparing Design in Nature with Science and Engineering* 87, (Southampton: WIT Press), (233–44).

Benyus, J. ([1997] 2002), *Biomimicry: Innovation Inspired by Nature*, New York: Harper Collins.

Biomimicry Institute, "What is Biomimicry?," https://biomimicry.org/what-is-biomimicry/ (accessed on August 15, 2021).

Brown, J. R. and M. T. Stuart (2020), "Thought Experiments," Oxford Bibliographies, https://www.oxfordbibliographies.com/view/document/obo-9780195396577/obo-9780195396577-0143.xml (accessed on September 12, 2021).

Brown, J. S., A. Collins, and P. Duguid (1989), "Situated Cognition and the Culture of Learning," *Educational Researcher* 18 (1): 32–42.

Cameron, A. (1998), "The Medium is Messy," *Eye* 30, http://www.eyemagazine.com/profile/author/andy-cameron (accessed on March 29, 2021).

Casakin, H. (2011), "Associative Thinking as a Design Strategy and its Relation to Creativity," in the Proceedings of the International Conference on Engineering Design, Iced11, https://www.designsociety.org/ (accessed on June 25, 2021).

Darwin, C. (1964 [1859]), *On the Origin of Species* [Reprint of 1859 edition], Cambridge: Harvard University Press.

Derman, E. (2010), "Metaphors, Models & Theories," https://www.edge.org/conversation/emanuel_derman-metaphors-models-theories (accessed on August 14, 2021).

Dubberly, H. and P. Pangaro. "Cybernetics and Design: Conversations for Action." *Cybernetics and Human Knowing* 22 (2–3): 73–82.

Dunne, A. and F. Raby (2013), *Speculative Everything: Design, Fiction, and Social Dreaming*, Cambridge: MIT Press.

Forlizzi, J. (2011), "Where is the Thinking in Systems Thinking?" *Interactions* 18:2, 49–51.

Frankel, F. (2003), "Richard Feynman's Diagrams," *American Scientist* 91:5, 450–1.

Friedman, K. (1997), "Design Science and Design Education." *The Challenge of Complexity.* Peter McGrory, ed.. Helsinki: University of Art and Design Helsinki UIAH, 54–72.
Gell-Mann, M. (1994), *The Quark and the Jaguar*, New York: Holt.
Goldschmidt, G. (1997), "Capturing Indeterminism: Representation in the Design Problem Space," *Design Studies* 18 (4): 441–55, https://www.sciencedirect.com/science/article/abs/pii/S0142694X97000112 (accessed on June 29, 2021).
Gregoire, C. and S. B. Kaufman (2015), "A Messy Mind Is Good for Creativity," *Time*, https://time.com/4115669/wired-to-create (accessed on August 21, 2021).
Hensel, M., A. Menges, and M. Weinstock, "Emergence in Architecture," (2004) *Architectural Design* 74 (3) Special Issue: *Emergence: Morphogenetic Design Strategies*: 6–9.
Hensel, M., A. Menges, and M. Weinstock (2010), *Emergent Technologies and Design: Towards a Biological Paradigm for Architecture*, London: Routledge.
Hester, P.T. and K. M. Adams (2013), "Thinking Systemically about Complex Systems," Complex Adaptive Systems Conference, Procedia Computer Science 20: 312–17, 10.1016/j.procs.2013.09.278.
Jones, K.D. and M. F. Platzer (2006), "Flapping-Wing Aerohydromechanics in Nature and Engineering," in C.A. Brebbia, ed. *Design and Nature III: Comparing Design in Nature with Science and Engineering* 87, (Southampton: WIT Press), (3–12).
Kolko, J. (2010), "Abductive Thinking and Sensemaking: The Drivers of Design Synthesis," *Design Issues* 26 (1): 15–28.
Lende, D. (2008), "Play and Embodiment," Neuroanthropology.net, https://neuroanthropology.net/2008/02/23/play-and-embodiment/ (accessed on July 29, 2021).
Leski, K. (2011), *The Storm of Creativity*, Cambridge: The MIT Press. Kindle Edition.
Lindgaard, K. and H. Wesselius (2017), "Once More, with Feeling: Design Thinking and Embodied Cognition," *She Ji* 3 (2): 83–92.
Love, A. C. (2010), "Darwin's 'Imaginary Illustrations': Creatively Teaching Evolutionary Concepts & the Nature of Science," *The American Biology Teacher* 72 (2): 82–9, https://online.ucpress.edu/abt/article/72/2/82/3237/Darwin-s-Imaginary-Illustrations-Creatively.
Markel, H. (2021), *The Secret of Life: Rosalind Franklin, James Watson, Francis Crick, and the Discovery of DNA's Double Helix*, New York: Norton.
Mathieson, L. (2016), "Synergies in critical reflective practice and science: Science as reflection and reflection as science," *Journal of University Teaching & Learning Practice* 13 (2.4).
Miller, A. (1996), "Metaphors in Creative Scientific Thought," *Creativity Research Journal* 9 (2) (3): 113–30.
Mukherjee, S. (2017), *The Gene: An Intimate History*, New York: Scribner.
Myers, W. (2012), *Bio Design: Nature, Science, Creativity*, New York: MoMA.
Polanyi, M. (2012; 1958), *Personal Knowledge: Towards a Post-Critical Philosophy*, Chicago: University of Chicago Press.
Roberts, M. A. J., R. M. Cranenburgh, M. P. Stevens, and P. C. F. Oyston, (2013), "Synthetic biology: biology by design," *Microbiology* 159 (Pt 7): 1219–20. doi: 10.1099/mic.0.069724-0, https://www.ncbi.nlm.nih.gov/pmc/articles/PMC3749723/ (accessed on August 17, 2021).
Rogers, H. S. (2022), *Art, Science, and the Politics of Knowledge.* Cambridge: MIT Press.

Rust, C. (2004), “Design Enquiry: Tacit Knowledge and Invention in Science,” *Design Issues* 20 (4): 76–85.

Santa Fe Institute, (2014), “Emergence: A Unifying Theme for 21st Century Science,” https://medium.com/sfi-30-foundations-frontiers/emergence-a-unifying-theme-for-21st-century-science-4324ac0f951e (accessed on August 4, 2021).

Schön, D. (1992), “Designing as reflective conversation with the materials of a design situation,” *Knowledge-Based Systems* 5 (1): 3–14.

Simon, H. (1982), *The Sciences of the Artificial*. Cambridge: MIT Press.

Smith, M. K. (2003), “Michael Polanyi and tacit knowledge,” *The Encyclopedia of Pedagogy and Informal Education*, https://infed.org/mobi/michael-polanyi-and-tacit-knowledge/ (accessed on August 21, 2021).

Stevens, V. (2014), “To think without thinking: The implications of combinatory play and the creative process for neuroaesthetics,” *The American Journal of Play* 7 (1): 99–119, https://files.eric.ed.gov/fulltext/EJ1043946.pdf (accessed on June 29, 2021).

Wahl, D.C. (2006), “Bionics vs. Biomimicry: From Control of Nature to Sustainable Participation in Nature,” in C.A. Brebbia, ed. *Design and Nature III: Comparing Design in Nature with Science and Engineering* 87, (Southampton: WIT Press, 289–9).

Ware, C. (2021), *Information Visualization: Perception for Design,* Amsterdam: Elsevier.

Watson, J. D. (1969), *The Double Helix*, New York: Signet (first published by Athenium in 1968).

Weinstock, M. (2010), *The Architecture of Emergence: The Evolution of Form in Nature and Civilization*, Chichester, UK: John Wiley and Sons.

Wiener, N. (1948), *Cybernetics: Or Control and Communication in the Animal and the Machine.* Cambridge: MIT Press.

Young, L., (2011), “Regen(d)erating Decoration: Cultural Narrative in Ornamented Fonts,” in L. Atzmon (ed.), 391–411, *Visual Rhetoric and the Eloquence of Design*, Anderson, SC: Parlor Press.

Part 1 Visual Metaphor, Conceptualization, and Modeling Ideas in Design and Science

1 Matters of Mathematics: Designerly Practices in Geometry

K. LEE CHICHESTER

In February 2014, UCLA neurobiologists Stan Schein and James Gayed reported the discovery of a new class of regular geometric bodies: equilateral Goldberg polyhedra (Fig. 1.1) (2014). Schein and Gayed's polyhedra are a striking example of design-based discovery in science, and relevant not only to the study of virus structure, but also to the architecture of dome-shaped buildings. Their finding was spectacular: the last time a new class of convex equilateral polyhedra with polyhedral symmetry had been discovered was in the early seventeenth century. At that time, the court mathematician and astronomer Johannes Kepler presented the full group of so-called rhombic polyhedra, along with the first-known depiction of all thirteen Archimedean solids (Kepler 1619). While Archimedean and Platonic solids consist of perfect equiangular and equilateral polygons, the angles of the Goldberg polyhedra's hexagonal faces vary.[1] Although they therefore lack the perfection of their ancient cousins, the Goldberg polyhedra nevertheless qualify as genuine polyhedra.

Asked about the role that models played in their discovery, Stan Schein points to the unique combination of computer-based and analog tools that he and his colleague employed: "My collaborator, James Gayed, and I have used some of the best physical 'toys' in our laboratory to build models of geometric objects."[2] He specifically names the molecular construction kits *Polydron, Zometool,* and *Molymods*, as well as a toy with magnetic struts and ball bearings that allows for unequal angles. In the press photograph accompanying the news reports, Schein and Gayed pose with a whole collection of stereometric bodies (Fig. 1.2). Stereometric bodies are solid figures including pyramids, prisms, and other polyhedrons. The photograph references the late-nineteenth-century iconography of mathematicians staged with models, in which the models point to the ideal and immaterial contents of the mathematicians' work. But, in the case of Schein and Gayed, the physical models obtain a higher relevance. As Schein explains:

> In recent times, most molecular modeling is done on the computer, with software like Wavefunction's Spartan chemistry software and UCSF's Chimera visualization software. James and I use both of those programs extensively, but we always start with the toys. I have long believed that if the Ancient Greeks and even

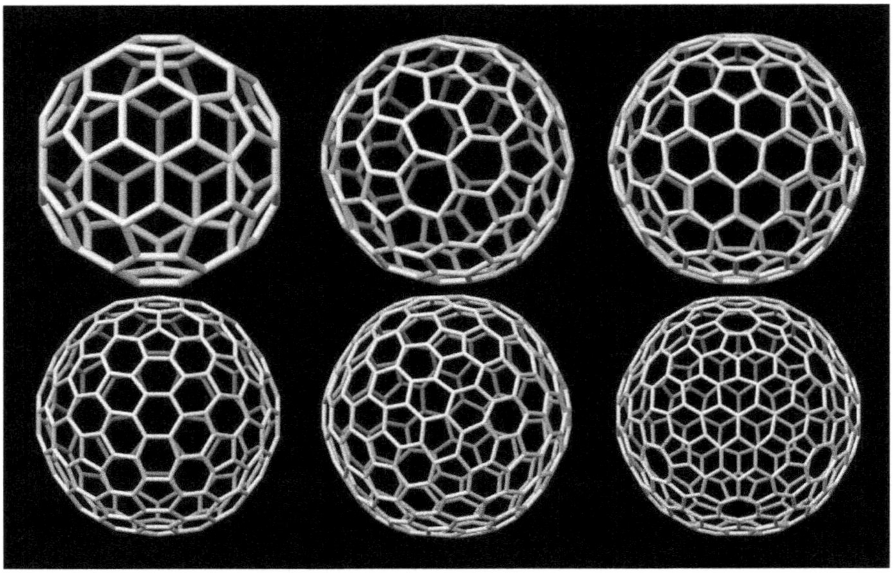

Fig. 1.1 Stan Schein and James Gayed, Goldberg polyhedra. In Stan Schein and James Gayed, "Fourth class of convex equilateral polyhedron with polyhedral symmetry related to fullerenes and viruses," *Proceedings of the National Academy of Sciences* 111.8 (February 2014): 2920–5. Image used with permission of the authors.

Fig. 1.2 Stan Schein and James Gayed posing with "toy" models. Photograph by Alex Yeh, 2014. In Dana Mackenzie, "Goldberg Variations: New Shapes for Molecular Cages," *Science News*, February 14, 2014, www.sciencenews.org/article/goldberg-variations-new-shapes-molecular-cages (accessed on February 25, 2018). Image used with permission of Stan Schein and James Gayed.

> Kepler had [had] better toys, they would have made even more progress than they did
>
> (email correspondence with the author), February 21, 2014

It is striking that despite the sophisticated software used for modeling complex geometric bodies, Schein underscores the creative potential of tactile models. He suggests that it is easier to come up with new ideas by toying around with a physical thing than by working with algorithms.[3] In an article on Goldberg polyhedra in *Science News*, the Boston mathematician Egon Schulte confirms this point of view: "You have to get your hands dirty to answer those questions" (Schulte quoted in Mackenzie 2014). He also concedes, though, that his own discipline has shamefully neglected this kind of hands-on research in past years. Physical models and visualizations in general were discredited in mathematics around 1900, as they were deemed inadequate to represent complex geometric shapes or non-Euclidean space. Biologists and chemists, on the other hand, held on to physical models because they used them to refer to real-life, "material" structures: molecules (de Chadarevian 2002).[4] It is therefore no coincidence that Schein and Gayed's discovery was made using models and software from chemistry and molecular biology "as a geometry engine" (2014).[5]

Schein acknowledges the influence of chemist Linus Pauling (1901–1994) and molecular biologists James Watson (*1928) and Francis Crick (1916–2004) on his working methods. While Pauling discovered the helical shape of peptides in 1948 by experimenting with folded strips of paper, Watson and Crick identified the double helix structure of the DNA molecule five years later using cut-out pieces of cardboard.[6] In a video-documented restaging of the original discovery (Fig. 1.3), Watson shows how he found the spatial shape of the DNA-molecule by fitting together the pieces of a self-made puzzle—or what he calls a "solid fiddling with models" (1968: 77). By pairing the cut-out cardboard nucleobases, Watson's cognitive processes were exteriorized, extending into the modeling material itself. The model thereby facilitated the mental task (Clark & Chalmers 1998: 7–19). As Reinhard Wendler has shown, instead of merely illustrating given facts, physical models function as epistemic tools, that is, as agents that participate in a thought process with unforeseen results (2013: 25ff.; 2015: 74–80). The material properties and spatial structure of the modeling components guide the researcher's actions (of movement and thought) within predetermined boundaries that function as embodied constraints. The choice and shape of the working materials—such as Watson's flat, polygonal cardboard nucleobases—becomes decisive for the way an epistemic thing—in this case the DNA-double-helix—is conceptualized, and for the resultant scientific insights.[7] Like Pauling, Watson and Crick did not forcefully "make the structure with an integer screw," but "let the models fold naturally into any screw they were comfortable with" (Crick 1988: 58, 60). The parts of the model function, then, as agents within what Bruno Latour terms actor-networks, in which human-made and natural things and experiences co-exist in a network of relationships (2005). These networked things impact each other. It is this ability of model structures to direct researchers to

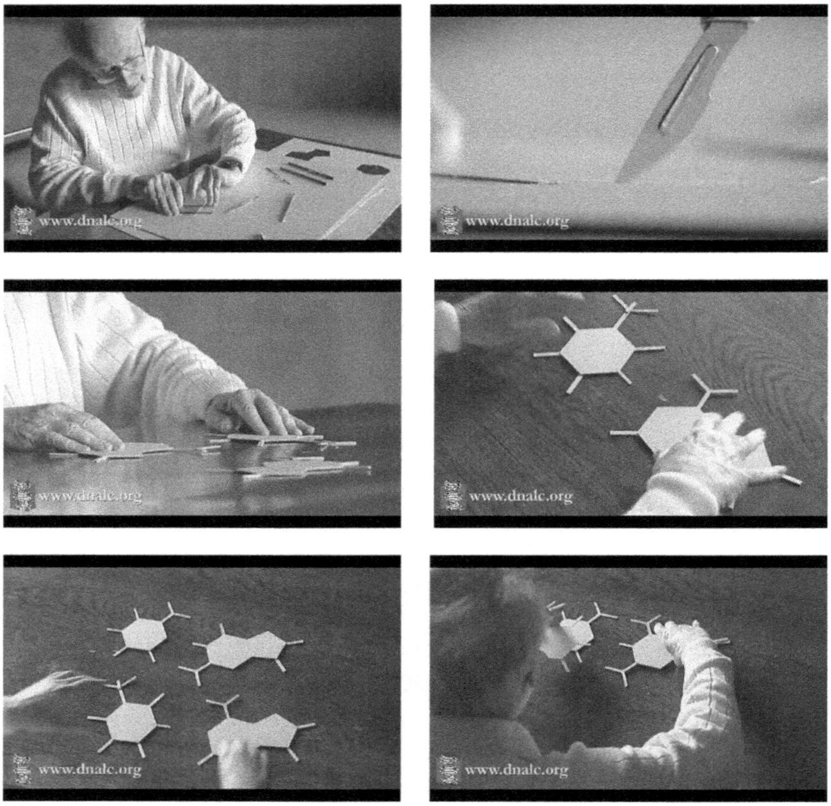

Fig. 1.3 Six video stills of James Watson restaging his 1953 discovery of the DNA-double-helix with cardboard models of the four nucleobases, DNA Learning Center. www.dnalc.org/view/15492-Discovering-the-double-helix-structure-of-DNA-James-Watson-video-with-3D-animation-and-narration.html (accessed on February 25, 2018).

(energetically) favorable configurations—the perfect fitting of all parts—that Schein seems to value for his own work.

Instead of placing himself in the tradition of model-based discoveries in molecular biology discussed above, however, Schein could have invoked the history of geometric research on polyhedra; embodied thinking through matter likewise marks the rediscovery of the (semi-) regular solids during the early modern period. But at that time—at least in the beginning when science was still largely a matter of scholastic learning—it was not scientists who were experimenting with models and visualizations, but *artisans*—the "designers" of the Renaissance.[8] The rediscovery of ancient solid geometry, or stereometry, between 1450 and 1600 demonstrates that 2D- and 3D-modeling practices in mathematics, as in science in general, originated in *artisanal* workshops. It was hence the result of *designerly* practice and tradition. Artisans' modeling methods were adopted by scholars, who recognized that, by "getting their hands dirty," artisans were more successful at envisioning complex shapes and understanding the laws of nature (e.g., demonstrated by how they solved mechanical problems) than were scholars studying ancient texts.

Artisans began to reflect the advantages of their empirical approach and made their insights accessible in vernacular treatises. They argued that the most important aspects of their work could not be put into words, but were implicit only in actions: a pragmatic "knowing how" as opposed to a merely theoretical or propositional "knowing that." In *The Body of the Artisan*, Pamela Smith has argued that this kind of knowledge-production, which she terms "artisanal epistemology," helped to shape the Scientific Revolution by inoculating science with *embodied* experimental methods (2004). In this essay, I show that this "knowing by doing" also propelled mathematical research from the fifteenth to the seventeenth centuries. It engendered important advances in mathematics—first in artisans' workshops, then through collaborations between artisans and scholars, and finally by scientists and mathematicians themselves. It is in this tradition that Schein and Gayed's model-based discovery can legitimately be located.

The final two parts of this essay deal with the aforementioned decline of visual and physical models in mathematics around 1900. This development paralleled the rise of non-mechanical explanations in physics, specifically in electromagnetics and quantum theory. Noting this paradigm shift, early historians of science were prompted to investigate the roles that artisan-engineers played in establishing the mechanical world-view that was so foundational to the Scientific Revolution. By analyzing artisans' contributions to early modern science, these historians and philosophers of science paved the way for the recognition of design-specific ways of knowing and artistic research. They also showed that, by integrating designerly methods, modern science is marked by bodily practices, visual sense-production, and tacit knowledge. Aware of the deepening divide between increasingly specialized scientific disciplines and art, these scholars called for the renewal of collaborations between design (or art in a broader sense) and science that had issued such fruitful results in the early modern period.[9]

This essay reveals historical predecessors, both in early modernity and in the first half of the twentieth century, for the practice and theorization of what design writer Nigel Cross has termed "designerly ways of knowing." Designerly ways of knowing concern design-specific methods; but they are also relevant to embodied, visual, experimental, and creative methods in the sciences that have their origin in design thinking and processes of making. The intention of this essay is to make past insights productive for present challenges in times when synthetic biology and biorobotics, and nano-physics and self-moving materials, are bridging the fields of nature and design, calling for a renewed collaboration between scientists and designers.

Artisanal Epistemology and the Rediscovery of Ancient Geometry

Piero della Francesca

Knowledge of the Platonic solids—so-called "regular" polyhedra—was preserved from Antiquity throughout the Middle Ages in Arabic, Greek, and Latin copies of Euclid's *Elements*. The pictorial tradition begun in these medieval codices survived well into the fifteenth century (Sorci 2001).[10] Medieval translators including Campanus of Novara (ca. 1220–1296), however, had already expressed doubts about the usefulness of the two-dimensional diagrams they employed to represent stereometric figures. The representation of three-dimensional ideal figures in two-dimensional space seemed to be a logical impossibility, and an irresolvable problem given medieval representational practices.

Although translators accompanied their texts with drawings, they admitted that these might be more confusing than helpful. While simple stereometric figures, such as the cube, could be presented in parallel or axonometric projection, polyhedra with higher numbers of faces, or solids inscribed within one another, presented greater difficulties. Illustrators would draw individual aspects of a figure—such as the shape of one of its faces—in the margins of the text and expect readers to mentally infer the complete spatial structure. In other places, illustrators employed orthogonal projections. They, however, chose viewpoints that made it close to impossible for readers to imagine the ideal body. Or they included the construction lines, causing the structure itself to be obscured. The diagrams ultimately presented ideal bodies that were askew, creating more obfuscation than elucidation (Sorci 2001: 149–52). Although the shapes of the simpler Platonic solids could be envisioned in the mind's eye, the more complex structure of the semiregular Archimedean solids remained a mystery. The Greek mathematician Pappus, who wrote about the semiregular solids, only mentioned the number and shape of their faces, but not their arrangement in space (*Synagoge*, book 5).

It was Tuscan painter Piero della Francesca (1415–1492) who presented the first drawings of Archimedean solids in his *Trattato d'abaco*, which was composed in the 1470s (Fig. 1.4). Piero was also the first person to write a complete and systematic treatise on perspective, the *De prospectiva pingendi* of ca. 1474/5. And he was the first post-antique author to dedicate an entire book to geometry, the *Libellus de quinque corporibus regularibus* (short book on the five regular solids), which was written sometime after 1482 and dealt specifically with the regular and semiregular solids (Field 1997a: 76).[11] Although he abstained from applying the new perspectival technique to geometry—which would have sacrificed the figures' measurability and absoluteness because they would be rendered from an individual viewpoint—he did come up with diagrammatic innovations that demonstrate his designerly acuity for problems of spatial representation. He eliminated

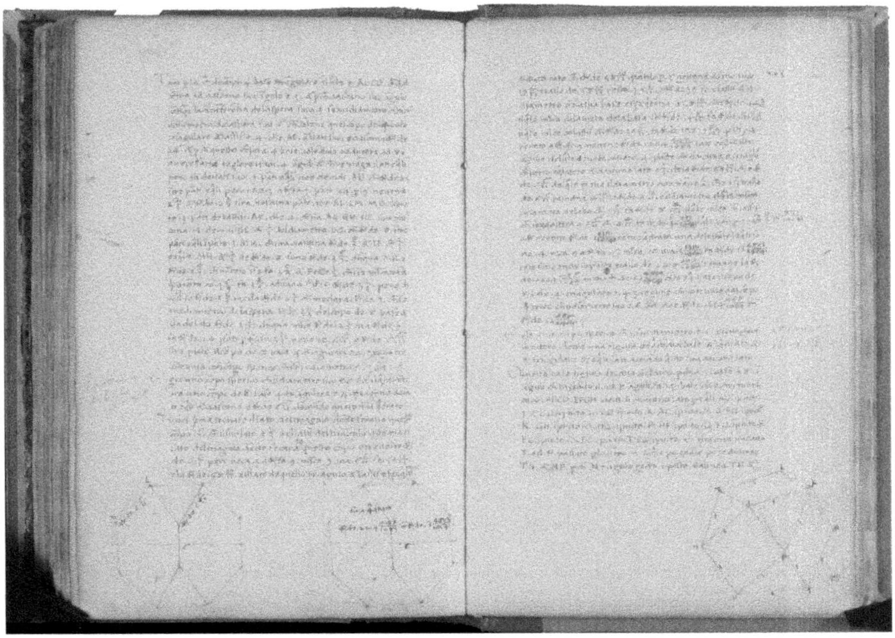

Fig. 1.4 Piero della Francesca, *Trattato d'abaco*, 1470s, Florence. Biblioteca Medicea Laurenziana, Ms. Ashb. 359*, ff. 107v–108r. In: Filippo Camerota (ed.), *Piero della Francesca. Il disegno tra arte e scienza, ex. cat.*, Palazzo Magnani (Milan: Skira, 2015), 134. Image used with permission of Biblioteca Medicea Laurenziana, Florence.

the lines of construction, showed geometric operations in series of drawings instead of conflating them in one drawing, and rendered his polyhedra opaque to heighten the clarity of his geometric figures. Consciously choosing from among orthogonal, oblique, or axonometric projections, Piero created stereometric bodies whose spatial properties are immediately evident to the eye (Sorci 2001: 164, 167; Field 1997: 75).

Historian of science Judith Veronica Field speculates that Piero's trained visual imagination and familiarity with the new art of perspective enabled his rediscovery of six of the in total thirteen Archimedean solids (1997: 73; 1997b: 242, 250). Although Piero rejected perspectival representation in his mathematical treatises—he referred to the perspectival depictions as *degradato* (degraded) as opposed to the *propria forma* (actual form) of the figure—Field suggests that he may have used the new pictorial technique in exploring stereometrical geometries, as a method of discovery (Field 1997a: 86).[12] Despite skewing the angles and measurements of a figure, perspective is the only manner of representation that conveys a full sense of a polyhedron's ideal properties and spatiality to the eye of the beholder. The image thus creates the impression of an object that can be worked with experimentally and epistemically, in the mind as well as on paper.

The fact that Piero introduced the concept of "cutting off" the corners of polyhedra—called "truncation" in geometric terminology—suggests that he may also have used physical models to literally "cut off" corners from geometrical bodies. He could have derived the Archimedean from the Platonic solids by experimenting with clay, wax, wood, or precious stone, which were common materials in artisanal workshops.[13] Decisively, it was an artist who initiated the rediscovery of ancient geometry by applying artisanal and designerly methods of experimental modeling, decoration, perspectival construction, and visual communication to the scholarly discipline of mathematics.

Luca Pacioli, Leonardo da Vinci, and Jacopo de' Barbari

One of the main disseminators of Piero's achievements in geometry was the Tuscan mathematician Luca Pacioli (1445–1517). Historian of science Leonardo Olschki (1885–1961) suggests that Pacioli's mathematics training with Piero (whom he referred to as his first teacher in their hometown of Sansepolcro), along with his friendships with artist-engineers including Leonardo da Vinci, exposed Pacioli to the virtues of applied mathematics and visualization (Olschki 1919: 160, 197f.). Although Pacioli is not remembered for any noteworthy innovations, his relevance to the history of science lies in his popularization of mathematical learning through vernacular publications, and his interweaving of theory with practical applications (Olschki 1919: 151–239). Pacioli made the results of an age-old science, so far reserved for an erudite elite, accessible to artist-engineers. He likewise offered mathematicians rich sources of unstudied material by bringing to their attention problems from artisanal and engineering practice.

Pacioli was also acutely aware of the importance of visualization for progress in mathematical research. In his introduction to *De divina proportione* (1498)—a treatise on the "divine proportion," the golden section in geometry, painting, and architecture—Pacioli remarks that "in vision knowledge has its origin," referring to the eye as the first gate to the intellect.[14] Inspired by Plato's *Timaeus* as well as by Aristotle, he compares the human head to a citadel that protects the body with its five senses, and informs the mind about the environment: for "nothing is in the intellect that was not first in the senses."[15] Already his major mathematical treatise, the *Summa de arithmetica, geometria, proportioni et proportionalità* of 1494, had begun with the aesthetic reflection that, upon seeing the Platonic solids, even the most simple-minded person could sense something of their "sweetness" (*dolcezza*) to the mathematician (Baader 2003: 195). It is visual enchantment, Pacioli writes, that offers the impetus for scientific inquiry; the beauty of these Platonic solids alleviates the difficulties of their analysis.[16] Even the most noble and exact science, according to Pacioli, depends upon the mimetic and affective abilities of visual representation to address the beholder. He deemed vision to be epistemically the most important of the five senses, and consequently considered perspective to be the eighth of the seven liberal arts, and the prerequisite for all the others.[17]

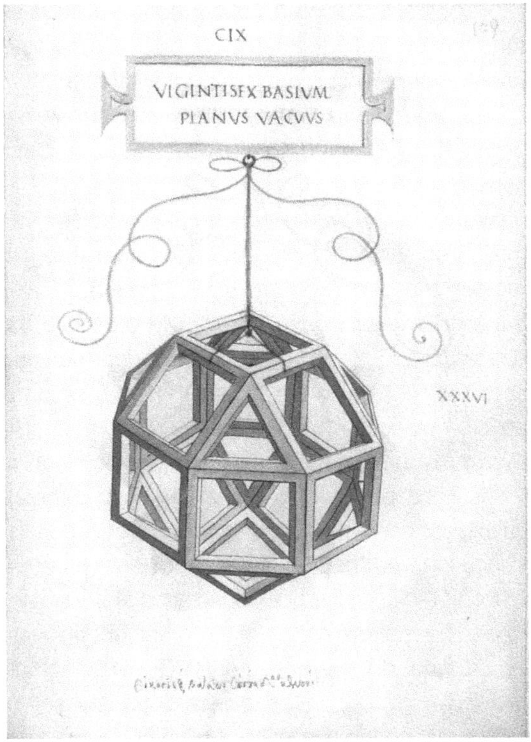

Fig. 1.5 Leonardo da Vinci, Regular rhombicuboctahedron, in Luca Pacioli, De divina proportione, ca. 1498, plate XXXVI, folio 109r. © Veneranda Biblioteca Ambrosiana – Milano/ De Agostini Picture Library. AKG-images / Mondadori Portfolio.

A major part of Pacioli's *Divina proportione* is, in fact, an Italian translation of Piero's *Libellus*, including Piero's figures, which Pacioli printed in the margins without mentioning his source.[18] For the appendix, he commissioned Leonardo da Vinci (1452–1519), who was living with him at the time, to produce a series of lavishly colored perspectival depictions of polyhedra (Fig. 1.5). Leonardo rendered them as "skeletal" constructions, which became a standard method, as it showed the front and back sides of a body simultaneously without confusing the viewer. Leonardo's technique was another design innovation in the visual communication of geometric bodies.

Among Leonardo's illustrations are two Archimedean solids whose shapes were then unknown: the icosidodecahedron and the rhombicuboctahedron.[19] Although they were discussed in Pacioli's text, Leonardo's images present information on the spatial configuration of the solids that the text could not offer. In fact, the text explicitly refers readers to the illustrations for additional information (Field 1997b: 253f.). Far from being mere addenda to a self-contained text, the images constitute the source as well as the ultimate aim of the geometric knowledge presented. Essentially, the text is worth next to nothing without the illustrations. The images function in the strictest sense of the Latin *illustrare*: to "bring to light" (Amelung 2018; Amelung 2019).

Pacioli also adopted Piero's notion of truncation, as he speaks of "cutting off" the corners of geometric bodies, which indicates that he may also have been experimenting with models. We know from the second chapter of the *Divina proportione* that he made models of the polyhedra "with his own hands," which he consigned to the dedicatee of the book, Ludovico Sforza, Duke of Milan (1452–1508) (Pacioli 1509: chap. 2).[20] He expresses his hope that their heavenly beauty would incite the Duke's thirst for knowledge as it had his own. This may also explain why the figures in the book seem to represent physical models made of wooden struts, bound up and suspended by a curling ribbon, rather than ideal bodies. Their vivid presence may have been a tribute paid to the epistemic role played by actual wooden models. Leonardo's figures—which Pacioli says are "shown to the eye in their proper *material* shape"—thus stand in for the physical objects, replacing them as valid mathematical evidence and functioning in a similar manner as epistemic tools ((italics mine) (1509: part 1, chap. 49).

As Field points out, the rhombicuboctahedron, which Pacioli is credited with having rediscovered, is one of four Archimedean solids that cannot be obtained by truncation, as they require an additional procedure today known as "distortion" to arrive at their regular polygonal faces after cutting off the corners (Field 1997b: 256–62).[21] The representation of a rhombicuboctahedron in the famous portrait of Pacioli by Jacopo de' Barbari (1460/70–1516) actually suggests another construction method (Fig. 1.6).[22] The ethereal, translucent body of the rhombicuboctahedron pending in space has been described by many scholars as a crystal body half-filled with water suspended from the ceiling by a red string. This interpretation has prompted questions regarding the mechanics of the construction and the optical accuracy of the painted reflections (see Fig. 1.6). A simpler explanation is that Pacioli constructed the rhombicuboctahedron from forty-eight identical thin wooden or metal struts, stabilized by a string around the perimeter, and dipped into soapsuds. This is the easiest way of obtaining a rhombicuboctahedron; it may also explain the odd surface reflections and the attachment of the string to the bottom of the structure for stability. Pacioli may have let these sticks guide his actions—much like Schein and Gayed did in the case of their modeling toys.

The painting by Jacopo de' Barbari is the first-known representation of this Archimedean solid in history. Pacioli only described it in writing three years later in his *Divina proportione*. The painting, which prominently shows the date of 1495 on the *cartellino* in the lower right-hand corner, thus functions as a scientific publication and is Pacioli's claim to priority. Pacioli probably even relied on the painter to present the solid in perfect perspective in order to prove his rediscovery—something for which words alone, as we have seen, would not have sufficed. The introduction of designerly methods to the scientific quest hence enabled new discoveries—and Pacioli was one of the first scholars to advocate for the artisan-engineers' empirical methods of problem-solving and visualization in the sciences.

Fig. 1.6 Jacopo de' Barbari (attr.), *Portrait of Luca Pacioli and a Gentleman (Guidobaldo da Montefeltro?)*, oil on panel, 120 x 99 cm, 1495, Museo e Gallerie Nazionali di Capodimonte, Naples. Public domain.

Albrecht Dürer

A further innovation in the field of polyhedral geometry was made by the Nuremberg artist, typographic designer, and goldsmith Albrecht Dürer (1471–1528)—another design thinker *avant la lettre*. Dürer may even have visited Pacioli during his travels to Italy to learn the secret laws of perspective. If so, it is likely that he was introduced to the mathematician by their common friend Jacopo de Barbari.[23] What is clear is that geometry, and especially polyhedra, were to take on a prominent role in Dürer's work. In his *Underweysung der Messung* of 1525, he depicts polyhedra as ground plans and elevations, but also as *nets*—a manner of visualization that seems to have been his own invention (Fig. 1.7) (Field 1997b: 266). Dürer recommends assembling the geometric bodies by pasting together two sheets of paper, cutting the nets out with a knife, and folding them along the lines. In this manner, he presents two Archimedean bodies that had never been published as images before—the truncated cuboctahedron and the snub cube. Since both of these solids, again, cannot be obtained by truncation, Field assumes that he used the nets as an epistemic tool (1997b: 267–9).[24] It is possible that the net idea was inspired by ornamental tiling.

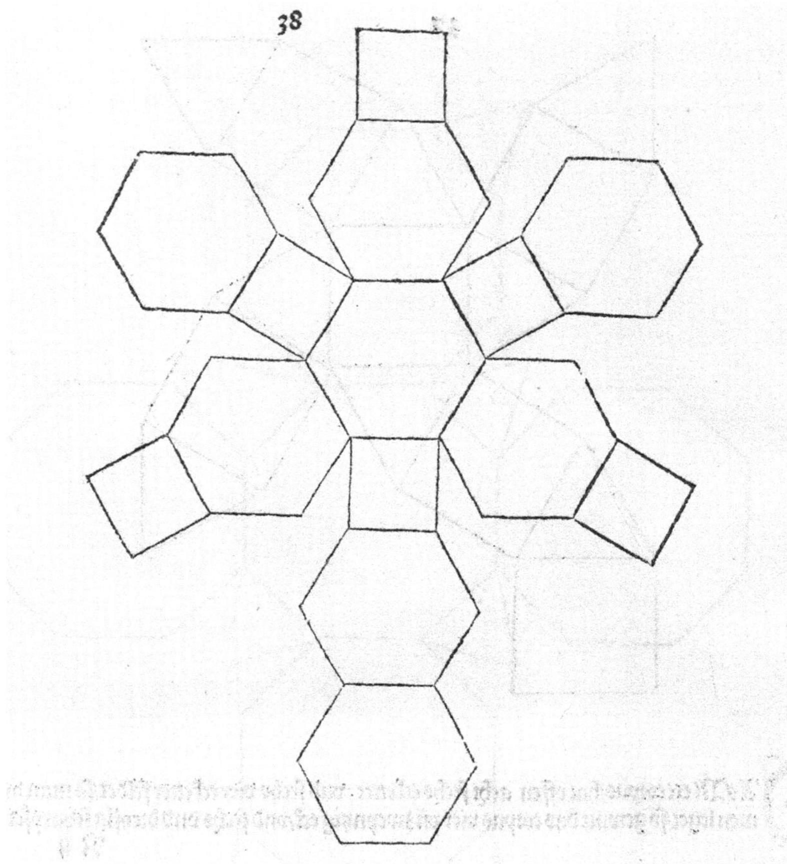

Fig. 1.7 Albrecht Dürer, Truncated Cuboctahedron Net, woodcut, from *Underweysung der Messung* (Nuremberg: H. Andreae, 1525).

Dürer likewise presented ornamental tiling in his *Underweysung* (1525: 61–6) as part of a classical workshop tradition based on the intricate organization of geometric figures in two-dimensional space.[25]

As Pacioli already noted in *De divina proportione*, instead of setting up equations, workshop masters such as Dürer were "calculating their works with level and compass, and bringing them to wonderful perfection" (Olschki 1919: 155, 210). Dürer remained loyal to the workshop tradition in his treatises. In his *Underweysung*, he does not mention Euclid's construction of the pentagon with the golden section, but instead shows an approximation employed by craftsmen. Although it was mathematically incorrect, Dürer uses this approximation for its *Behendigkeit*—its "handiness."[26] For this reason, Dürer's craft-based approximation was adopted by scholars, such as Kepler, and his method became an object of study in its own regard in mathematics and physics. As Olschki has shown, the originality of Dürer's treatises lay in their making available techniques and problems from artisanal and engineering practice for exact mathematical analysis and explication. Translated into Latin in three editions, as well as into French and Italian, the *Underweysung* became a highly popular book, known to scholars throughout Europe and cited by scientists Galileo Galilei, Gerolamo Cardano, Tycho Brahe, and Johannes Kepler, among many others (Olschki 1919: 414, 433–44). Using workshop methods, Dürer brought important advances to geometry and was inspirational to a century of mathematical research.

Johannes Kepler

The astronomer and mathematician Johannes Kepler (1571–1630) was the first modern author to publish all thirteen Archimedean solids with illustrations. Kepler also described these solids in perfect mathematical terms and offered proof that there are no more than thirteen (Olschki 1919: 275f.). Field presumes that Kepler recognized the potency of Dürer's method, and may have used the artisan's craft-based nets as an "invention machine" (1997b: 269). Kepler produced his own models. Indeed, a poem in his magnum opus, the *Astronomia nova* of 1609, describes his scientific quest as manual labor, claiming that he "put his hands to work in the hope of succeeding in supporting the vault of the heavens with new beams / Finest building material having been provided by the five [Platonic] bodies" (Kepler 1937: 3:14). This self-description marks a paradigm shift, beginning around 1600, from contemplative thinking to an empirical, mechanical, and ultimately *designerly* approach to the exploration of nature.

A model of the cosmos composed of Platonic solids, nesting one within the other, is famously depicted in a print from Kepler's *Mysterium cosmographicum* of 1596 (Fig. 1.8). Deploying a paper model, Kepler convinced Duke Frederick I of Württemberg to have a *Kunstkammer* version of the cosmos fabricated in silver. This proposed silver model represented the planets using inlayed gemstones on the edges of hemispherical orbits, and moved them by mechanical clockwork. The final piece was to function both as a test model for Kepler's hypothesis and a fancy goblet—each orbit to be filled with a different alcoholic beverage (Cromwell 1997: 147f.; Bredekamp 1995: 37; Chojecka 1967: 56, n. 6). To heighten the visual clarity of his silver model construction in the etching done of it (see Fig. 1.8), Kepler adopted the skeletal, wooden-looking polyhedra that had been introduced by Leonardo. Kepler may well have encountered Leonardo's rendering style in an opulent book by the Nuremberg goldsmith Wenzel Jamnitzer, the *Perspectiva corporum regularium* of 1568.

Although the silver model was never completed, its epistemic relevance should not be underestimated. Working with tactile, mobile models—and closely collaborating with artisans in the princely workshops—made it possible for Kepler to conceive ways to apply the laws of mechanics to celestial bodies. This intellectual leap had been previously unthinkable because the heavens were conceived of as an ideal sphere with laws distinct from worldly physics. Inspired by the idea that as the first geometer, God had made the cosmos according to aesthetic principles[27]—in the sense of the Greek word *κόσμος*, which means "order," but also "ornament"—

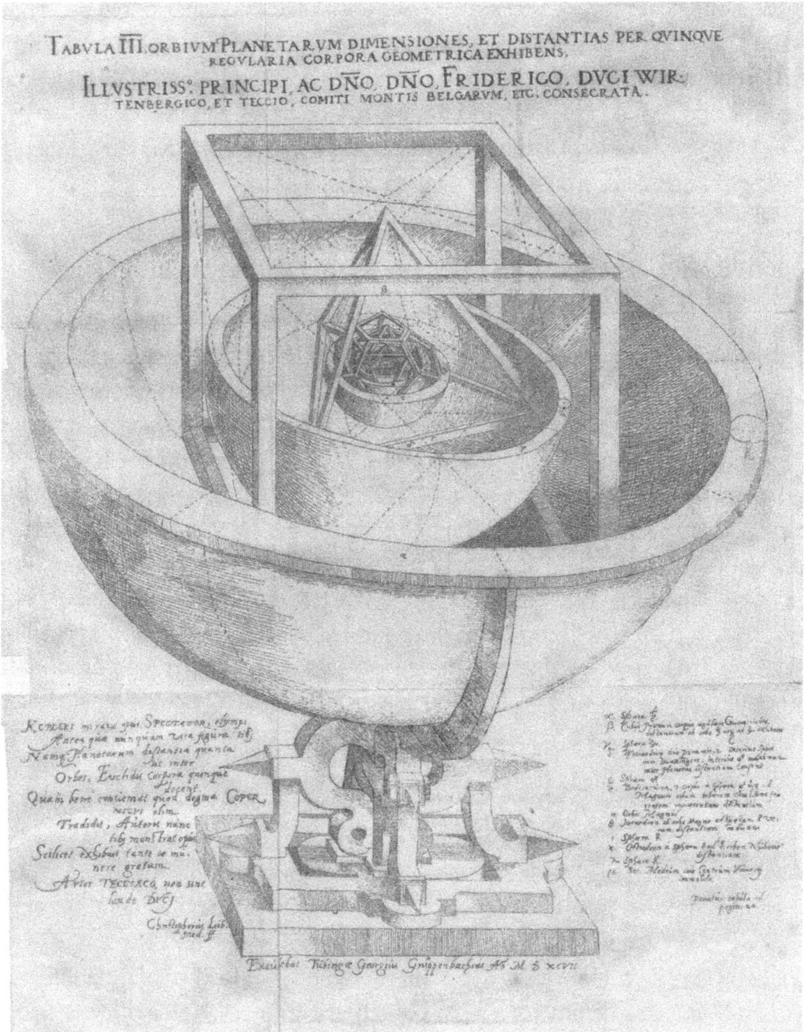

Fig. 1.8 Johannes Kepler, Diagram of the Planetary Spheres, etching, in *Mysterium Cosmographicum* (Tübingen: Georgius Gruppenbachius, 1596), plate 3. ETH-Bibliothek Zürich, RAR 1367: 1, http://doi.org/10.3931/e-rara-445 / Public Domain image from ETH-Bibliothek Zürich / Library of the ETH Zürich.

Kepler's polyhedral model of the cosmos was also intended as a compelling aesthetic argument for the Copernican heliocentric system (Martens 2000). If God, the original craftsman, had made the world as a mechanical clockwork, artisanal reasoning could be legitimately applied to its analysis, and "reverse-engineering" became the method of choice for gaining insight into its workings.

Steeped in the German tradition of instrument-making, Kepler's hemispherical, mechanically-turning model of the cosmos propagated the idea of the world as a machine, open to the eyes of the inquiring beholder (Bredekamp 1995: 37). In 1956, the art historian Erwin Panofsky (1892–1968) suggested that Kepler's inundation in a mannerist visual and material culture ultimately enabled his discovery of the elliptical shape of the planetary orbits. Kepler's point of view was repudiated by his contemporary Galileo, who, for equally aesthetic reasons, preferred the classicist ideal of circles (Panofsky 1956: 3–15; Bredekamp 2019: 41–59). Kepler's discovery of the elliptical planetary orbits was fortified by his adoption of the artisanal language that was introduced into mathematics by Dürer. Dürer had translated Latin technical terms into common words to make them comprehensible to artisanal readers. Kepler used Dürer's term the "Eierlini," the "egg-line"—which was uncommon in astronomy but common in mannerist designs—to describe the orbits of celestial bodies. By thus adopting artisinal, vernacular language, Kepler found the shape of one of the most mundane objects replicated in the heavens.[28]

Tacit Knowledge in Craftsmanship and Science

Realizing the relevance of their work to scholars, certain early modern artisans began to reflect the superiority of a knowledge that is involved and expressed in making as opposed to verbalized or formalized as theory (Smith 2004). As Dürer claimed, "practice [Gebrauch] and understanding must go hand in hand."[29] He used the German word *Gebrauch* to refer to practical knowledge or skill, and *Kunst*, meaning "art," to refer to intellectual understanding or theory. Dürer writes: "For verily, art [i.e., theory] is embedded in nature; he who can (with)draw it has it" (1528: 4:198).[30] Through this notion, Dürer fixed the then new idea that knowledge could be obtained by physically drawing from nature—the transformation of contours into lines and incisions by means of close observation, technique, and manual skill. As a result, he framed art and design as embodied theory: knowledge was demonstrated by the precision and perfection of the finished object, and this kind of knowledge could not necessarily be expressed in words. In his *Underweysung*, Dürer called upon students to prove that they understood Euclidean geometry through their exact use of the dot and the line.[31] In the end, the visual work itself would reveal the student's geometric proficiency—his or her implicit knowledge—and the *Gewalt*, or power, of the maker (Panofsky 1955: 164, 273; Smith 2004: 67–74).

This concept of knowledge through practice was picked up by humanists such as Paracelsus (1493/4–1541), Walther Hermann Ryff (1500–1548), and Petrus Ramus (1515–1572), who visited artisans' workshops to learn from their experience. Ramus subscribed to the idea that, since craft was older than science, the latter had evolved from the former (Hooykaas 1958: 21).[32] He came to regard *use* as the ultimate aim of science, concluding that mathematics should be defined as a practical art, since measurement, architecture, fortification, and navigation were its final purpose.[33] What was decisive for the advancement of science, Ramus writes, was essentially technological progress: that one "put the hand to work, and . . . worked well."[34] He found that this principle was better demonstrated in the products of craftsmen than in the precepts and rules of the mathematicians.[35] Significantly, Ramus was one of the first humanists to acknowledge the value of an embodied, habitual knowledge that cannot be verbalized but can only be evidenced in *doing* and its effects—"knowing how" as opposed to a merely theoretical "knowing that."

The proposition that artisans played a prominent role in instigating the Scientific Revolution by introducing mechanical models and anticipating the methods of modern empirical science, began with the establishment of the History of Science around 1920. The three-volume work by the literary scholar Leonardo Olschki on the history of early modern scientific language (*Geschichte der neusprachlichen wissenschaftlichen Literatur* 1919–1927) is one of the first in-depth analyses of artist-engineers' contributions to advancement in science, especially in the field of mathematics. At the Second Congress for the History of Science in London in 1931, the Russian physicist Boris Hessen also propagated the notion that the Scientific Revolution had been the result of mechanics, not cosmology (Hessen 2009: 41–101). His lecture inspired a series of publications, by mostly Marxist historians and scientists, who took a so-called "externalist" view of the development of science through technological innovation.

The author most closely connected to this approach is the Viennese philosopher Edgar Zilsel (1891–1944) (for whom the "Zilsel-thesis" was named). Zilsel, whose ideas sprang from Marxist materialism, was involved with the Vienna Circle and its logical empiricism in the 1920s and 1930s. He explored how the importance of mechanical instruments and machines in early capitalist society had instigated a mechanical world-view that regarded nature as a "gigantic but lawfully functioning mechanism" (Zilsel 1941a; Raven & Krohn 2000: 178).[36] Zilsel concluded that the artisan-engineers, the makers of the instruments, would have been first to deal with questions of cause and effect that were central to experimental science (Zilsel 1941a; Raven & Krohn 2000: 176ff.). Although universities taught their students contempt for manual labor—which was considered inferior to the "liberal arts"—artisans were carrying out experiments in their workshops, dissecting bodies, measuring the land and the heavens, and developing "considerable theoretical knowledge in the fields of mechanics, chemistry, metallurgy, geometry, anatomy, and acoustics" (Zilsel 1939; Raven & Krohn 2000: 4). Their measuring instruments, Zilsel claims, are the precursors of the modern physical apparatus, their "quantitative rules of thumb are the forerunners of the physical laws of modern science" (Zilsel 1942; Raven & Krohn 2000: 13f.).

Zilsel observed that early-sixteenth-century artisans realized the value of their practical knowledge and wrote treatises in which they attempted to describe in words what they knew from experience. Academic scholars of that time, such as Pacioli, Ramus, and Kepler, were interested in this work, setting aside prejudice against manual labor that was inherited from Antiquity, and integrating the artisans' knowledge into their own scholarly treatises. Insights that emerged from craft workshops were published in the scientific *lingua franca* of Latin. At the same time, scholars also began to present their own research in vernacular language—

creating what Pamela Long refers to as new "trading zones" between practitioners and the learned (2011: 9).[37] Modern science emerged within these trading zones. It did so through the adoption of artisanal "hands-on" methods used by rationally trained university scholars who systematized workshop knowledge, as well as by craftsmen and engineers who synthesized scholarly knowledge into their practical works.[38]

In a 1952 paper entitled "Artist, Scientist, Genius: Notes on the *Renaissance-Dämmerung*," art historian Erwin Panofsky links the "anticlimax between two peaks" in the sciences between 1400 and 1600—which the historian of science George Sarton had ascertained—to the climax in the arts (including craftsmanship and illustration) of the same period.[39] He suggests that the major achievements in the arts and crafts—even if they were non-propositional and contained in objects—should be considered to be "vital contributions to the progress of the sciences" (Panofsky 1962: 126ff.). He dismisses the idea of a "clean-cut separation between artistic and scientific activities" in the Renaissance, and instead considers how "such a science and such an art advanced, as it were, on a united front"—a development he describes as "decompartmentalization" (1962: 127). Panofsky examines how various kinds of artisans influenced the development of the observational and descriptive sciences, such as zoology, botany, paleontology, anatomy, and physics. In these sciences, Panofsky remarks, "illustration is not so much the elucidation of a statement as a statement in itself" (1962: 147).[40] Essentially, he believes that an observation that was not visually recorded was lost for scientific inquiry since words could never achieve the clarity visualizations have in describing an observation.

Panofsky uses this idea to critique Zilsel and Olschki, who had suggested an opposition or parallelism—and cross-fertilization—between "purely theoretical speculations" and "purely operational procedures" (1962: 136). A reference he makes to the philosopher Ernst Cassirer shows Panofsky's skepticism about the assumption that practice could ever be separated from reflection, or form from meaning. As Cassirer wrote in his *Philosophy of Symbolic Forms*: "By symbolic pregnance we mean the way in which a perception as a sensory experience contains at the same time a certain non-intuitive sense [*nicht-anschaulichen Sinn*] which it immediately and concretely represents."[41] In other words, since form always conveys meaning, engagement with matter and its arrangement into patterns and shapes is in itself a non-verbal act of reflection and sense-production. Designerly practices were not merely enriched by theory in the early modern period, they were already theory. It was *science* that, by adopting experimental methods and practices of visualization, incorporated a new approach to theory-production. This approach was based on the tactile manipulation and visual contemplation of forms and matter. The relevance of skill and connoisseurship to knowledge—which is laid out in Michael Polanyi's 1950s concept of "personal" or "tacit" knowledge—is still rarely acknowledged today.[42] This is especially true in science. Polanyi explained in his Aberdeen lecture of November 1952: "The arts of doing and of knowing are indeed but two aspects of a single art, which may be called the art of making sense."[43]

Modern science's empiricism could only come about through the acceptance of this "knowing by doing"—of tacit knowledge as true, legitimate knowing. Although this was readily acknowledged in the early modern period, with time, the significance of tacit or personal knowledge was suppressed by the self-stylization of science as an "objective," impersonal endeavor (Daston & Gallison 2007; Latour 1993).[44] A trace of this history may, however, still be found in the use of the word "*labor*atory" to refer to the scientific *work*bench.[45]

The Modern *Anschauungskrise* and Computational Geometry

Zilsel's thesis demonstrating the relevance of designerly practices to the seventeenth-century Scientific Revolution emerged at a time when the sciences and the arts were drifting apart, and the mechanical world-view was being replaced by the immaterial concepts of electromagnetism, relativity theory, and quantum physics. By 1920, mathematics had similarly turned away from physical modeling as a research method or a means of insight. The Weierstraß-Function and so-called "monster-curves" had revealed the limits of visualization in geometry; they were impossible to visualize until the advent of computers. The metallic zinc physical models that were common in nineteenth-century mathematics—models that the famous geometrician Felix Klein (1849–1925) had understood to be the objective of his research—fell victim to suspicions regarding the imprecision of embodied mathematics. This phase in the history of science has come to be known as the *Anschauungskrise*, the crisis of visualization or intuition.[46] In 1920, the German mathematician Richard Baldus (1885–1945) wrote that with the end of the early modern period of discovery and the onset of modern analytic critique, the need for visualization (*Anschauung*) had been overcome. Models, according to Baldus, "show more than one would wish for altogether, and too little in each detail" (1921: 9f., 6). They represent special cases, while mathematicians require abstract generalization, and they induce logically unfounded suppositions based on mere visual suggestions. Exact logical thinking, Baldus concluded, was not possible with purely visual means (1921: 10). He instead advocated the use of formulas in combination with the more flexible and less concrete mind's eye.[47]

In contrast to mathematics, biologists, chemists, and engineers—fields through which "practical mathematics" developed—continued to use models. The mathematically inspired Scottish biologist D'Arcy W. Thompson (1860–1948) published a paper in 1925, entitled "On the Thirteen Semi-Regular Solids of Archimedes," in which he described a new method for obtaining the Archimedean solids by mechanically converting a plane net of paper polygons into a series of semi-regular polyhedra. The similarity to Dürer's method is no coincidence: Thompson knew of and drew from the artisan's works in his famous transformation diagrams published in *On Growth and Form* in 1917. Thompson also embraced design techniques that he encountered in the Scottish Arts and Crafts movement to explain form processes in nature, which he interpreted as fundamentally mechanical, and therefore mathematically describable (Chichester 2019).[48]

The novelty of Thompson's paper—which was the justification for its publication in the peer-reviewed *Proceedings of the Royal Society*—was his new mechanical method of constructing already-known geometric bodies. The fact that both of the paper's peer reviewers doubted that it was publishable, reveals just how marginalized modeling had become in mathematics by the 1920s.[49] Thompson was a biologist who applied math to understanding actual forms in nature. He was, therefore, comfortable using visualizations and designerly techniques of open-ended, experimental modeling to advance his research.[50] As a biologist, he understood imprecision to be a fundamental condition of the physical world, and not a contradiction to scientific exactitude and the formulation of natural laws.[51]

Only relatively recently, with the rise of computational geometry, has visualization, or *Anschauung*, returned to mathematics. The virtual model unites the seemingly irreconcilable poles of concretization and abstraction. Like Baldus' 1920 description of the advantages of using the mind's eye to visualize, virtual modeling allows zooming in to check specific details, offers different modes of representation for different aspects of complex figures, and makes it possible to examine these details and representations in temporal succession. In an article from 1995, the mathematician and philosopher Brian Rotman even referred to "virtual reality" as the reality of mathematical objects and procedures (1995: 414). His idea, however, only reveals to what extent the design of the virtual interface has become transparent. Mathematical figures appear so naturally in their algorithmic environment that the design decisions about coloring or light-effects, spatial organization, and interactive options become invisible.

Mathematicians Faniry Razafindrazaka and Konrad Polthier at the Free University in Berlin are working on the computer-based tessellation of large genus surfaces to produce so-called "regular maps" (Fig. 1.9). These are knotty figures with regular tilings that make them distant relatives of the ancient polyhedra. These mathematicians explicitly state that the production of good images is a central aim of and challenge in their work: "regular maps are intriguing surfaces and having a nice visualization of them remains an interesting and unsolved problem" (2014: 250). At the same time, the aesthetic qualities of these surfaces, which they note, go hand-in-hand with complexity and symmetry, are crucial—they become part of the research task. Razafindrazaka and Polthier describe their aim as being "not only to generate some genus *g* surface but also a surface with rich topological structure and nice looking shape" (2014: 243). The beauty of the figure becomes integral to the mathematical problem and is a guide to new insights that cannot be read in the formulas.

The early modern collaborations that I presented earlier in this essay suggest that designers could help to create even better visualizations—either physical or virtual—that enable as yet unimaginable geometric bodies to be represented and studied. In his 1960 lecture "The Fear of Knowledge," the Art Historian, and former student of Panofsky and Cassirer, Edgar Wind (1900–1971) criticized the widely held conviction that intellectual ideas were detrimental to the artistic imagination. Much to the

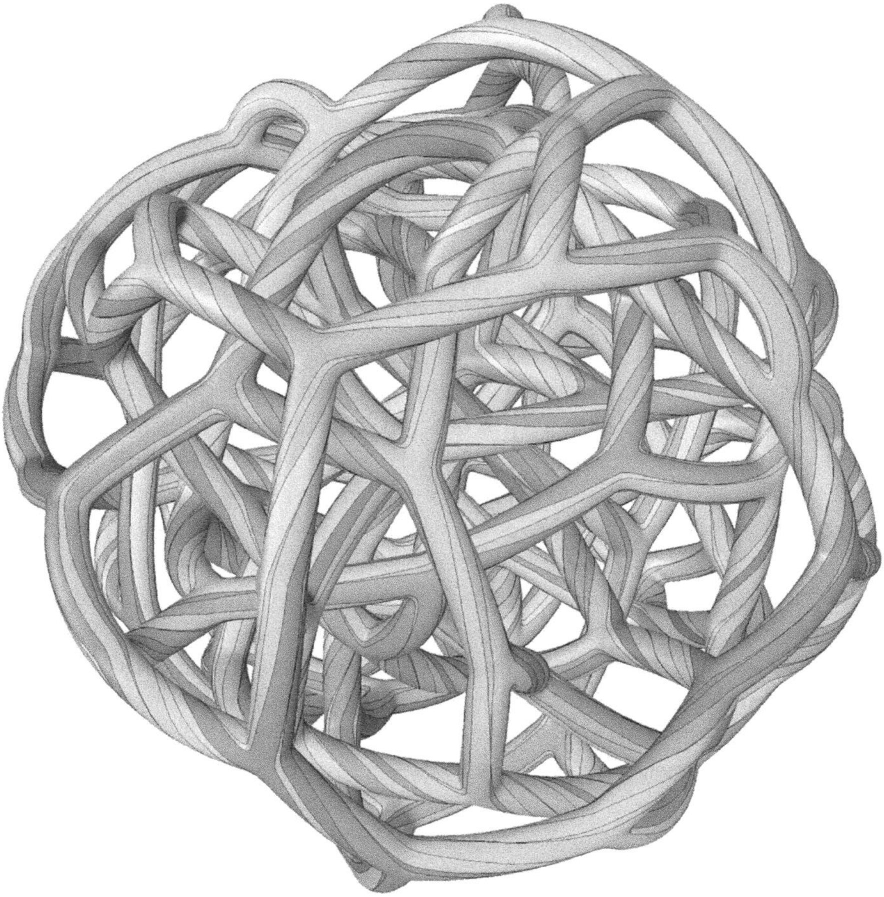

Fig. 1.9 Faniry Razafindrazaka and Konrad Polthier, "Visualization of a genus 61 surface tiled with 480 hexagons following the regularity of the map R61.1" {6,4}, 2014, Freie Universität zu Berlin. Image used by permission of Faniry Razafindrazaka and Konrad Polthier, Freie Universität zu Berlin.

contrary, he points out, some of the best works of art have been inspired by scientific thought—the rift between artistic disciplines and science being essentially the modern-day result of the Romantic critique of reason (Wind 1979: 52).[52] He argues that this conceptual divide leads to designs that imitate science from a distance, without being truly involved in the research process.[53] Wind argues instead that intellectual precision and pictorial fantasy go hand in hand, as clear representation sharpens scientific concepts (1979: 61f.).

As an example, he refers to a contemporary book on *Mathematical Models*, published by H. M. Cundy and A. P. Rollett in 1956. Quoting the poet Samuel Taylor Coleridge (1772–1834), the authors describe their aim as being "to assist reason by the stimulus of the imagination" (Wind 1979: 157, n. 110). The authors' approach prompts

Wind to compare and contrast their illustrations of Archimedean solids with Luca Pacioli's *Divina proportione*, since Pacioli likewise understood images as key to the intellect. Cundy and Rollett's modern illustrations, however, do not fare well in Wind's comparison: the lack of eloquence, or *Anschaulichkeit*, that marks Cundy's and Rollett's figures, Wind maintains, was by no means needed for greater mathematical precision. In Leonardo's images, all characteristics of the regular and semi-regular polyhedra are evident at first sight. These same kinds of traits, however, have to be inferred intellectually from Cundy's and Rollett's diagrams, which have confusing shading and visual distortions. In a sense, the modern authors have reverted to the medieval problems of representing three-dimensional bodies in two-dimensional space.

Wind is not asking contemporary scientists to adopt the graphic and modeling methods of the Renaissance, instead he hopes to encourage modern illustrators of scientific and mathematical structures and phenomena to "penetrate the machinery at their disposal with some of the imaginative force and persuasive elegance that Leonardo applied to the more unwieldy mechanics of his time" (1979: 59). Sixty years after Wind's call for a closer collaboration between scientists and designers, contemporary science does more work from and with visualizations and modeling. Wind's recommendation for intellectual precision combined with pictorial fantasy is even more relevant today. Interdisciplinary considerations are crucial to this kind of progress in both science and design. Contemporary designers should apply to the processes of scientific inquiry their understanding of shapes and materials, the organization of space, as well as new imaging or modeling technologies. In doing so, they could reintroduce scientific inquiry to their methods of open-ended experiment and visual reasoning. Applying their specialized training in *designerly ways of knowing* to scientific research could lead to critical discoveries—just one of which may be new and yet more complex geometric bodies.

Notes

1. The angles may vary, for instance between 124,9 and 110,2 degrees as opposed to the 120 degrees typical for a regular hexagon, but the faces are all flat, as required for regular polyhedra. Stan Schein and his collaborators more recently published a fifth class of polyhedra derived from the truncated tetrahedron, in which neither the hexagons nor the pentagons are regular (Schein et al. 2016: 1–10).
2. Email correspondence with the author, February 21, 2014.
3. Algorithms tend to require users to already have some idea of what it is they are looking for, and also to have a firm grasp of a figure's spatial properties.
4. A mathematician who actively reintroduced model making to her work in the early-twentieth century while participating in bio-molecular research was Dorothy Wrinch (1894–1976) (Senechal 2013).
5. The discovery of the Goldberg polyhedra was prompted by the observation of similar shapes in cells of the human retina, where large protein structures called clathrin cages act as transport vehicles for molecules (Liebelson 2014).

6. Schein writes: "Although I am a neuroscientist, my PhD was in molecular biology. I have long been impressed with the revolution in studies of molecular structure started by Linus Pauling's use of physical models to understand protein structure. That work deeply influenced James Watson and Francis Crick, who adopted use of physical models to figure out DNA structure" (email correspondence with the author, February 21, 2014). For a detailed account of these model-based discoveries, see for instance H. Judson, (1996), *The Eighth Day of Creation. Makers of the Revolution in Biology*: 63f. and 149f.; J. D. Watson (1968), *The Double Helix: A Personal Account of the Discovery of the Structure of DNA*; T. Hager (1998), *Linus Pauling and the Chemistry of Life*: 47 and 86; S. Schindler (2008), "Model, Theory, and Evidence in the Discovery of the DNA," *British Journal for the Philosophy of Science* 59: 610–58, esp. 636f.; and R. Wendler (2013), *Das Modell zwischen Kunst und Wissenschaft*, Munich: Fink: 23ff. and 98ff.
7. Hans-Jörg Rheinberger coined the term "epistemic thing" to describe the as-yet-unknown, but anticipated, object of scientific discovery that is in the process of being materially defined by means of an experimental setup (Rheinberger 1997).
8. The term "design" in the modern sense dates to the early-nineteenth century and is closely linked to the introduction of industrial production. When tracing the history of designerly practices as embodied epistemologies related to making, it becomes necessary to exchange "design" for "artisanry," "craftsmanship," and "engineering." No fundamental distinction was made in the early modern period between art, artisanry, or craftsmanship and engineering, as they were all considered related fields of invention, experiment, and visualization. They are hence treated as forerunners to "design" in this essay.
9. Contemporary studies dealing with artistic or design research, as well as with designerly practices in the sciences and humanities, have their origin in these debates of the 1930s to 1960s. Instances for current collaborative research projects in Berlin are the DFG Research Training Group "Knowledge in the Arts" at the Berlin University of the Arts (UdK, 2012–2021), and the Cluster of Excellence "Image Knowledge Gestaltung" at Humboldt University (2012–2018), a group which is now called "Matters of Activity" since 2019. See also W. Schäffner, (2010), "The Design Turn. Eine wissenschaftliche Revolution im Geiste der Gestaltung," in ed. C. Mareis, G. Joost, and K. Kimpel (eds.), *Entwerfen—Wissen—Produzieren. Designforschung im Anwendungskontext*, 33–46, Bielefeld: transcript.
10. See esp. Chapter 5, "La rappresentazione dei solidi regolari e semiregolari nel *Libellus*," 145ff.
11. Leon Battista Alberti had been the first to publish a book on perspective (*De pictura* 1435/6), which contained only two construction methods, however, omitting perhaps the most relevant one using ground plan and elevation. Piero in turn applied a strict mathematical methodology to an artistic problem, and thereby essentially founded the subdiscipline of projective geometry. For a detailed discussion of his contributions to the progress of science see L. Olschki, (1919), *Die Literatur der Technik und der angewandten Wissenschaft vom Mittelalter bis zur Renaissance*, vol. 1, *Geschichte der neusprachlichen wissenschaftlichen Literatur* (Heidelberg: Olschki): 138ff.

12. In his treatise, Piero explains how to construct three-dimensional bodies from ground plans and elevations—the very method involved in the perspectival depiction of polyhedra.
13. Leonardo Olschki suspects that Piero may have adopted this practice from decorative stone cutting and the grinding of gemstones (Olschki 1919: 218f).
14. "Che dal vedere avesse inizio el sapere . . . E de li nostri sensi per li savii el vedere più nobile se conclude. Onde non immeritamente ancor da vulgari fia detto l'ochio esser la prima porta per la qual lo intelletto intende e gusta." (Pacioli 1509: 17); see also E. Tiller (1509), "Perochè dal corpo umano ogni mesura con sue denominazioni deriva." Luca Paciolis *De divina proportione* und die mathematische Aneignung des Körpers,' *Kunsttexte.de* 3 (2011): 1–21; and H. Baader, 'Das fünfte Element oder Malerei als achte Kunst. Das Porträt des Mathematikers Fra Luca Pacioli,' in V. von Rosen, K. Krüger, and R. Preimesberger (eds.), *Der stumme Diskurs der Bilder*, (Munich and Berlin: Deutscher Kunstverlag, 2003): 177–203, 194.
15. "Perochè, commo la massima filosofia canta, 'nihil est in intellectu quin prius sit in sensu."' (Pacioli 1509: 64). Pacioli is citing Thomas Aquinas' interpretation of Aristotle. Through John Locke the phrase became the motto of British empiricism.
16. In a poem at the beginning of the *Divina porportione* he writes: "El dolce fructo vago e si dilecto / construise gia i philosophi cercare / causa di noi che pasci l'intelleto" (It is a pleasure to construct this sweet and lovely fruit; as it pleases the mind, also philosophers devote themselves to our affairs.) (Pacioli 1509). Pacioli believed that it was the unbroken, ideal beauty of the regular polyhedra that first inspired the idea of universal principles hidden behind their perfect shapes.
17. Pacioli added *prosectiva* to Seneca's seven liberal arts (grammar, rhetoric, dialectic, music, arithmetic, astronomy, and geometry), advocating its placement next to music, the fourth art, as its visual equivalent (Pacioli 1509: chap. 3; Baader 2003: 178).
18. Piero's *Libellus* may have been written in the Italian vernacular first and later translated. However, only the Latin manuscript survives today (Field 1997b: 252).
19. Their form was unknown even though they were described by Pappus.
20. It is known that already in 1489 Pacioli was commissioned to produce models of polyhedra for the young Guidobaldo da Montefeltro, the dedicatee of Piero's *Libellus*, as well as for clerical patrons. He is said to have made more than sixty models of polyhedra, which became his main interest and occupation (Olschki 1919: 172, 223). The fact that Leonardo made changes to the drawings in the Geneva manuscript, which can be interpreted as changes of viewpoint, suggests that he was working with physical models, which he may have drawn using a window as a drawing instrument (Field 1997b: 262).
21. Field discusses the complex process of "distortion" in greater detail. She comes to the conclusion that Leonardo was the true discoverer of the rhombicuboctahedron, as Pacioli apparently was not aware of the need for this extra step in construction.
22. The young man in the background is presumably Pacioli's patron Guidobaldo da Montefeltro, who had been tutored by the mathematician as a child and would have been twenty-three years old when the painting was made. Some authors, including Leonardo Olschki (1919) and Hannah Baader (2003), have surmised that it showed the painter, Jacopo de' Barbari, himself, underpinning the relevance of artistic knowledge to

mathematics and perspective. However, it is doubtful that the painter would have been in a social position to represent himself wearing fur and gloves, and raised above the portrayed mathematician, even though this would make the point.

23. In 1506, Dürer wrote in a letter to Willibald Pirckheimer that he would be leaving Venice for Bologna to stay with someone who would introduce him to the secret laws of perspective. Many authors have discussed the controversial identity of Dürer's teacher in Bologna (Casotti 2005: 187–98; Kemp 1990: 55). Be this as it may, it is probable that Dürer knew Pacioli's *Divina proportione*, and hence the results of Piero's work, as almost all the information he gives on polyhedra could be derived from the treatise (Field 1997b: 266).
24. Field even assumes that Dürer knew all thirteen Archimedean solids, as the ones that are missing could easily have been obtained by this technique.
25. For a discussion of the relevance of artisanal tilings for the history of geometric research see K. Lee Chichester (2021).
26. The advantage of this technique is that it allows for the figure to be drawn without readjusting the compass. The construction of polygons was of practical relevance to craftsmen and architects of the time as it was needed for fortifications (Olschki 1919: 426f., 432).
27. This idea was originally stated in Plato, *Timaeus*, 31b–36d. In the Christian tradition it is encapsulated by the statement: "You, however, ordered all things by measure, number, and weight." *Book of Wisdom*, 11, 20. See also M. Folkerts, E. Knobloch, and K. Reich (eds.), (1989) *Maß, Zahl und Gewicht. Mathematik als Schlüssel zu Weltverständnis und Weltbeherrschung*, ex. cat. (Weinheim: VCH); and F. Ohly (1982), "Deus Geometra. Skizzen zu einer Geschichte der Vorstellung von Gott," in N. Kamp and J. Wollasch (eds.), *Tradition als historische Kraft*, (Berlin: De Gruyter): 1–42, 6–14.
28. Kepler used the formula "Umbkreis einer Eierlinien" (circumference of an egg-line) in his *Auszug auß der Uralten Messe-Kunst Archimedis* (1616, §7), after having used the Latin form *ovalis* with reference to Dürer in the *Astronomia nova* (1609, vol. 4, chap. 46): "Sequitur jam, viam hunc . . . vere esse ovalem, non elliptiam, cum mechanici nomen ab ovo ex abusu collocant (Durerus)," as well as in a letter to Fabricius: "Jam igitur hoc habeo, Fabrici; viam planetae verissimam esse ellipsin (quam Durerus itidem ovalem dixit), aut certe insensibili aliquo ab ellipsi differentem" (Pfeiffer 2000: 86, n. 28; Whiteside 1974: 19, n. 26a).
29. " . . . geprawch vnd verstand mus pey einander seyn" (Dürer 1969: 272; Gluch 2009: 105–14).
30. "Dann warhafftig steckt die kunst in der natur; wer sie herauß kann reyssen, der hat sie." Dürer plays with the multiple meanings of *reißen*, connoting drawing as well as tearing (Dürer 1528: 4:198). On the materiality of the concept of *reißen* and embodied practices of precision, see K. Lee Chichester, 'Von Tupfen, Rissen und Fäden. Präzision als verkörperte Praxis in der Frühen Neuzeit,' *Bilder der Präzision. Praktiken der Verfeinerung in Technik, Kunst und Wissenschaft*, ed. Matthias Bruhn and Sara Hillnhütter (Berlin: De Gruyter, 2018), 137–52.
31. "Auff das die unsichtig Lini/ durch den geraden ryß im gemut verstanden werd/ Dann durch solche weyß muß der innerlich verstand im eussern wreck angetzeigt werden" (Dürer 1525: 41).

32. This idea was explicated in the twentieth century by Cyril Stanley Smith (1970).
33. ". . . ex suo fine defineretur mathematica ars" (Ramus 1569a: 108; Hooykaas 1958: 24, 21f.).
34. ". . . il falloit mettre la main à l'œuvre, e . . . bien labourer" (Ramus 1567; Hooykaas 1958: 21.).
35. ". . . cette fin apparait avec beaucoup plus de splendeur dans l'usage et dans les œuvres géométriques que dans les préceptes et les règles, comme on voit les astronomes, les géographes, les géodètes, les navigateurs, les méchaniciens, les architectes, les peintres, les statuaires faire usage de rien d'autre que de la géométrie dans la description et le mesurage des astres, des pays, de la terre, des machines, des mers, des édifices, des tableaux . . ." (Ramus 1569b: 1; Hooykaas 1958: 25).
36. Edgar Zilsel's essay "Problems of Empiricism" was originally published in 1941 in the *International Encyclopedia of Unified Science*, edited by Otto Neurath, Rudolf Carnap, and Charles Morris. The essay was reprinted in Diederick Raven and Wolfgang Krohn (eds.), *The Social Origins of Modern Science* (Dordrecht: Kluwer, 2000), 171–99. Zilsel further expanded on his ideas in his essay "The Sociological Roots of Science," which originally appeared in the *American Journal of Sociology* 47 (1942): 1–32, and is also reprinted in Raven and Krohn, *The Social Origins of Modern Science*, 7–21. Zilsel's first paper on the topic, "The Social Roots of Science," was presented at the 5th International Congress for the Unity of Science held at Harvard University on September 3–9, 1939, reprinted in Raven and Krohn, *The Social Origins of Modern Science*, 3–6. Page numbers are given for the reprints.
37. See esp. chap. 1, "Artisan/Practitioners as an Issue in the History of Science," 10–29.
38. "On the whole, the rise of the methods of the manual workers to the ranks of academically trained scholars at the end of the 16th century is the decisive event in the genesis of science" (Zilsel 1942; Raven & Krohn 2000: 17, see also 14f; Zilsel 1939; Raven & Krohn 2000: 5).
39. A revised version of this essay was printed in *The Renaissance: Six Essays* (1962). Panofsky refers to Zilsel's "The Origins of Gilbert's Method" (1941b) and "The Genesis of the Concept of Scientific Progress" (1945); as well as to Olschki (1919).
40. See also pp. 145ff. Fritz Saxl argues similarly (1957).
41. "Unter 'symbolischer Prägnanz' soll also die Art verstanden werden in der ein Wahrnehmungserlebnis, als 'sinnliches' Erlebnis, zugleich einen bestimmten nicht-anschaulichen 'Sinn' in sich faßt und ihn zur unmittelbaren konkreten Darstellung bringt" (Cassirer 1954: 235; Cassirer 1957: 202). I have changed "meaning" to "sense" in the English translation to emphasize the etymological relationship between "sensory" and "sense" as in *Sinn* and *sinnlich*.
42. Decisively, Michael Polanyi did not restrict personal knowledge to empirical science, but claimed its equal importance for theory (Polanyi 1952; Polanyi 1958). His book on *Personal Knowledge* (1958) is based on the Gifford Lectures Polanyi delivered at the University of Aberdeen in 1951–2. Kepler's polyhedral model of the cosmos is one of the examples Polanyi gives for faith-based research inspired by beauty.

43. Although Polanyi polemicized against Boris Hessen's utilitarian call for applied as opposed to pure sciences, he did develop his ideas against the backdrop of the latter's accentuation of practitioners' relevance for the Scientific Revolution. Citation from an unpublished manuscript of Polanyi's lecture as delivered in Aberdeen, November 1952, "Skill and Connoisseurship," p. 3, Michael Polanyi Papers, University of Chicago. I thank Rebekka Ladewig for this reference.
44. One of the first considerations of tacit knowledge in modern science was given by Patrick M. S. Blackett in "The Craft of Experimental Physics" (1933).
45. I thank Matthias Bruhn for drawing my attention to modern science's Freudian slip in the use of the word "laboratory."
46. As Felix Klein wrote in his *Erlanger Programm*, "the task is to grasp the spatial figures in their full figurative reality [*gestaltliche Wirklichkeit*], and (which is the mathematical side) to understand the relations valid for them as obvious consequences of the principles of spatial intuition [*Anschauung*]. For this geometry, a model—be it realized and observed or only vividly imagined—is not a means to an end but the thing itself" (Klein 1872; Wussing 1974: 75). See also Klaus Thomas Volkert, *Die Krise der Anschauung. Eine Studie zu formalen und heuristischen Verfahren in der Mathematik seit 1850* (Göttingen: Vandenhoeck & Ruprecht Gm, 1986); Herbert Mertens, 'Mathematical Models,' in *Models. The Third Dimension of Science*, ed. Soraya de Chadarevian and Nick Hopwood (Stanford: Stanford University Press, 2004), 276–306; and Eberhard H.-A. Gerbracht, "Wie die Gottheit erschaubar wird—die Figuren Felix Kleins," in *Mathematische Formeln.* Bildwelten des Wissens 7,2, ed. Wladimir Velminski and Gabriele Werner (Berlin: Akademie Verl., 2010), 90–107.
47. In the mind's eye, according to Baldus, the confusing simultaneity of a complex figure's properties can be resolved into a manageable sequence of individual aspects, as relevant for a specific research question (Baldus 1921: 7ff).
48. A dissertation on D'Arcy Thompson, organicist biology, and the design reform movement is in preparation by the author.
49. Letter from James Hopwood Jeans to D'Arcy W. Thompson of October 28, 1924, St. Andrews University Library, Special Collections, D'Arcy W. Thompson Manuscripts Collection, ms42744(2).
50. Thompson is known to have used soap bubbles and rubber bands for his experiments, but he also referred to pottery and glass-blowing to explain natural form processes (Thompson 1917; Jarron 2015: 160–71).
51. This also holds true for his contemporaries as shown by Deborah Coen for the biologists of the Vienna Vivarium (2006).
52. The lecture was part of the famous 1960 BBC Reith lecture series "Art and Anarchy." Wind was the first to introduce Charles S. Peirce's concept of *embodiment* to art history to denote the unity of sensible matter and spiritual force.
53. "Even the most agile imagination is not at its best when it apes a science from which it is in fact cut off" (Wind 1979: 59).

This essay was written in 2018, based on my master's thesis.

References

Amelung, K. M. (2018), "Illustratio(nen)," In M. Lauschke and P. Schneider (eds.), *23 Manifeste zu Bildakt und Verkörperung*, 97–103, Berlin: De Gruyter.

Amelung, K. M. (2019), "Illustration: On the Epistemic Potential of Active Imagination in Science," in A. Male (ed.), *A Companion Guide to Illustration: Art and Theory,* 330–53, New Jersey: Wiley Blackwell.

Baader, H. (2003), "Das fünfte Element oder Malerei als achte Kunst. Das Porträt des Mathematikers Fra Luca Pacioli," in V. von Rosen, K.Krüger, and R. Preimesberger (eds.), *Der stumme Diskurs der Bilder*, 177–203. Munich and Berlin: Deutscher Kunstverlag.

Baldus, R. (1920), "Mathematik und räumliche Anschauung. Inaugural lecture held at the University of Karlsruhe," in *Jahresbericht der deutschen Mathematiker-Vereinigung*, vol. 30, edited by A. Gutzmer. Leipzig: B.G. Teibner, 1921.

Blackett, P. M. S. (1933), "The Craft of Experimental Physics," In H. Wright (ed.), *University Studies, Cambridge 1933*, 76–96, London: Ivor Nicolson and Watson.

Bredekamp, H. (1995), *The Lure of Antiquity and the Cult of the Machine: The Kunstkammer and the Evolution of Nature, Art, and Technology*. Princeton: Wiener.

Bredekamp, H. (2019), *Galileo's Thinking Hand: Mannerism, Anti-Mannerism, and the Virtue of Drawing in the Foundation of Early Modern Science*. Berlin: De Gruyter.

Casotti, M. W. (2005), "Un episodio controverso del soggiorno di Dürer a Venezia: Il viaggio a Bologna." *Arte Veneta* 61: 187–98.

Cassirer, E. (1954), *Philosophie der symbolischen Formen.* 1923–1929. Vol. 3. Berlin: Bruno Cassirer. Translated as *Philosophy of Symbolic Forms*. New Haven: Yale University Press, 1957.

Chadarevian, S. de. (2002), *Designs for Life: Molecular Biology after World War II*. Cambridge: Cambridge University Press.

Chichester, K. L. (2018), "Von Tupfen, Rissen und Fäden. Präzision als verkörperte Praxis in der Frühen Neuzeit," in M. Bruhn and S. Hillnhütter (eds.), *Bilder der Präzision. Praktiken der Verfeinerung in Technik, Kunst und Wissenschaft*, 137–52, Berlin: De Gruyter.

Chichester, K. L. "Form als Spur von Kraft und Prozess bei D'Arcy Thompson." Talk held at the workshop "Imaginarien der Kraft: Kunst, Literatur, Wissenschaft," Center for Advanced Study "Imaginaries of Force," Universität Hamburg, Warburghaus (October 2, 2019).

Chichester, K. L. (2021), "'Snowflake Generation' – Die Kristallisierung kosmischer (Un-Ordnung)," in M. Bruhn (ed.), *Kältebilder. Erkenntnis und Ästhetik am Gefrierpunkt*, Berlin: De Gruyter.

Chojecka, E. (1967), "Johann Kepler und die Kunst. Zum Verhältnis von Kunst und Naturwissenschaften in der Spätrenaissance," *Zeitschrift für Kunstgeschichte* 30: 55–72.

Clark, A. and D. J. Chalmers (1998), "The Extended Mind," *Analysis* 58:1: 7–19.

Coen, D. (2006), "Living Precisely in Fin-de-Siècle Vienna," *Journal of the History of Biology* 39: 493–523.

Crick, F. (1988), *What Mad Pursuit: A Personal View of Scientific Discovery*, New York: Basic Books.

Cromwell, P. R. (1997), *Polyhedra. One of the Most Charming Chapters of Geometry*, Cambridge: Cambridge University Press.
Daston, L. and P. Galison (2007), *Objectivity*, Cambridge: MIT Press.
Dürer, A. (1525), *Underweysung der Messung mit dem Zirkel und Richtscheyt*, Nuremberg: Formschnyder.
Dürer, A. (1528), *Vier Bücher von menschlicher Proportion*, Nuremberg: Formschnyder.
Dürer, A. (1969), H. Rupprich (ed.), *Schriftlicher Nachlass* 3: *Ästhetischer Exkurs*, Berlin: Dt. Verein für Kunstwiss.
Field, J. V. (1997a), *The Invention of Infinity: Mathematics and Art in the Renaissance*. Oxford: Oxford University Press.
Field, J. V. (1997b), "Rediscovering the Archimedean Polyhedra: Piero della Francesca, Luca Pacioli, Leonardo da Vinci, Albrecht Dürer, Daniele Barbaro, and Johannes Kepler," *Archive for History of Exact Sciences* 50 (3/4): 241–89.
Folkerts, M. E. Knobloch, and K. Reich (eds.), (1989), *Maß, Zahl und Gewicht: Mathematik als Schlüssel zu Weltverständnis und Weltbeherrschung*. Ex. cat. Weinheim: VCH.
Gerbracht, E. H. A. (2010), "Wie die Gottheit erschaubar wird—die Figuren Felix Kleins," in W. Velminski and G. Werner (eds.), *Mathematische Formeln. Bildwelten des Wissens Band* 7-2, 90–107, Berlin: Akademie Verlag.
Gluch, S. (2009), "Geometria practica und ars pictoria theoretica. Albrecht Dürer zwischen Theorie und Praxis," in Ulrich Großmann (ed.) *Buchmalerei der Dürerzeit. Dürer und die Mathematik. Neues aus der Dürerforschung*. Dürer-Forschungen 2, 105–114, Nürnberg: Verl. des Germ. Nationalmuseums.
Hager, T. (1998), *Linus Pauling and the Chemistry of Life*. New York: Oxford University Press.
Hessen, B. (2009), "The Social and Economic Roots of Newton's Principia," reprinted in G. Freudenthal and P. McLaughlin (eds.), *The Social and Economic Roots of the Scientific Revolution: Texts by Boris Hessen and Henryk Grosmann*, 41–101, Berlin: Springer.
Hooykaas, R. (1958), *Humanisme, Science et Réforme. Pierre de la Ramée, 1515–1572*, Leiden: E.J. Brill.
Jarron, M. (2015), "Independent and Individualist," *Art in Dundee 1867–1924*, Dundee & Perth: W&G Baird.
Jeans, J. H. Letter to D'Arcy W. Thompson, October 28, 1924. St. Andrews University Library, Special Collections, D'Arcy W. Thompson Manuscripts Collection, ms42744(2).
Judson, H. (1996), *The Eighth Day of Creation: Makers of the Revolution in Biology*, New York: CSHL Press.
Kemp, M. (1990), *The Science of Art: Optical Themes in Western Art from Brunelleschi to Seurat*. New Haven: Yale University Press.
Kepler, J. *Astronomia nova* (1609; 1937), in *Gesammelte Werke* 3 (14) Munich: C.H. Beck.
Kepler, J. (1617–1621), *Epitome Astronomiae Copernicanae*, 7 vols. Linz: Johannes Plancus.
Kepler, J. (1619), *Harmonices mundi libri V*. Frankfurt/M.: Gottfried Tambach & Johann Planck.
Klein, F. (1872), "Das Erlanger Programm. Vergleichende Betrachtungen über neuere geometrische Forschungen," Edited by Hans Wussing. Leipzig: Akademische Verlagsanstalt, 1974.
Latour, B. (1993), *We Have Never Been Modern*. Cambridge: Harvard University Press.

Latour, B. (2005), *Reassembling the Social: An Introduction to Actor-Network-Theory*. Oxford: Oxford University Press.

Liebelson, D. (2014), "These New 3D Shapes Could Help Us Combat Herpes and Fight Climate Change," *The Week*, http://theweek.com/articles/450389/new-3d-shapes-could-help-combat-herpes-fight-climate-change (accessed on February 9, 2018).

Long, P. (2011), *Artisan/Practitioners and the Rise of the New Sciences, 1400–1600*. Corvallis: Oregon State University Press,

Mackenzie, D. (2014), "Goldberg Variations: New Shapes for Molecular Cages," *Science News*, www.sciencenews.org/article/goldberg-variations-new-shapes-molecular-cages (accessed on February 25, 2018).

Martens, R. (2000), *Kepler's Philosophy and the New Astronomy*, Princeton: Princeton University Press.

Mertens, H. (2004), "Mathematical Models," In S. de Chadarevian and N. Hopwood (eds.) *Models. The Third Dimension of Science*, 276–306, Stanford: Stanford University Press.

Ohly, F. (1982), "Deus Geometra. Skizzen zu einer Geschichte einer Vorstellung von Gott," in N. Kamp and J. Wollasch (eds.), *Tradition als historische Kraft*, 1–42, Berlin: De Gruyter.

Olschki, L. (1919), *Die Literatur der Technik und der angewandten Wissenschaft vom Mittelalter bis zur Renaissance*. Vol. 1. *Geschichte der neusprachlichen wissenschaftlichen Literatur*, Heidelberg: Olschki.

Pacioli, L. (1509), *De divina proportione*, Venice: A. Paganius Paganinus.

Panofsky, E. (1952; 1962), "Artist, Scientist, Genius: Notes on the *Renaissance-Dämmerung*," revised and reprinted in *The Renaissance: Six Essays*, New York: Harper and Row.

Panofsky, E. (1955), *The Life and Art of Albrecht Dürer.* Princeton: Princeton University Press.

Panofsky, E. (1956), "Galileo as a Critic of the Arts: Aesthetic Attitude and Scientific Thought," *Isis* 47: 3–15.

Pappus of Alexandria, *Synagoge* [Collection], book 5.

Pfeiffer, J. (2000), "La creation d'une langue mathémathique allemande par Albrecht Dürer. Les raisons de sa non reception," in *Sciences et langues en Europe. Une conference organisée par le Centre Alexandre Koyré*. Luxembourg: Office for Official Publications of the European Communities.

Polanyi, M. (1958), *Personal Knowledge. Towards a Post-Critical Philosophy*. Chicago: University of Chicago Press.

Polanyi, M. (1952), "Skills and Connoisseurship," *Atti del Congresso di Metodologia*, Torino.

Ramus, P. (1567), *La Remonstrance de Pierre de la Ramée, faite au conseil privé, en la chamber du roy au Louvre, le 18 de janvier 1567, touchant la profession royalle en mathématique*. Paris: André Wechel.

Ramus, P. (1569a), *Scholarum mathematicarum libri unus et tringinta*. Book 1. Basel: Esebius Episcopius,

Ramus, P. (1569b), *Arithmeticae libri II et Geometriae XXVII* . . . Basel: Esebius Episcopius.

Raven, D., R. S. Cohen, and W. Krohn (eds.) (2000), *The Social Origins of Modern Science*. Dordrecht: Kluwer, 171–199.

Razafindrazaka, F. and K. Polthier, (2014), "Realization of Regular Maps of Large Genus," *Mathematics and Visualization* 38 (January): 239–52.

Rheinberger, H-J. (1997), *Toward a History of Epistemic Things: Synthesizing Proteins in the Test Tube*. Stanford: Stanford University Press.

Rotman, B. (1995), "Thinking diagrams: Mathematics, writing and virtual reality," *South Atlantic Quarterly* 94: 389–415.

Saxl, F. (1957), "Science and Art in the Italian Renaissance," in idem. *Lectures*, 111–24, London: Warburg Inst.

Schäffner, W. (2010), "The Design Turn. Eine wissenschaftliche Revolution im Geiste der Gestaltung," in C. Mareis, G. Joost, and K. Kimpel (eds.), *Entwerfen – Wissen – Produzieren. Designforschung im Anwendungskontext*, 33–46, Bielefeld: Transcript.

Schein, S. and J. M. Gayed (2014), "Fourth Class of Convex Equilateral Polyhedron with Polyhedral Symmetry Related to Fullerenes and Viruses," *Proceedings of the National Academy of Sciences* 111 (8): 2920–5.

Schein, S., et al. (2016), "Decoration of the Truncated Tetrahedron—an Archimedean Polyhedron—To Produce a New Class of Convex Equilateral Polyhedra with Tetrahedral Symmetry," *Symmetry* 8 (82): 1–10.

Schein, S. (2014), Email correspondence with the author, February 21, 2014.

Schindler, S. (2008), "Model, Theory, and Evidence in the Discovery of the DNA," *British Journal for the Philosophy of Science* 59: 610–58.

Senechal, M. (2013), *I Died for Beauty: Dorothy Wrinch and the Cultures of Science*. New York: Oxford University Press.

Smith, C. S. (1970), "Art, Technology, and Science: Notes on Their Historical Interaction," *Technology and Culture* 11(4): 493–549.

Smith, P. (2004), *The Body of the Artisan: Art and Experience in the Scientific Revolution*. Chicago: University of Chicago Press.

Sorci, A. (2001), "La forza de le linee," *Prospettiva e stereometria in Piero della* Francesca, Florence: SISMEL-Edizioni del Galluzzo.

Tiller, E. (2011), "Peroché dal corpo umano ogni mesura con sue denominazioni deriva," Luca Paciolis *De divina proportione* (1509) und die mathematische Aneignung des Körpers'. *Kunsttexte.de* 3: 1–21.

Thompson, D. W. (1917), *On Growth and Form*. Cambridge: Cambridge University Press.

Thompson, D. W. (1925), "On the Thirteen Semi-regular Solids of Archimedes," *Proceedings of the Royal Society of London*. Series A, Containing Papers of a Mathematical and Physical Character 107 (742): 181–8.

Volkert, K. T. (1986), *Die Krise der Anschauung: Eine Studie zu formalen und heuristischen Verfahren in der Mathematik seit 1850*. Göttingen: Vandenhoeck & Ruprecht Gm.

Watson, J. (1968), *The Double Helix. A Personal Account of the Discovery of the Structure of DNA*. New York: Atheneum.

Wendler, R. (2013), *Das Modell zwischen Kunst und Wissenschaft.* Munich: Fink.

Wendler, R. (2015), "Thinking with Models. On the Genesis of James Watson's Molecular Biology of the Gene," in H. Bredekamp, V. Dünkel, and B. Schneider (eds.), *The Technical Image: A History of Styles in Scientific Imagery*, 74–80, Chicago: University of Chicago Press.

Werner, G. (2002), *Mathematik im Surrealsimus. Man Ray—Max Ernst—Dorothea Tanning*, Marburg: Jonas Verl.

Whiteside, D. T. (1974), "Keplerian Planetary Eggs, Laid and Unlaid, 1600–1605," *Journal for the History of Astronomy* 5: 1–21, 19 n. 26a.

Wind, E. (1979), "The Fear of Knowledge," in E. Wind and G. Duckworth and Co., Ltd. (eds.), *Art and Anarchy. The Reith Lectures 1960 Revised and Enlarged*, London: Faber & Faber.

Zilsel, E. (1939), "The Social Roots of Science," paper presented at the 5th International Congress for the Unity of Science, Harvard University, 3–6, printed in Raven and Krohn, *The Social Origins of Modern Science*.

Zilsel, E. (1941a), "Problems of Empiricism," in O. Neurath, R. Carnap, and C. Morris, *Foundations of the Unity of Science: Towards an International Encyclopedia of Unified Science* 2 (8): 53–94, Chicago: University of Chicago Press.

Zilsel, E. (1941b), "The Origins of Gilbert's Method," *Journal of the History of Ideas* 2: 1–32.

Zilsel, E. (1942), "The Sociological Roots of Science," *American Journal of Sociology* 47: 1–32.

Zilsel, E. (1945), "The Genesis of the Concept of Scientific Progress," *Journal of the History of Ideas* 6: 325–49.

2 Agile Artifacts—Designing the Incomplete

JAN ECKERT AND DANIEL ECKERT

Introduction

In 1993, design strategist and industrial designer Adam Richardson investigated the dialogue between designers and the people who use their designs. Applying ideas from French literary theorist Roland Barthes' essay "La mort de l'auteur" (The Death of the Author) (Barthes 1968), Richardson characterizes designers as "locked into a controlled, predictable dialog" with users and other stakeholders (1993: 36). This controlled designer–user–stakeholder dialogue, according to Richardson, leads to an identity crisis that he calls "The Death of the Designer" (1993). Richardson argues that "industrial designers . . . have much less control over the process of product development" (1993: 34) than is commonly thought to be true. He continues:

> How users and cultures respond to the products which designers help create is not well understood. Most conventional theories tend to exaggerate the designer's influence over these interactions, and exactly what the designer's responsibilities are toward the culture as a whole must be given closer attention
>
> 1993: 34

Before the introduction of contemporary open-design models, and cooperative-design approaches in Scandinavian countries in the late 1960s and early 1970s, the main objective of designers was to deliver a functioning solution (which includes aesthetic functions) with specific predefined uses or objectives. The user's participation in design was limited to working within this range of specific functions. Nearly thirty years after Richardson's essay, this locked-up dialogue among designers, stakeholders, and users continues to open up. Participatory design (Schuler & Namioka 1993) and co-creation processes (Sanders & Stappers 2008), incorporate non-designers into the design process by involving them in analysis, testing, or evaluating prototypes. Furthermore, new manufacturing technologies and the spread of computer-aided design tools have led to a democratization of design practices and an increase in design literacy among non-designers. Contemporary approaches such as open design (Avital 2011) or the maker movement (Anderson 2012) attract communities of people who design and then share their designs and "blueprints" widely. Consequently, one might ask whether Richardson's death of the designer has indeed come to pass, or whether, thanks to the openness of some contemporary design processes, designers have begun to leave room for incompleteness in artifacts and the involvement of others in their making and use.

This essay explores ideas about incompleteness in artifacts through a collaborative project carried out by the two authors, designer Jan Eckert and molecular biologist Daniel Eckert. Both of us work on artifacts that are characterized by various levels of incompleteness, which open up areas of interaction, interpretation, and mutation. We decided, therefore, to investigate the phenomenon of incompleteness in artifacts from a design perspective, using the concept of *Handlungsspielraum* (Eckert 2012), as well as from a molecular biology perspective, using genetically engineered knockout cell lines as objects of investigation.

Handlungsspielraum (German: *Handlung* = action; *Spiel* = play/game; *Raum* = space/range or latitude), a concept that comes out of Jan Eckert's doctoral studies in design science, considers what happens when certain models of design (e.g., open design), and the artifacts they employ, deliberately incorporate gaps or conceptual spaces. *Handlungsspielraum* demonstrates that, from a user's point of view, these gaps can become a "stage" for future interpretations and modifications, and for the development of both the design process and the "final" artifact itself.

Daniel Eckert's work on genetically modified knockout cell lines—cells that deliberately carry a defect (e.g., a gap) in their genetic code—likewise deals with deliberate gaps. Scientists use the genetic defects in knockout cells to investigate the role that a missing gene plays in an organism. By providing precisely designed genetic gaps, knockout cell lines serve as a sort of platform or "stage" for their designers' experiments.

The aim of the comparison of *Handlungsspielraum* and knockout cell lines is to see how the incompleteness that is part of both approaches opens up "dialogue between designers and users" (Richardson 1993). We begin our discussion in this essay by reviewing research on incompleteness in design processes and artifacts. Our investigation unearthed three different types of artifacts: products, prototypes, and triggers. These three types of artifacts are used in the design process in different ways. In the following sections, we discuss how the three types of artifacts give rise to distinct sorts of dialogue among designers, artifacts, and their users.

We then consider how different episodes of *Handlungsspielraum* provide various levels of incompleteness, and compare *Handlungsspielraum* episodes with the gene-editing process of knockout cell lines. Our research revealed a new variety of *Handlungsspielraum* episode, one that combines both constraints and openness for both makers and users. This episode provides stages for interpretation and mutation, but it also incorporates opportunities for further development on the part of users—that is, a stage for user experimentation. This specific kind of incompleteness encourages an open approach in science and design, and produces what we call an agile artifact, an artifact that "simultaneously embraces [openness in] both process and outcome" (Garud et al. 2008). The aim of our investigation is to gain a better understanding of how, and under which circumstances, agile artifacts can be more successful than completed ones.

Products vs. Prototypes vs. Triggers

In "Different models of Designing and Consuming: Objects, Practices and Processes" (2007), design engineer Jack Ingram, sociologist Elizabeth Shove, and human geographer Matthew Watson divide design processes into three different models. Using their research, we subdivide the artifacts we consider in our investigation into three types (Fig. 2.1): first, (A), the design of products that are meant to be solely consumed products, second, (B), the design of products that open up new design opportunities through their consumption prototypes, and third, (C), the design of products that are purposely designed to open up new opportunity triggers. In the following sections, we elaborate the differences among products (A), prototypes (B), and triggers (C), and describe their relationship to *Handlungsspielraum* and knockout cell lines (Fig. 2.1).

Type A—Products: Artifacts and their Further Use and Interpretation

Type A artifacts, or products, are the most common types of artifacts that we encounter; for example, our kitchenware or a tablet computer. In Type A artifacts, the design process is finalized before they are available to users. Options for modifying these artifacts are mostly non-existent. In some cases, designers of Type A artifacts even try to extend control over the final design beyond the design process. For instance, ten years ago Apple laptops were more flexible; users were able to make upgrades (e.g., RAM or battery) themselves. Today, it isn't even possible for users to open their computers without special tools.

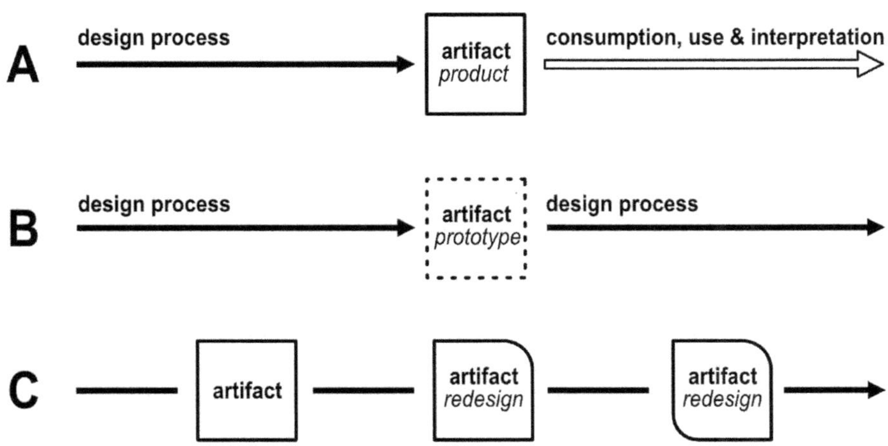

Fig. 2.1 Products vs. prototypes vs triggers: Artifacts that carry different properties and functions in the design process. Eckert & Eckert, 2018.

But, research shows that users still attempt to adjust “finalized” designs. Both literary critic and semiotician Umberto Eco and literary theorist Roland Barthes argue that the authorship of an artwork or text shifts when the beholder reads, perceives, performs, or interprets them (Eco 1962; Barthes 1968). The character of these shifts is dependent upon situation, context, and audience. In design these shifts are executed by users, even in Type A artifacts.

In the 1970s, psychologist James J. Gibson investigated these kinds of shifts as part of his concept of affordances (Gibson 1977), an investigation that was carried on by cognitive scientist Donald Norman (1988, 2007) in the 1980s. Affordances represent transactions that occur between individuals and their environments, including artifacts. Both Gibson and Norman contend that artifacts invite a range of possibilities to act. As political scientist Herbert Simon explains, “an artifact can be thought of as a meeting point—an ‘interface’ in today’s terms—between an ‘inner’ environment, the substance and organization of the artifact itself, and an ‘outer’ environment, the surroundings in which it operates” (Simon 1981: 9). In *Design as Interface*, design critic Gui Bonsiepe (1999, 2009) characterizes Simon’s “interface” (German: *Interaktionsraum*) as a conceptual space where users interact with an artifact.

Brandes’ ideas in “Non-Intentional Design” (2008) and “Design by Use” (2009) extend the notion that artifacts remain “open” even after the end of the design process. According to Brandes, unintended interactions with artifacts—such as changing the original function for entertainment—are a kind of follow-up design that happens after the designer finalizes and leaves the initial design process. Clearly, even Type A artifacts are not fully finalized and closed.

Type B—Prototypes: Artifacts as Part of the Design Process

In this section we discuss Type B artifacts, which can be described as open prototypes. Prototypes are crucial to a wide range of analytic, planning, or thinking processes in design. In the late 1920s, economic historian Abbott Payson Usher wrote that working with incomplete patterns in prototypes can precipitate insights and innovations during the design process (1929). Fifty years later, Tom Markus (1969) and Tom Maver (1970) argued that sketch plans and working drawings function in the same way, giving rise to novel ideas and outcomes (RIBA Royal Institute of British Architects—Plan of Work).

Architects Bill Hillier, John Musgrove, and Pat O’Sullivan, in fact, coined a term for incomplete and open prototypes: primary generators (1972). A primary generator is a “single unifying concept, idea, analogy, or configuration” that “aids the designer in creating the concept, [but] not the mechanism [or the design] of the solution” (Taylor 2011). Architect Jane Darke interviewed other British architects to uncover the role that primary generators played in their design processes (1978). Darke’s professor, architect Bryan R. Lawson, contended that developing rough designs and prototypes, helps to reveal the problem focus during a design process: “first decide what you think might be an important aspect of the problem, develop a crude design on this basis, and then examine it to see what else you can discover about the problem" (1990: 45). Finally, Call Briggs and Spencer W. Havlick (1976) used Hillier and Darke's definition of primary generators to claim that a hypothesis is itself the central artifact of a problem-solving process like design.

In the early 2000s, innovation became a major buzzword in design research processes. This new interest in originality and ingenuity in design further associated prototypes with innovation. Designers Clement Mok and Keith Yamashita, in fact, describe prototypes as central elements of innovation in their article "AIGA Process of Designing Solutions" (2003), in which they maintain that rapid prototyping speeds up decision making in innovation. Psychologist Willemien Visser (2006) built on these ideas about prototypes and innovation. He asserts that cognitive artifacts, which are physical objects such as prototypes that enhance cognition, allow preconceptual ideas such as hypotheses or scenarios to facilitate designing in complex stakeholder ecosystems. Designers commonly use prototypes in their design processes. So, designers are used to dealing with Type B artifacts as prototypes, which are characterized by their capacity to serve as open and incomplete drafts of future artifacts that have yet to be negotiated.

Type C—Triggers: Artifacts and their Further Mutation in Time

While Type A products and Type B prototypes are well documented in design research, Type C artifacts that function as triggers—those that have the potential to induce an ongoing cycle of interpretation and mutation during design—are less well represented. Although some ideas discussed in Brandes' "Non-Intentional Design" (2008) and "Design by Use" (2009) include re-creations and new creations based upon existing artifacts, there has been little discussion about which properties of artifacts trigger the act of re-designing.

In his famous metaphor of the "cathedral and the bazaar," though, software developer Eric Raymond (1999) considers exactly these properties. He compares a metaphorical closed "cathedral" to a metaphorical "bazaar," which he describes as an environment that represents both openness and negotiation. Raymond uses this metaphorical comparison in his characterization of open-source software, an agile, purposely open platform. In the early 2000s, Raymond's ideas about open and negotiated processes that intentionally aim for innovative outcomes began to interest designers. Designers began investigating openness as a concept for design that involves a collection of participants—including users and other non-designer stakeholders—throughout the creation process. This re-thinking of the design process promoted agility and openness in both the design process and in design outcomes.

Involving various stakeholders at all stages of a design process promotes incompleteness. In "Incomplete by Design and Designing for Incompleteness," Raghu Garud, Sanjay Jain, and Philipp Tuertscher conclude that "incompleteness allows participants to engage with design in a way that is meaningful to them" (2008). This statement also connects incompleteness to the acceptance and success of incomplete designs. Because an incomplete design provides space for designers to collectively negotiate its final form, the design is better accepted by its users. Together with Garud and Tuertscher, business professors Jel Gehman and Arun Kumaraswamy conclude that the incompleteness of the artifact sets the stage for a process that goes beyond the artifact itself (Garud et al. 2017). Incomplete artifacts, they contend, act as triggers that facilitate subsequent designs and design processes.

The notion of artifacts as triggers also relates to sociologist Susan Leigh Star and philosopher James R. Griesemer's concept of boundary objects. These researchers investigated the role that shared materials—for example, specimens, field notes, and maps of species' territories—play in traversing the viewpoints of multiple stakeholders—such as amateur collectors and museum professionals—at Berkeley's Museum of Vertebrate Zoology. They named these items boundary objects, which they describe as

> Plastic enough to adapt to local needs and the constraints of the parties employing them, yet robust enough to maintain a common identity across sites. They are weakly structured in common use, and become strongly structured in individual-site use. These objects may be abstract or concrete. They have different meanings in different social worlds, but their structure is common enough to make them recognizable, a means of translation.
>
> Star, Griesemer 1989

Boundary objects "allow coordination without consensus" according to Professor of Management and Organizations Beth Bechky (2003). The concept of boundary objects demonstrates how closely use and further development are tied up with the artifact itself. Bo L. T. Hedberg, Paul C. Nystrom, and William H. Starbuck, however, looked beyond the artifact to the entire design process, that is, to Type C artifacts. They describe Type C artifacts as triggers: "Designs can themselves be conceived as processes—as generators of dynamic sequences of solutions, in which attempted solutions induce new solutions and attempted designs trigger new designs" (1976).

The notion of Type C artifacts as triggers has a lot in common with Type B artifacts as prototypes. But there is a fundamental difference: while prototypes are always followed by a finalized design, Type C artifacts as triggers embrace incompleteness throughout the design process, which triggers cycles of further development. Type C artifacts as triggers can also be understood as actors in future change and development in Actor-Network Theory (ANT), in which people (and other living organisms), objects, ideas, and processes together create social situations.

Building on this notion of artifacts as triggers for mutation and re-design, in the following sections of this essay we explore the model of *Handlungsspielraum* (Eckert 2012) and compare it to Daniel Eckert's work on knockout cell lines. We find that both

Handlungsspielraum and knockout cell lines are characterized by intentional incompleteness as the primary quality of an artifact that leads to continuous updating of the artifact itself.

Handlungsspielraum—a Concept of incompleteness

Handlungsspielraum (Eckert 2012) involves artifacts that are deliberately incomplete in order to open up conceptual space for interpretations or mutations of the artifact. *Handlungsspielraum* thus relates most closely to Type C trigger artifacts. While in English the word *Handlungsspielraum* could be best translated by latitude or wiggle room, we prefer the German word (German: *Handlung* = action; *Spiel* = play/game; *Raum* = space/range). In particular, the double meaning of *Spiel*/play is significant to the concept, since it can mean to play a game, but also describes a little space or gap that allows a mechanism to move (e.g., a bolt has a little play). Indeed, the qualities of incompleteness of artifacts and playful interactions connect *Handlungsspielraum* to both Eco's discussion of the open (art)work – or "Opera Aperta" (1962) and to Gibson and Norman's concept of affordances discussed earlier.

Affordances, as we have already noted, offer ways for those involved "to access a range of possibilities to act" with an artifact (Bonsiepe 2009). Bonsiepe's interface, or *interaktionsraum* (interaction space), is a space in which the users' actions unfold in conjunction with the artifact. The actions in Bonsiepe's model and the concept of affordances mainly refer to information that is provided by the artifact itself. *Handlungsspielraum* further extends the interaction space by considering all contextual information, such as the reason that the artifact is being used, where and when the artifact gets used, which constraints might come with these circumstances, and the cultural or social frameworks that are part of the interaction. In addition, *Handlungsspielraum* encompasses users' deliberate interpretation and mutation of artifacts. Besides intended uses and interactions, therefore, *Handlungsspielraum* includes unintended ones (Brandes 2008; Brandes et al. 2009) and extends them with playful interactions that lead to modifications of the artifact's physical properties as well as its meaning and context. An example of *Handlungsspielraum* are "life hacks" that are shared on the internet: most of them are based upon very ordinary artifacts that are used in new ways in a specific contexts. Some of these hacks are useful, others are playful, and some are even silly. These properties connect *Handlungsspielraum* to Eco's description of the open (art)work that is revived by the beholder's interaction with the work.[1]

In this essay, we use a simplified model of *Handlungsspielraum* that focuses on two episodes (Fig. 2.2). In the first one, which is called a negotiating episode, a rough artifact (similar to the Type B prototype) has a low level of constraints on its use. This situation allows the artifact to function in a variety of contexts, which still allows it to be negotiated by the user. Contemporary furniture, for instance, may be designed in a manner that allows clients to customize according to their needs, and according to

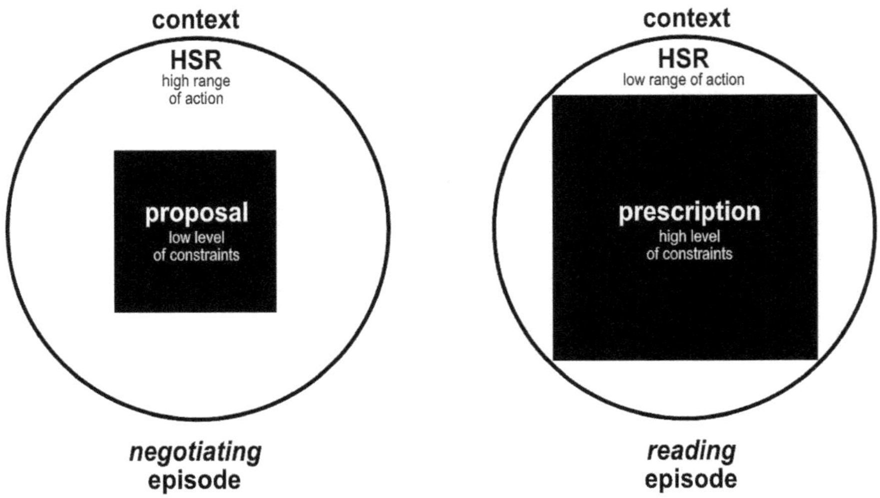

Fig. 2.2 Negotiating and reading episodes illustrating low-defined artifacts with a high range of action, and highly defined ones with a low range of action. Eckert & Eckert, 2018.

the context in which it is used. Websites for ordering customizable furniture often provide modifiable 3D images: a prototype of the real piece of furniture. The user can modify and re-modify this 3D prototype in a process of negotiation.

The second episode, a reading episode, on the other hand, offers a well-defined artifact with constraints, a finished product like the Type A artifact. The number of contexts into which such an artifact may fit is limited. Its *Handlungsspielraum*, therefore, provides a smaller range of possible actions. This kind of artifact is less agile—like a piece of furniture that needs to be assembled following precise instructions, or a musical score that must be read and performed by a musician without interpretation. In both examples the instructions or the score represent a high level of constraints and, therefore, the range of possible interpretations or actions is limited.

Development of open-source software, which was mentioned earlier, the shared development of 3D-manufactured artifacts in online platforms, and the collaborative environments in fab labs are examples of design that function as negotiating episodes. Crowd-funded designs likewise provide future users the opportunity to back projects during their development, and also include users as early adopters who test designs and feed back into the future design process. On the other hand, designer-driven industrial design mostly includes artifacts that fit into the reading episode. A classic "jump and run" video game is typical of a reading episode. Players may choose between a variety of actions such as jumping and moving, but there are certain parameters over which players have no influence (e.g., a constant timeline of the game, or the fact that the player might only move in one direction); these parameters act as constraints on players' behavior.

Most design artifacts fit into one of the A, B, C types discussed previously—products, prototypes, or triggers—and also into one of these two *Handlungsspielraum* episodes—negotiating or reading.

In the next section we explore both the design and function processes of so-called knockout cell lines in the context of products, prototypes, or triggers, and negotiating or reading *Handlungsspielraum* episodes. Knockout cell lines are cell populations that are deliberately designed to be incomplete and serve as platforms for future cell designs. In a fashion similar to the negotiating episode of *Handlungsspielraum*, these cell lines incorporate a genetic gap that is deliberately designed by biologists to enable researchers to conduct experiments and produce mutations in the cells.

Knockout Cell Lines—Perfectly Incomplete Artifacts

Knockout cell lines are designed artifacts. They are specifically created cell populations in which a targeted gene and its function have been purposely switched off or knocked out. Genes are part of codes that carry out specific functions in the cell (or the body). If a gene is knocked out, its accorded cell function will not take place. One of the possible applications of knockout cell lines, then, is to investigate the role of the knocked-out gene and its interaction in cellular mechanisms. This information could be used to produce new health products, such as pharmaceutical drugs or therapies, which target the investigated gene's function to block or enhance the gene's function (for example, protein expression). Knockout cell lines can also provide cell growth assays, demonstrating, for example, that the knocked-out gene plays a crucial role in tumor development. In a case where a gene produces a protein that suppresses tumor growth, a knockout of the gene would not produce that protein. The knockout would therefore allow tumor development as compared to the healthy control group.

Creating knockout cell lines is limited by several factors (Fig. 2.3) that—in a similar fashion to the reading episode of *Handlungsspielraum*—act as constraints when designing or using these cell lines (Fig. 2.3). At one end of the spectrum there is the healthy cell, which doesn't offer the openness of knockout cells, and at the other end is the limitation called gene essentiality (Blomen et al. 2015). Gene essentiality means that the targeted gene is essential for cell survival, and a knockout of this gene would lead to cell death. These constraints limit the design of knockout cell lines, just as *Handlungsspielraum* can be limited by the predefined function of an artifact, the context its functionality requires, or even ethical or cultural constraints that check the usability of an artifact.

In addition to the limitations generated by healthy cells and gene essentiality, designing knockout cell lines has two other constraints. First, healthy human cells have one copy of each gene on each of its two chromosomes, which means there are two copies of each gene in a cell. This constraint means that both copies of the gene must be knocked out to create a knocked-out gene. Even when a researcher "hits" the gene on one chromosome, the gap produced might be masked by the remaining second copy of this gene on the other chromosome. To overcome this limitation, scientists generate a haploid parental cell line, that is, a cell line that carries only one chromosome of each of the twenty-three pairs (Essletzbichler et al. 2014). This haploid cell line is the perfect starting material for knockout cell lines.

Another constraint when designing knockout cell lines are so-called off-targets. Off-targets are unintended genetic modifications that happen as a result of using a short molecule of guide RNA (ribonucleic acid) in the knockout process. DNA and RNA are made up of sequences of four nucleotide bases—cytosine (c), guanine (g), Adenine (A) and thymine (T) or Uracil (U) in RNA—that line up in a specific order along a strand of sugar/phosphate that functions as a backbone. The order of these bases determines which protein the gene produces.

Beyond creating a pattern of bases along a sugar/phosphate strand, the bases can pair with other bases to create a second backbone parallel to the first one. Cytosine bonds with guanine, and adenine bonds with thymine or uracil to create a kind of ladder structure with the bases as the rungs. The bases of guide RNA line up along the bases of the DNA strand to produce a complementary "fit" to the DNA sequence of the target gene to be knocked out.[2]

The guide RNA sequence of bases identifies the particular sequence of bases in the DNA to be cut in a knockout cell. Since there are only four possible base pairs—G-G, G-C, A-U, U-A—the pattern of bases in a guide RNA sequence is likely to be found in different genes elsewhere in the DNA. The guide RNA that is used to knockout a gene, therefore, has to be as specific as possible to make sure that no other genes (off-targets) get hit and knocked out (Fu et al. 2013).

The Cell Design Process

Designing a knockout cell line consists of multiple carefully planned and executed steps, some of which relate to *Handlungsspielraum* episodes. In the next section we will focus on key steps in the cell design process (Fig. 2.3): the creation of an artifact prototype, the selection of cells for prototyped cell lines, and upscaling the incomplete cell lines to what is called knockout cell lines.

Vectors—Creating Knockout Prototypes

Considering the constraints posed by healthy cells and gene essentiality, and of off-targets as well, an elegant way to introduce a specific mutation into a gene is to deliver an already mutated DNA template into the cell. In this first step, a defective version of the target gene—a circular piece of DNA called a vector—that carries the altered version of the target gene is designed and produced. The process of developing knockout cell lines is strongly tied to the biochemical CRISPR/Cas-method. CRISPR/Cas (Clustered Regularly Interspaced Short Palindromic Repeats/CRISPR-associated) is a protein complex that cuts DNA at very precise positions in order to edit the disrupted site of the DNA. Using CRISPR/Cas protein complex and the specific guide RNA, the vector is transferred to the healthy, parental so-called

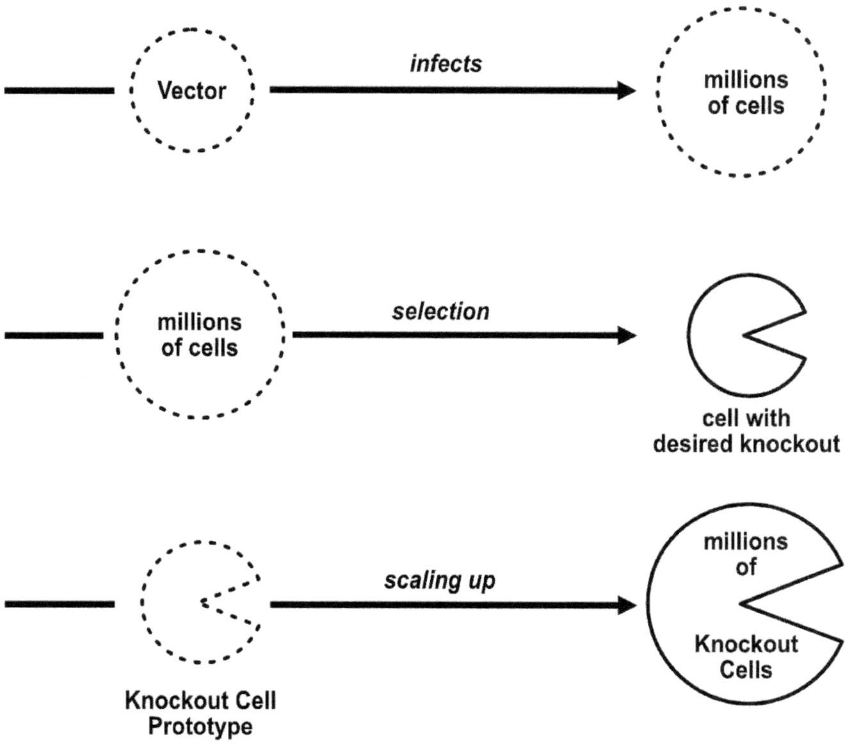

Fig. 2.3 Key steps in the cell design process. Eckert & Eckert, 2018.

wild-type cells. In these "infected" cells, CRISPR/Cas cuts and breaks the DNA at the target site. As discussed above, prototypes are a core element in design processes; in knockout cell lines, the vector functions as the prototyped version of the desired knockout in the cell design process. If the designer/scientist determines that the vector works properly, they allow the cell to incorporate the vector's mutated gene into its own mechanism. The cell uses the vector as a template to rebuild the disrupted piece of DNA. The new piece of DNA will then be passed down to all of the cell's descendants, which multiplies the prototypes for the knockout cell line.

Selecting Knockout Cells

The first step of the selection process for cells separates edited from non-edited cells within the cell population. In the next step, this population is limited to single isolated cells, so-called clones, from which colonies are grown. This process ensures that each colony consists of genetically identical clones with exactly the same mutation.[3] The aim of this selection phase is to narrow down the population of prototyped cells to one single cell sub-population of the desired knockout.[4] The selection and cloning of relevant cells corresponds to evaluating different prototypes in design or negotiating properties of a design artifact, as in the negotiating episode in *Handlungsspielraum*.

Upscaling, Quality Control, and Distributing Knockout Cells

In this step, colonies of identical clones are selected for those that carry the mutated version of the target gene without any additional, random mutations that could lead to off-target effects. These clones are then grown to create populations of millions of cells in a stable knockout cell line (Fig. 2.3). The knockout cell line, the corresponding healthy wild-type cell line, and the entire cell line's documentation—e.g., the used guide RNA and the new, mutated version of the target gene's DNA sequence—are all sent to the client. In design, we would call this the rollout of a product or service whose design process has been finalized. As in the reading episode discussed earlier, their further use mostly is constrained by specific parameters or design choices.

Interestingly, though, this is where knockout cell lines are different from finished design artifacts: even after elaborate selection and multiplication, the final cell lines still carry a carefully designed gap. This particular observation about the cell-design process revealed a new property that we decided to call perfectness of incompleteness. By perfectness of incompleteness, we mean that an artifact is precisely designed, completed, and yet it is open to mutation at the same time. Understood in the context of types of artifacts or *Handlungsspielraum* episodes, this kind of incompleteness suggests a new type of artifact or design episode, one that embraces perfectness, incompleteness, and the artifact's future mutation.

Mutation—A Core Element of Design

Our comparison of *Handlungsspielraum* and the knockout cell line design process revealed some striking similarities. On the one hand, as in some design, molecular biologists intentionally design incomplete cell lines, which, thanks to their agility, easily adapt to the user's contexts by providing them a platform—or stage—for their own experiments and future products. Also, as in design, the procedure for creating cell lines is constrained by certain restrictions (e.g., diploid cell lines or issues with the guide RNA). And, like many design products, the design of knockout cell lines requires prototyping, selection, and multiplication (production). We also found that the selection process in cell design corresponds to the negotiating episode in *Handlungsspielraum*, while the final step of upscaling and delivering knockout cell lines has some similarities with the reading episode.

Our main finding, though, is that while most design processes that include prototyping, selection, and production result in artifacts that can hardly be changed by users, knockout cell lines embrace incompleteness as one of the main properties of a final product. Compared to the two *Handlungsspielraum* episodes discussed earlier, this form of incompleteness is new. Because knockout cell lines can be redesigned or reconfigured during the design process *and during use*, they can be characterized as having a kind of perfect incompleteness.[5] To us, this perfect incompleteness suggests a third episode of *Handlungsspielraum* that we call the staging episode (Fig. 2.4).

Inspired by the process by which knockout cell lines are designed, this third staging episode includes both a high level of control over the artifact's quality and properties, and an extension of the range of action that we call the stage. This stage provides an intentional gap for future mutations (or patches, as in open source; comp. Raymond, 1999). Although both the negotiating and reading episode in *Handlungsspielraum* are based upon more or less complete artifacts, they do not necessarily incorporate future mutations as part of the design process or the artifact itself. In the negotiating and reading episodes, mutations are mostly driven by context and the user's decision to re-interpretate or re-design the artifact. In the staging episode, the artifact *itself* is characterized by a carefully designed permeability and, thus, incorporates its future mutations by staging them before they even take place. Think of it like subscribing to a car-sharing service instead of buying a car: while buying your own car might provide you with a number of options that can be negotiated, once bought, the car itself will mostly remain the same and may suit most of your needs. The car-sharing service, instead, is characterized by the fact that you do not know which car you will get to use. It might depend on availability or upon your specific need in a specific situation. Consequently, the car-sharing

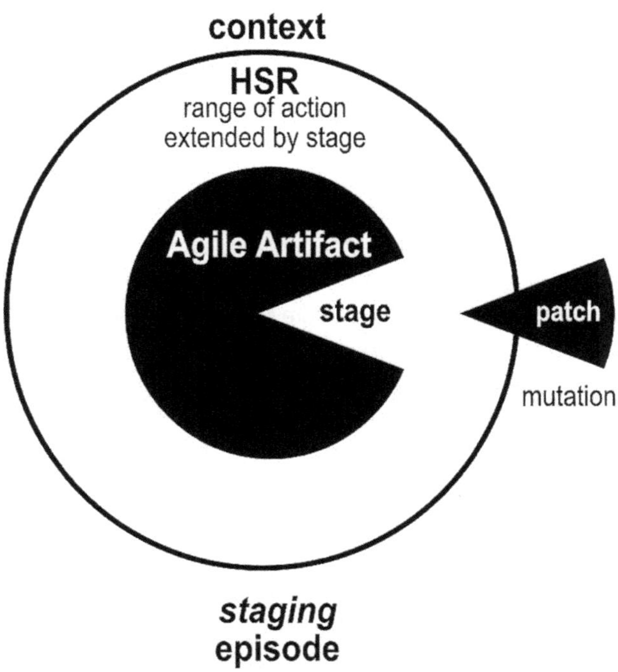

Fig. 2.4 The staging episode and the agile artifact. Eckert and Eckert, 2016.

service incorporates the fact that you expect an agile solution that can adapt to your changing needs over time. With regard to their future use and interpretation, artifacts of the staging episode come with a high level of agility; this is why we have named them agile artifacts.

Agile Artifacts—An Ongoing Journey

The comparison of different types of design artifacts and the cell design process of knockout cell lines unveiled a number of similarities, as well as the discovery of a new specific quality of incompleteness. This sort of incompleteness might help designers to produce final artifacts that incorporate change and mutation by providing their future users with specific stages upon which to engage with the artifact. Although our literature review earlier in this essay mostly showed that this sort of engagement can be driven by context or an incompleteness in terms of a raw design or prototype, the agile artifact's incompleteness is characterized by its *intentionally incomplete design and resulting agility*. Even if the idea of designing in agile or iterative episodes is not new to design, the agile artifact could extend concepts such as iterative design (Nielsen 1993), agile software development (Beck et al. 2011), or scrum methodology (James 2012) through the notion of incorporating intentional incompleteness in a final object. The agile artifact does so not only during the design process, as we have seen, it also produces agile design outcomes in which "incompleteness acts as a trigger for action" rather than a threat to the design artifact (Garud et al. 2008). Incorporating incompleteness that triggers an ongoing circle of future mutations of the artifact is the biggest strength of the agile artifact (Fig. 2.5).

This perspective also adds a new aspect to incompleteness. Garud et al., write about this value: "Any outcome is but an intermediate step in an ongoing journey, representing both the completion of a process as well as its beginning" (2008: 367). They conclude: "This observation underscores the value of incompleteness" (Garud et al. 2008: 363). The agile artifact embodies their idea of an ongoing process that foregrounds continuous updating of the artifact in its use contexts and with its users. Agile artifacts are designed to promote future adaptations and mutations by users. In our opinion, they have the potential to "unlock the controlled and predictable dialogue between designers and users" described by Richardson (1993).

Agile artifacts include the designer, the users, and the artifact in a circle of continuous creation where all agents act autonomously and embrace the imperfect

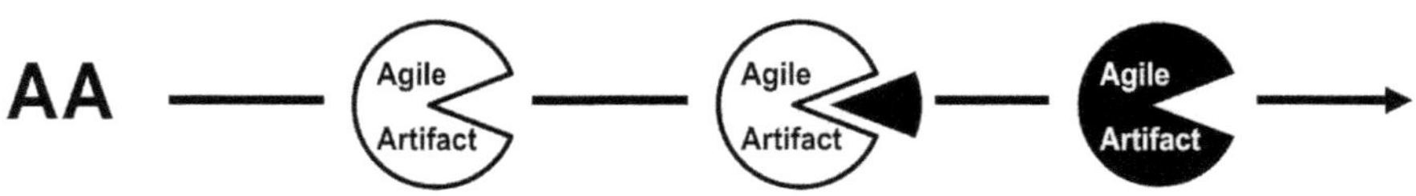

Fig. 2.5 The ongoing process of the agile artifact. Eckert and Eckert.

as part of an ongoing journey. Italian composer Luciano Berio describes this relationship between continuity and creativity:

> I think a work always stays there. Because I don't believe that in any form of creativity—be it science, in music, in literature, in painting—there are separate entities, objects. Certain works take shape because something else happened before. There is a continuity. It is like a journey. And these works simply are signals in the moment of this journey—but the journey continues.
>
> Berio 2000

This journey belies Richardson's notion of the death of the designer. Rather, the stunning correspondences between *Handlungsspielraum* and knockout cell line design point to a profound continuity among creators, their creations, and those who use them.

Notes

1. Furthermore, it connects to Bo L. T. Hedberg, Paul C. Nystrom, and William H. Starbuck's notion of designs that may trigger new designs and design processes (Hedberg et al. 1976).
2. Guide RNA determines precisely where the gene is cut by CRISPR/Cas to create a mutation.
3. Doing so is important as it is still possible that random mutations have occurred throughout the cutting and repair process.
4. In cases where using antibiotics during the selection process is not efficient enough, further detection of the edited cells might be necessary. A popular option is tagging the edited cells with a detectable cell surface marker or a *green fluorescent protein (GFP)* for example. This step enables scientists to screen the edited cell population for cells producing the surface marker or *GFP* using *fluorescence assisted cell sorting (FACS)*.
5. Whereas today's standard *knockout cell lines* already come with a switched off gene, the next level of modified cell lines will give users the choice over which gene is being knocked out, but also the ability to decide when and in which context. This binds the knockout to a given condition and enables users to control its induction, broadening the scope for experimentation—in other words, it widens the range of possible actions for the users and interpretations of the cell artifact itself. Possible mutations that are available while designing and using the cell line is likewise an extraordinary feature of *knockout cell lines*.

References

Anderson, C. (2012), *Makers: The New Industrial Revolution*. New York: Crown.

Avital, M. (2011), "The generative bedrock of open design," in: *Open Design Now: Why Design Cannot Remain Exclusive*. BIS Publishers.

Barthes, R. (1968), "La mort de l'auteur." Manteia 5, 1968. repr., In *Le Bruissement de la langue*, Paris: Seuil, 1984, pp 61–7.

Bechky, B. A. (2003), "Sharing meaning across occupational communities: The transformation of understanding on a production floor," *Organization Science* 14 (3): 312–30.

Beck, K., M. Beedle, A. Van Bennekum, A. Cockburn, W. Cunningham, M. Fowler, J. Grenning, J. Highsmith, A. Hunt, R. Jeffries, J. Kern, B. Marick, R. C. Martin, S. Mellor, K. Schwaber, J. Sutherland, D. Thomas (2011), *Manifesto for Agile Software Development*. http://agilemanifesto.org (accessed on December 31, 2016).

Benington, H. (1956), *Waterfall Model.* Presented at: United States, Navy Mathematical Computing Advisory Panel (June 29, 1956), *Symposium on advanced programming methods for digital computers*, [Washington, D.C.]: Office of Naval Research, Dept. of the Navy.

Berio, L. (2000), *A portrait of the Italian composer Luciano Berio*. 59 min. video. Directed by Reuven Palgi-Hecker.

Blomen V.A., P. Májek, L.T. Jae, et al. (2015), "Gene essentiality and synthetic lethality in haploid human cells," *Science* 350 (6264):1092–6.

Bonsiepe, G. (1999), *Interface – An Approach to Design.* Jan Van Eyck Akademie.

Bonsiepe, G. (2009), *Entwurfskultur und Gesellschaft.* Basle, Boston, Berlin: Birkhäuser.

Brandes, U. (2008), "Non-Intentional Design," in Michael Erlhoff and Tim Marshall (eds.), *Wörterbuch Design: Begriffliche Perspektiven des Design*, Basle, Boston, Berlin: Birkhäuser.

Brandes, U. Sonja Stich, Miriam Wender (2009), *Design by Use – The Everyday Metamorphosis of Things.* Basle, Boston, Berlin: Birkhäuser.

Briggs, C. and S. W. Havlick (1976), "Scientific Problem-Solving Process," link defunct, in *How Do You Design? A Compendium of Models*, H. Dubberly (ed.), Dubberly Design Office, www.dubberly.com.

Caillois, R. (1958), *Les Jeux et les Hommes*. Paris: Gallimard.

Darke, J. (1978), "The primary generator and the design process," in W. E. Rogers and W. H. Ittelson (eds.) *New Directions in Environmental Design Research*: Proceedings of EDRA 9, Washington: EDRA, 325–37.

Eckert, J. (2012), "New Playdecks for Location Based Services," Doctoral Thesis submitted to the School of Doctoral Studies in Design Science at the IUAV University of Venice.

Eckert, J. & D. Eckert, (2018), "Designing the Incomplete: The Agile Artefact." Talk given at Beyond Change: Questioning the role of design in times of global transformations, Swiss Design Network Research Summit, Basel.

Eco, U. (1962; 2009), *Opera Aperta*, VIII edition, Florence: Tascabili Bompiani.

Engelbart, D. (1995), *Boosting our Collective IQ*, http://dougengelbart.org/pubs/books/augment-133150.pdf (accessed on May 2019).

Eppinger, S.D. and K.T. Ulrich (1995), *Product Design and Development*. New York: McGraw-Hill Education.

Essletzbichler P., T. Konopka, F. Santoro, D. Chen, B.V. Gapp, R. Kralovics, T.R. Brummelkamp, S.M. Nijman, and T. Burckstummer (2014), "Megabase-scale deletion using CRISPR/Cas9 to generate a fully haploid human cell line," *Genome Res*earch 24: 2059–65.

Fu, Y., J.A. Foden, C. Khayter, M.L. Maeder, D. Reyon, J.K. Joung, and J.D. Sander (2013), "High-frequency off-target mutagenesis induced by CRISPR-Cas nucleases in human cells," *Nature Biotechnology* 31 (9): 822–6.

Garud, R., S. Jain, and P. Tuertscher (2008), "Incomplete by Design and Designing for Incompleteness," *Organization Studies* 29 (03): 351–71.

Garud, R., J. Gehman, A. Kumaraswamy, and P. Tuertscher (2017), "From the Process of Innovation to Innovation as Process," in A. Langley and H, Tsoukas (eds.) *The Sage Handbook of Process Organization Studies*, London: Sage.

Gibson, J.J. (1977), "The Theory of Affordances," in R. Shaw and J. Bransford, *Perceiving, Acting, and Knowing: Toward an Ecological Psychology*, Hillsdale, New Jersey: Lawrence Erlbaum Associates Publishers.

Hedberg, B.L.T., P.C. Nystrom, and W.H. Starbuck (1976), "Camping on Seesaws: Prescriptions for a Self-Designing Organization," *Administrative Science Quarterly* 21 (1): 41–65.

Hillier, B., J. Musgrove, and P. O'Sullivan (1972), "Knowledge and design," in W. J. Mitchell (ed.) *Environmental Design: Research and Practice*, EDRA 3: University of California.

Ingram, J. E. Shove, and M. Watson (2007), "Products and Practices: Selected Concepts from Science and Technology Studies and from Social Theories of Consumption and Practice," *Design Issues* 23 (2): 3–16.

James, M. and L. Walter (2012), "Scrum Reference Card," www.collab.net/sites/default/files/uploads/CollabNet_scrumreferencecard.pdf (accessed on December 31, 2016).

Lawson, B. (1990), *How Designers Think: The Design Process Demystified*. London: Butterworth Architecture.

Markus, T. A. (1969), "The role of building performance measurement and appraisal in design method," in G. Broadbent and A. Ward (eds.) *Design Methods in Architecture*, London: Lund Humphries.

Maver, T. W. (1970), "Appraisal in the building design process," in G. T. Moore (ed.) *Emerging Methods in Environmental Design and Planning*, Cambridge: M.I.T. Press.

Mok, C. and K. Yamashita (2003), "Process of designing solutions," in H. Dubberly, *How do you Design? A Compendium of Models*, www.academia.edu/9821063/How_Do_You_Design (accessed on June 13, 2021).

Nielsen, J. (1993), "Iterative User-Interface Design," *IEEE* Computer 26 (11): 32–41.

Norman, D. A. (1988; 2002), *The Psychology of Everyday Things*. New York: Basic Books.

Norman, D. A. (1988; 2007), *The Design of Everyday Things*. New York: Basic Books.

Norman, D. A. (2009), *The Design of Future Things*. New York: Basic Books.

Raymond, E. (1999), "The Cathedral and the Bazaar," *Knowledge, Technology & Policy* 12 (3): 2–49.

Richardson, A. (1993), "The Death of the Designer," *Design Issues* 9 (2): 34–43.

Sanders, E. and P. J. Stappers (2008), "Co-Creation and the New Landscapes of Design," *CoDesign* 4 (1): 5–18.

Schuler, D. and A. Namioka (1993), *Participatory Design: Principles, and Practices*. Hillsdale, N.J.: Erlbaum.

Simon, H. A. (1981), *The Sciences of the Artificial* (2nd ed.). Cambridge: MIT Press.

Star, S.L. and J.R. Griesemer (1989), "Institutional Ecology, 'Translations' and Boundary Objects: Amateurs and Professionals in Berkeley's Museum of Vertebrate Zoology, 1907–39," *Social Studies of Science* 19 (3): 387–420.

Taylor, P. (2011), "Inductive design and the 'primary generator"' on Designerly Thinking: Design Thinking, Enterprise Architecture, Systems Thinking, Theory, and Practice, https://designerlythinking.wordpress.com/2011/04/10/inductive-design-and-the-%E2%80%98primary-generator%E2%80%99/ (accessed on September 3, 2020).

Usher, A.P. (1929), *A History of Mechanical Inventions*. New York: McGraw Hill.

Visser, W. (2006), *The Cognitive Artifacts of Designing.* Hillsdale, N.J.: L. Erlbaum Associates.

3 Diagramming ArtScience: Designing at Knowledge Intersections

CLARISSA AI LING LEE (WITH MATT CORNELL)

This essay is produced as a contrapuntal dialog between dancer and choreographer Matt Cornell and the author. His four images: the utility diagram, naval gazing statues, the Schrödinger black box theatre, and the Big Bounce, represent his conversation with, and interpretation of, the processes involved in this essay.

Introduction

In June 2020, the journal *East Asian Science, Technology, and Society* dedicated a special issue to thinking with diagrams. The purpose of the special issue was to illustrate the cultural and historical lineage that shaped the lexicon of diagrammatic thinking that predated modern science, and that had been cultivated through a long durée of knowledge histories. This special issue also acknowledges the roles that thinking through diagramming plays in science and culture. Diagrams can help generate knowledge that spans the social, material, scientific, and technical spheres. This capacity makes room for so-called emergent diagrams, which link known ideas with emerging concepts and, by doing so, can reveal shortcomings in extant knowledge.

Emergent diagrams can look conventionally diagrammatic, but they can also be 2D or 3D, and immersive, environmental, tactile, sonic, or uncategorizable. They can even be fictional, with no obvious connection to reality, because they represent aspects of the physical world that we cannot fully conceive without the help of a diagram. Emergent diagrams are not meant to predict the inevitable; rather they anticipate and exhibit unexpected possibilities. Making (and interpreting) emergent diagrams reveals unforeseen ideas through a two-pronged process of speculative experimentation and defamiliarization. Speculative experimentation challenges the experimental setting itself and forces consideration of new possibilities. Defamiliarization reframes common ideas in unfamiliar and uncomfortable ways.

Artistic and designerly processes and scientific methodologies all utilize speculative experimentation and defamiliarization. ArtScience, an integrative practice in which ideas and methodologies from art or design and science come together, makes it possible to share this speculative experimentation and defamiliarization. ArtScience, though, can operate in both active and passive modes. In the diagrammatic thinking that I discuss here, active-mode diagramming deploys common diagrammatic practices from both design and the sciences simultaneously. Active-mode emergent diagrams are hybrid forms in which the design and science aspects are not easily separable. Final forms of active ArtScience diagrams can sometimes be interpreted by users without them understanding the active incorporation of the visual vocabularies of design and science. In addition, there are passive ArtScience diagram modes, which are mostly utilized in sciences that don't "discipline" their methods of knowledge generation. These scientific knowledge traditions unknowingly incorporate design thinking and form making—whether they use design thinking and visualization for knowledge production, or they incorporate them as some ingredient in the development of scientific knowledge. In other words, in passive-mode diagramming, ArtScience—designerly thinking and making—is part of the knowledge process.

In this essay, I therefore consider how ArtScience diagramming practices, in particular, can shape emergent knowledge in both design and science by making the invisible visible, generating new perspectives, and suggesting new ways of asking questions. Collaboration between scientists and artists or designers can take place through the mediation of emergent diagrams. Diagrams translate information from different fields into a shared visual language, revealing commonalities among various kinds of knowledge, and bridging disparate information grammars. This knowledge is intermingled in a shared grammar space, as well as in a shared thinking and making space. Together, all of these aspects of emergent diagrams foster new perspectives and ways to see the world differently. And this capacity to perceive the world differently is the staple of design fiction.

In the following sections, I discuss how emergent diagrams function. In this essay, I use the term "diagram" broadly to mean imagery that reflects emergent knowledge in a visual form. I begin with a consideration of how producing geometric forms and computational imaging can disrupt the status of current knowledge by making the invisible visible. Then I show how science diagramming benefits from being understood instead as ArtScience diagramming. To do so, I use examples from geometry, computational imaging, astrology and astronomy, and biology.

Finally, I consider several thought-experiment, design-fiction exhibitions that utilize diagrams to express emergent scientific thinking and ideas. Design fiction is a practice that explores speculative scenarios through designed artifacts. Design fiction has direct and indirect counterparts in creative science prototyping, and can be used in conjunction with ArtScience diagramming to reveal novel perspectives on familiar ideas. These novel perspectives are often emergent ideas. Design fiction also offers users a "playful" approach for comprehending novel perspectives. I use precisely these kinds of emergent, playful processes to suggest unforeseen connections between quantum physics and human biology in the exhibitions that are discussed in the final section of this essay.

Emergent Scientific Diagrams

Scientific diagrams can be emergent when they act as primary illustrations of relationships and connections in novel scientific thinking. These emergent diagrams function as scientific arguments in their own right, and there are several ways that such diagrams may be deployed:

1 A scientist could use a diagram to help themselves or other people to picture a stubbornly abstract or difficult-to-imagine scientific object in space and time, for example, as in Fig. 3.1.

2 A scientist could use a diagram to simulate calculations of abstract phenomena before or after an experiment, as in Figs. 3.2 and 3.3. These diagrams simulate, on paper, what could potentially come out of an actual experiment.

3 A scientist could use a diagram to present data that describe social conditions—for example, heat maps of particular 2020 "hotspots," such as climate disasters (e.g., the Australian bush fires) and COVID-19 infection networks. This kind of diagram can also demonstrate the impact of human activities (or lack of activities) on the Earth, such as satellite mapping of clear skies, and the decline in production of pollutants.

4 A scientist could use a diagram to develop philosophical arguments relating to the science they are working on, or even for sketching out logical arguments

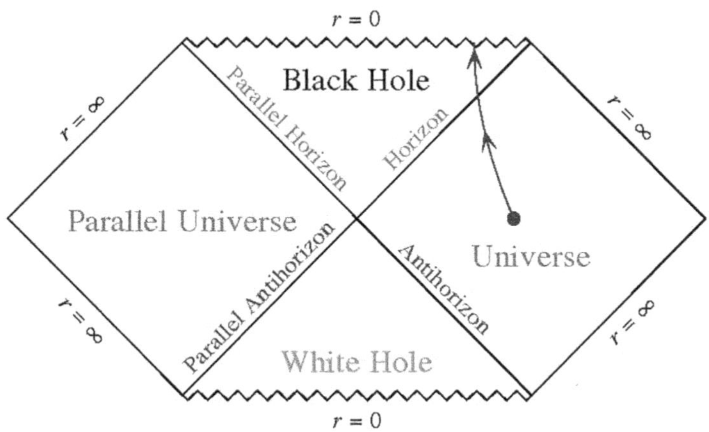

Fig. 3.1 This is the Penrose diagram representing an illusory dimension of a black hole (a very heavy dead star that had collapsed into itself). It is shown in relation to its "fictive" counterpart, the white hole—which no one may access from the outside—along with the hypothetical/real spacetimes of the universe. Even though these phenomena seem like science fiction to the general public, they actually have mathematical and scientific significance to astrophysicists. Penrose's work describing blackhole formation provides a robust prediction of Einstein's General Relativity theory, which won him half of the 2020 Nobel Prize in physics.

for visual or conceptual clarity. Newton's *Principia Mathematica*, which introduced original versions of physicist Isaac Newton's laws, featured geometric drawings of thought experiments—which are experiments carried out only in the mind—for imagining how mechanics could be played out in the real world. A good example is Newton's Rotating Bucket experiment.[1]

Because they embody and communicate ideas and narratives, diagrams in these four categories are typically relevant across a range of human endeavors.

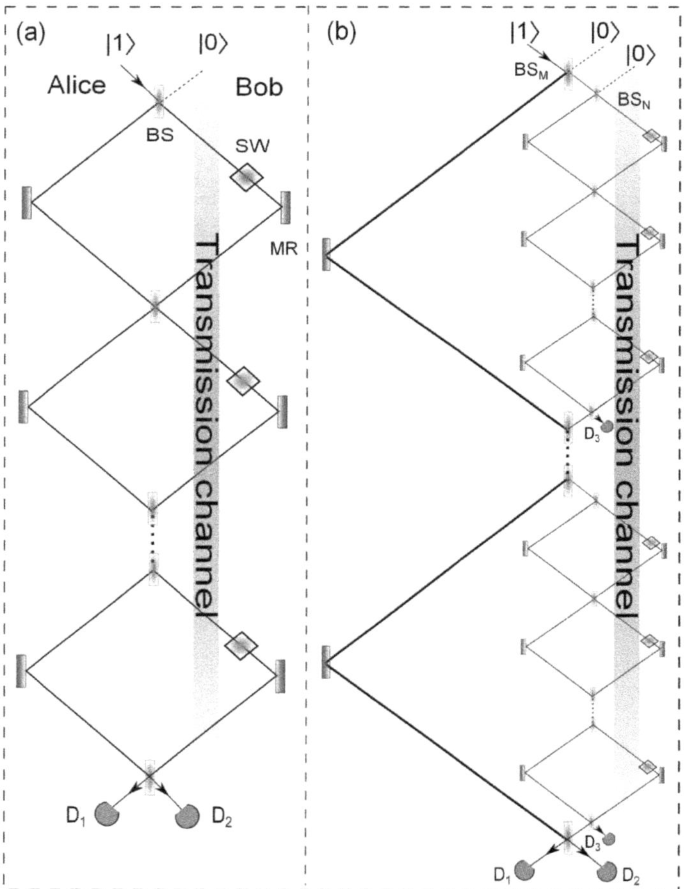

Fig. 3.2 This diagram illustrates how quantum information bits could be transmitted between two parties, eponymously named Bob and Alice. These two parties represent how quantum computers would communicate in various futures, and the diagram charts the logical system of their communication. This diagram could also be viewed as a conceptualization of the beginnings of the quantum internet. Figure printed with permission from Hatim Salih et al. (2013), *Physical Review Letters* 110 (6). © 2022 by the American Physical Society.

Fig. 3.3 This diagram by physicist Richard Feynman is his visualization of his own complicated calculations. The diagram represents the interactions among various subatomic particles that embody gravitational interaction, weak interaction, strong interaction, and electromagnetic interaction. These four interactional forces represent how the matter that makes up our physical universe is structured, organized, and connected. Image courtesy of Mark C. Kruse, Department of Physics, Duke University.

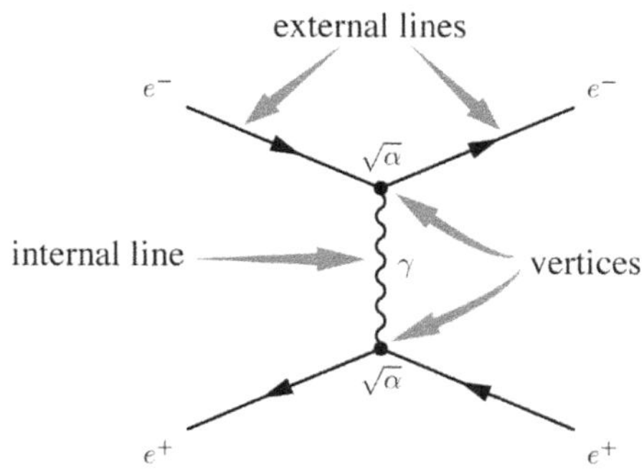

Diagramming in the Performance of Knowledge Discoveries

Science utilizes the scientific method, a knowledge system that measures, replicates, and applies rules and evaluative mechanisms to assess the physical world. As narrative artifacts, scientific diagrams integrate these analyses of the physical world with physical, social, or cultural themes. In the following sections, I will discuss examples of these kinds of themes as part of diagrammatic narratives in geometry, computational imaging—in which diagrammatic images are created indirectly from measurements using algorithms generated by computers—astronomy/astrology, and biology.

Geometric Diagrams

The physical laws that geometry represents are based on the association between an object and its environment.[2] Geometric diagrams typically use a minimalist visual vocabulary of lines and shapes to depict these associations and the interactions and movements of the invisible worldly functions they describe. In other words, geometric diagrams depict a visual narrative of theoretical notions: the lines and shapes in a geometric diagram come together to tell a story about how objects function in the physical world.

The theoretical "stories" revealed are often fascinating. Geometric diagrams can be used for computation, for example, to reveal how a billiard ball moves around a table. The Feynman diagram above (Fig. 3.3) is used for figuring out particle physics interactions (Lee 2013). Other geometric diagrams demonstrate how manipulation of one physical system could affect another system.[3] Some other Feynman diagrams, for instance, depict difficult-to-visualize, science-fiction-like realities, such as the idea that space is curved at the speed of light, or that there is a "color" micro-entity (gluon) that holds other micro-entities (quarks). Yet other geometric diagrams plot extraordinarily abstract information about the physical world. The Penrose diagram presented in Fig. 3.1, for instance, plots the curvature of spacetime containing

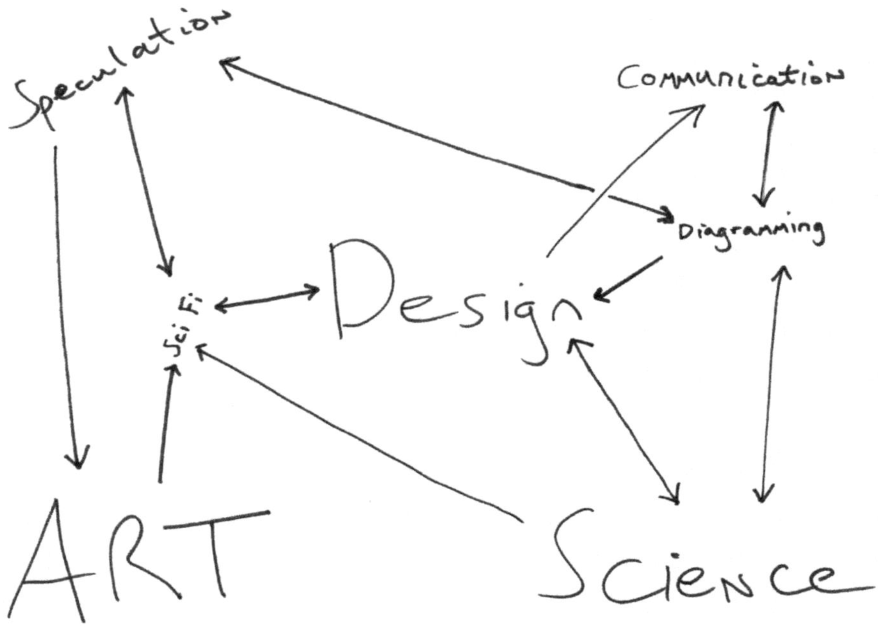

Fig. 3.4 I have tried to use diagramming as a tool to distill my thinking into a clear signal path of relationships. Instead, what I found staring back at me was a provocation questioning causation, and social value systems, and hierarchy. To attempt to know which skills and mediums are in service (only to be utilized by the others), has provoked conversation on utilitarian value vs. inherent value. This is especially true when considering how to allocate resources to these institutions/frameworks of thinking/models for explaining the world. One example on how one might read this diagram: Science can fuel speculation directly, but speculation must pass through the design process. Doing so allows design to become experiments that may function as science. It is an asymmetrical information flow. Even if art or design fuels speculations which are then designed into SciFi (as art form), that must still be designed by scientists into experiments or by science communicators into public awareness and discourse. This is discreet from the job of art. Neither art nor science have an obligation to be legible so as to be valid as good art or science (especially not to everyone immediately.) Are there direct links between all these things in the diagram or do the relationships play out in an almost noun-adjective hierarchy? Action seeking validation . . . This offer is an invitation to try your own diagramming so that your mental model stares back at you with all the questions that would otherwise remain conveniently invisible to our own cognitive dissonance red-flags and warning systems. Image courtesy of Matt Cornell.

infinity-type properties, which is not easily expressible (or even imaginable) through available mathematical conventions (Wright 2013).

Geometric diagrams can also reveal connections between the macro world of classical physics and the micro world of quantum physics. After all, quantum laws also apply in the perceptible, physical world (Plotnitsky 2006, Dieks 2016).[4] The engineering drawings of Leonardo Da Vinci, for example, exemplify theories of classical mechanical design (Laurenza 2006). But, surprisingly, they also happen to inadvertently reveal quantum theoretical interactions, such as oscillations, scattering, translation, and energetic transitions. These kinds of quantum interactions actually originated as geometric depictions in classical mechanics.

Computational Imaging

Although geometric diagrams use lines and shapes to represent the physical world, in computational imaging (sometimes called scientific computing), diagrammatic images are created indirectly from measurements using algorithms generated by computers. Computational diagrams can look like traditional images. In contrast to traditional imaging, though, computational imaging systems integrate sensing systems and machine learning affordances to create a holistic image from a series of images. A single X-ray image, for example, "does not reveal the precise location of a fracture, but a CT scan, which works by combining multiple X-ray images, can determine the precise location of one in 3D" (March 2012). And, in a second example, a typical camera cannot image around corners, but a recently developed system

Fig. 3.5 Naval gazing statues. Omphaloskepsis, or navel gazing, is contemplation of one's navel as an aid to meditation. Meditation to reach deeper truth or enlightenment . . . but how to be rigorous in this pursuit? Add in technology that augments the senses so we can see more! Seriously though, this image is just a playful interrogation of promise that there is actually a "bottom of things" that we can get to. Image by Matt Cornell.

> overcomes this problem by capturing ultra-fast time-of-flight information—that is, how long each photo has taken to reach the camera. This information is then decoded by a reconstruction algorithm conceived by team member Andreas Velten. "We are all familiar with sound echoes, but we can also exploit echoes of light," says Ramesh Raskar, head of the Camera Culture Research Group at the MIT Media Lab that carried out the study.
>
> Marsh 2012

By designing a set-up that involves sending fast pulses of light, and recording the received signal and using an algorithm, a novel idea emerged: a light-based system that sees around corners. The diagrammatic computation imagery that this device produces likewise communicates new kinds of information that might otherwise be imperceptible.

In computational imaging, an image—and a narrative of what is happening in the physical world it depicts—is produced when a selected object is measured over time and space. Then, its content is coded to emphasize important narrative information that is not necessarily visible. Although some computational images are easily accessible to the naked eye, others require customizable detectors to be seen. Some of these detectors communicate images from the production source to human observers, while others require further interpolation before a coherent image is available. Computational images are situated where visual information meets utility, and where abstruse scientific information transforms into a graphic rendition. The image narrative and content can be derived from a known object and environment, as in the X-ray example above, or from an invisible, theoretical, or fictional object and environment.

Astronomy and Astrology

Astronomy likewise visualizes and communicates narratives about the function of the visible and invisible physical world. Surprisingly, the modern science of astronomy also assimilated cultural aspects of astrology from ancient Asian civilizations, including the Middle East and West Asia. Astrological knowledge systems and diagrams in these regions revolved around heavenly, hellish, and earth-bound deities. These deities inhabited alternate dimensions and produced seen and unseen physical phenomena on earth.

Although today astrology is considered to be a fate-forecasting pseudoscience, some of the traditional astrological knowledge systems and diagrams were actually incorporated into modern astronomy (Rutkin 2006). The mathematical system of Chinese antiquity, for example, utilizes geometric, astrological/cosmological, and biological diagrams both to explain what was known and to reveal the unknown. The Chinese system was preoccupied with the day-to-day affairs of survival and maintenance of a dynasty's pre-eminence. Political concerns, though, did not preclude the development of sophisticated methods for computing the abstract universe that structures Chinese cosmology.[5] The Chinese developed a system that interlaced astronomy with astrology in order to conduct geomancy and divination—and it related astronomy and astrology to anatomy and physiology.

The ancient Chinese medical system is associated with the vital functions of the body, particularly how the body manages energy levels that affect its health. This pre-modern medical system was connected to astrological charts that represent how the body responds to its changing environment. Astrological beliefs were rendered in these detailed diagrams, which used systems thinking to map out the different vital points of the body. There is logic in these charts, even in those astrological practices that modern medicine equates with superstition.

Attempts have been made to sync pre-modern medical practices with those of modern medicine. Western medicine knowledge systems and practices may have some fortuitous connections with Chinese medicine, but there is definitely a strong relationship between how both use diagrams to chart their knowledge systems. The technology for charting these systems has evolved over time. In contemporary Chinese medicine, classical Chinese medicinal practices that drew upon astrological charts located in human anatomy (see Fig. 3.6) have been appropriated into modern-day visualization techniques that connect real bodies to diagrammatic representations. This includes the use of photographic and sonic devices for amplifying pain points and mapping out areas of pathology.

Despite the perception that it lacks scientific rigor, astrology actually provided science with a form of cosmological systematization. Indeed, a similar speculative spirit is echoed by contemporary scientists working to decipher the indiscernible phenomenon of dark matter. Today divination is carried out by radio telescopes and

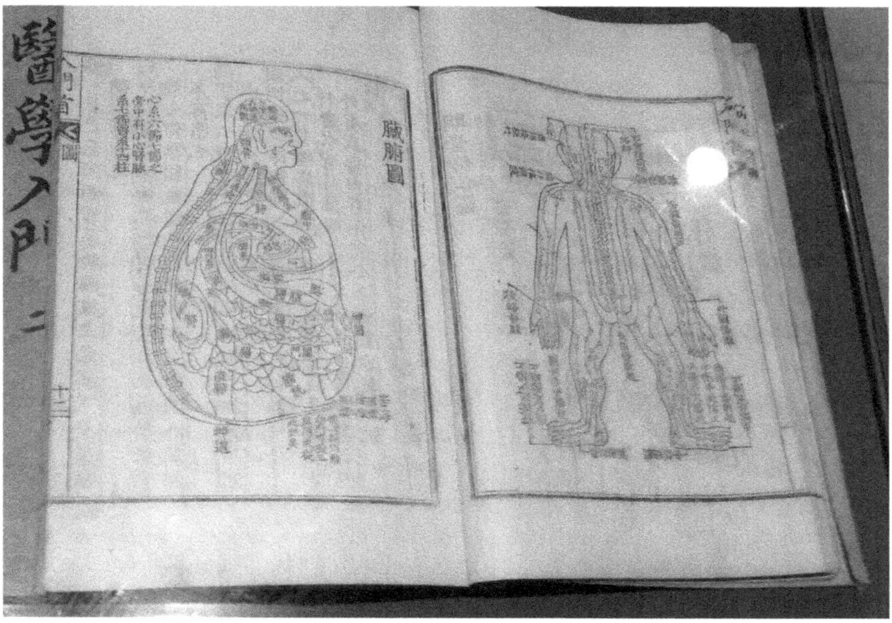

Fig. 3.6 Classical Korean medical book, traditional medicine museum in Seoul, Korea, Joseon Dynasty (c.1392–1897). There are many similarities between Korean medicine and Chinese medicine. Image by Clarissa Ai Ling Lee.

cosmic scanners.[6] These instruments are used to collect data from entities that cannot be observed with the naked eye; the dark matter data used in dark-matter diagrams is as sonic as it is visual.[7] Here, a diagram, as in Chinese medicinal astrology, is the platform for charting the connection between the known and the unknown, and the visible and the invisible.

Diagramming Biology: Theory and Applications Across Knowledge Cultures

Diagrams can be both poetic and speculative. As I've already noted, they can also serve as representations of the physical world or the body that merge invisible and visible aspects of those worlds. In biology, dissection techniques and imaging instruments, which increased in sophistication over the years, likewise revealed what couldn't otherwise be perceived. Anatomical diagramming that was based on dissections, vivisections, and microscopic slicing processes was, and still is, integral to biology education. Observations from anatomical dissections were sometimes reproduced in anatomical flap books. Anatomical flap books were "used to show what's hidden under our skin as early as the 1500s," according to Susan Isaac (2019). They are called flap books because of "the layers of movable paper flaps that are lifted to reveal the layers below, making the viewer an active participant who 'dissects' the body by opening the flaps" (Isaac 2019). Flap books also allow designers to diagram, slice-by-slice, the different layers that exist within a single anatomical structure, and to present functional and structural details that are not possible on a single two-dimensional page. Sophisticated medical illustrations produce life-like representations, but these images aren't diagrams. What makes flap book images diagrammatic representations are the design decisions about which aspects of the anatomy to emphasize, and how and why to do so. Little attention has been given to the role that these decisions play in understanding how anatomical structures have been understood over history. The reasons anatomical flap book diagrams look and function the way they do poses ontological questions about how we have perceived the body's structure and functions over time.

Some biological diagrams represented what couldn't be seen except in the minds' eye. Other diagrams attempt to explain what was unexplainable at the time they were created. Diagrams in biology can also be speculative, in which data and information reveal certain concepts or ideas that are without precedent at the time. Gregor Mendel, the nineteenth-century founder of modern genetics, carried out research on genes and the principle of inheritance in a pre-electron microscope age. Mendel diagrammed his findings, which revealed for the first time how observed genetic traits in organisms are inherited during the then-unobservable process of cross hybridization (Miko 2008). Other biological diagrams attempt to explain phenomena with unobservable causes in a fantastical fashion. In the sixteenth and seventeenth centuries, scientists couldn't figure out how sex produced a fetus. To explain this unseen process, they developed versions of the homunculus—a miniature human being within a human sperm that enters the female womb upon fertilization. There were various ideas about homunculi, so there were also diverse diagrammatic representations: some were depicted as fetuses and others as a fully grown person.

Finally, some biological diagrams are based on nature-inspired design. Natural historian Charles Darwin and theoretical biologist D'Arcy Wentworth Thompson's diagrammatic studies of beehive structures, for instance, led to novel architectural structures and also presaged the contemporary field of biomimicry. Diagrammatic (geometric) "thinking" in nature that is appropriated by scientists and designers is one form of biomimicry. Darwin and Thompson wrote about the diagrammatic properties of beehives—an environment that enabled bees to survive and thrive for millennia under various environmental conditions. The two scientists measured and reconstructed these hexagonal-based geometric structures, a process that also helped to lay the groundwork for the way contemporary biomimicry design is practiced.[8] In biomimicry, designers "mimic" natural structures, forms, and materials in order to reproduce their functions in design. Designers adapt the "diagrammatic" features of beehives in architectural design, and sometimes also base their choice of raw materials for these constructions on the materials that make up real beehives. Indeed, British architect Barry Jackson's HiveHaus design allows "people to build an entire home by connecting each individual honeycomb room to the other until the house is just as big as you want it" (Norris 2014). "The honeycomb is a masterpiece of engineering," according to National Public Radio science writer Robert Krulwich (2013).

Emergent diagrams like these, and those discussed in the previous sections, are more than just illustrations of information. Diagrammatic thinking is more than just exploring how to present data. Emergent diagrams offer ways both to provoke and model new ideas and innovative concepts. Watson and Crick produced two- and three-dimensional diagrams of the double helix, based on data and observations, in order to figure out how DNA works. They knew that "DNA was a geometric problem best understood by three-dimensional modeling" (Rust 2004). "Given the three-dimensional complexity of their problem," argues design critic Chris Rust, "it was only by constructing and, arguably, dwelling in their model that Watson and Crick could make the mental connections needed to complete the Puzzle" (2004). Their diagrams served as a kind of speculative design, or design fiction, a form of design that "thrives on imagination and aims to open up new perspectives" and acts "as a catalyst for collectively redefining our relationship to reality" (Dunne and Raby 2013, 2).

Design Fiction and Participatory Thought-experiment Exhibitions

As noted in my introduction, design fiction explores and presents speculative scenarios using designed artifacts. In this section, I first give some necessary background on the behavior of subatomic particles. I then present three speculative design-fiction exhibitions that have science and diagrams at their core.

Design fiction and speculative design relate to the philosophical concept of the thought experiment, a cognitive enterprise that combines logical meaning-making with make believe. Philosophical explanations of thought experiments, in fact, relate them to literary fiction (Meynell 2014), and, I would argue, to design fiction. Scientific thought experiments often forgo accepted knowledge about scientific outcomes—which is constrained by physical principles—to embrace inventive imaginings about scientific outcomes. As physician Michael Chandler notes in his essay for this volume, thought experiments are "performed in the laboratory of the mind . . . they involve mental manipulations . . . [and] are often (but not always) impossible to implement as real experiments." In scientific design fiction, the boundary for representing the causes of ideas and outcomes is inventive and porous. Despite being inventive and porous, however, scientific thought-experiment observations typically still infer a causal source, and there is a limit to how far causal actions can stray from acceptable physical reality in these thought experiments.

Inventiveness and porousness also make thought experimentation a productive process for participatory knowledge generation between designers and scientists, as well as between them and other stakeholders, such as users. Participatory thinking often takes place within a series of iterative loops between the creation of a research idea or problem statement, and the prototyping of solutions, often using diagrams in response to the problem. Both science and design processes are frequently iterative in this way. Transdisciplinary knowledge—including that expressed in scientifically based design fiction—emerges when the object, subject, and outcomes of the research are "questioned, reworked, and reinvented through sustained, deep, and long-term mutual collaboration and where new forms of material and social objects are invented" (Salter, Burri, and Dumit 2017, 152).

The projects presented here are thought-experiment exhibitions whose design content involves interactions among imaginary scientific materials or entities, emergent diagrams, and participants or citizen scientists. Citizen scientists are non-scientists who are involved in scientific research. Design fiction allows citizen scientists to inject a lay understanding of the scientific matter at hand into the process, and to participate in design-fiction scenario building with trained scientists. Collaborative design fiction, therefore, can be a platform in which non-scientist lay experts have a generative impact on the character of science (Davies et al. 2008). Transdisciplinary design fiction that is carried out with citizen scientists also encourages trained scientists to recalibrate both their roles as primary communicators of scientific knowledge, and their own relationships to their science.

Users or participants are critical to design fiction. Indeed, in their groundbreaking book *Speculative Everything*, Anthony Dunne and Fiona Raby argue that speculative design depends upon engagement with participants (2013, 139). Although visitors to science museums are often able to engage with exhibits, they must typically operate from a pre-defined "script," which allows only limited interactions with science knowledge making.[9] The scientific design-fiction exhibitions that I discuss here, though, feature dynamic diagrammatic environments that encourage participants to intervene in imaginative ways.

Diagramming the Design Fiction of an ArtScience Experience

Earlier in this essay, we discussed examples of emergent scientific diagrams and diagrammatic thinking across scientific traditions. The designs of these diagrams suggest, present, and assimilate stories around novel scientific theories or world views. Some of the diagrams portray these stories through abstract visual forms, while others use recognizable visual forms from everyday life. The imaginary exhibitions we discuss below likewise consider a spectrum of diagrammatic thinking. The differences among these exhibitions lies in how the diagrams are fleshed out into experiential settings for human visitors. Scientific processes, which are combined to create an imaginary sensual information experience, are partitioned into three theme-based rooms that contain specially curated content. Even though the exhibitions are fully fleshed out three-dimensional experiences in space and time that yield particular diagrams, visitors can immerse themselves into the exhibitions and also produce their own related diagrams and ideas.

Imagine visitors to these exhibitions walking from a room with an interactive, immersive space-time presentation into a room that is partitioned into a cavernous *Wunderkammer* (cabinet of curiosities). Changing the scale of the space-time experience, the first room allows visitors to imagine themselves participating in the process as a subatomic particle maneuvering through its environment, or as a human trying to make logical sense of a quantum-mechanical world. The cavernous *Wunderkammer* room connects the particle-wave duality of the previous room with a third room that offers an exploration of bees. These three rooms involve visitors in an ArtScience-infused fictional re-interpretation of the relationship between the quantum and biological worlds. In reality, quantum effects are part of complex biological systems although they are almost impossible to observe. Design fiction that utilizes emergent diagrams, however, can be used to produce an unconstrained immersive environment that overcomes the constraints of everyday experience.

Physical Indeterminism in Subatomic Particles and the Double-slit Experiment

In order to understand the thought-experiment design-fiction exhibitions discussed below, it is first necessary to provide a little background on the behavior of subatomic particles, in particular, the way they behave in the classic double-slit experiment. The classic double-slit experiment (Fig. 3.7) is centered on the idea that the physical state of the quantum object cannot be determined with certainty.

The double-slit experiment begins with electrons, which are negatively charged subatomic particles. In the double-slit experiment, a beam of electrons, which is

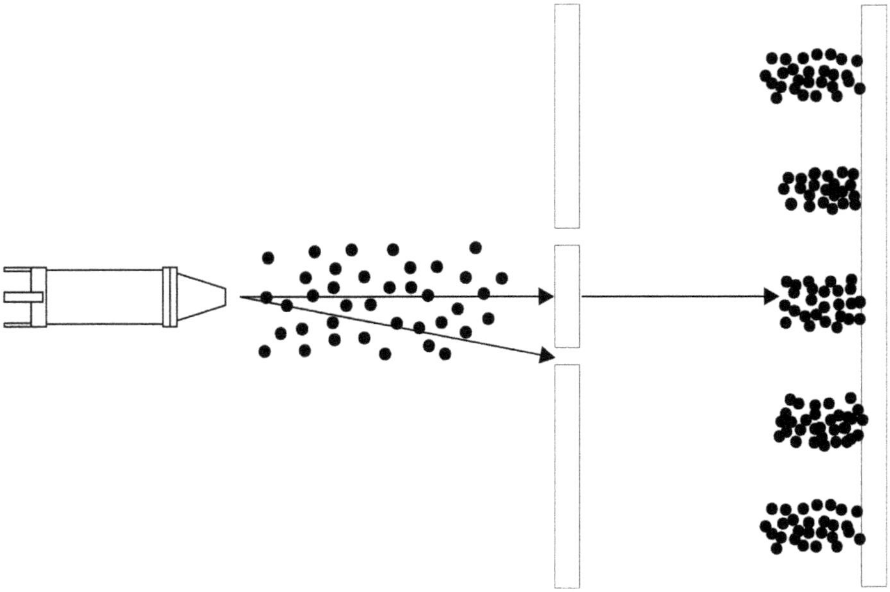

Fig. 3.7 The double-slit experiment, in its classic or quantum form, is concerned with the indeterminacy of an electron as it moves through obstacles. In this diagrammatic rendition of the experiment, miniature basketball hoops create a playful demonstration of how difficult it is to project classic double-slit experiment misses accurately—that is, which of the particles might go through the hoops. Illustration by Aaron Lee.

equivalent to a laser ray, is directed through a flat rectangular obstacle with two narrow slits carved into it. Some of the electrons go through the slits and some do not. The electrons that pass through the slits collide with a baffle in their path. This collision produces a row of shadow and light fringes.

This simple setup is similar to a real-world experiment that was conducted by a group of Japanese physicists (Tonomura et al. 1982). That experiment—which consisted of a negatively charged electron moving through a very weak electromagnetic field—was used to collect data from the moving electrons' interactions with photons in the field. Photons are particles of light with zero charge; they are found within the electromagnetic spectrum.[10] This experiment revealed the qualities of electromagnetic force by illustrating how the charged particles interact with this force.

In a related thought experiment using the double-slit experiment, the thought experimenter can count the probabilistic distribution of electrons that arrive at the baffle (which functions as a detector), as the electrons enter either the first slit with the second slit stopped up, or vice versa (Feynman 1965). The probabilistic distribution of electrons is important for our purposes. The other issue that is important for our discussion here is that the double-slit experiment demonstrates the problem of physical indeterminism of these electrons. That is, although the distribution of

Fig. 3.8 The Schrödinger black box theatre. Schrödinger's black box theatre is a space where a dancer takes on a superposition of being alive and dead until observed. The dance itself can also be assumed to be alive and dead until the dancing body collapses infinite potential and the dance becomes danced. Image by Matt Cornell.

electrons can be assigned a probability, the physical state and location of these particles cannot be determined with certainty.

Whether thought experiment or real world, the double-slit experiment is usually perceived by an observer situated at some distance from the experiment. This observer barely interacts with the setup or the interactions taking place inside. Imagine, though, if it was possible to make some adjustments to the role the observer plays. What happens if we bring the outside observer into the experiment and consider where they could be positioned within the experiment? If their position could be changed around rather than remain fixed, the observers could then diagram their experience of seeing the described interactions among particles and baffle within the experiment as an "insider." What would the diagrams of this process look like from a dynamic inside perspective?[11] I present the first exhibition in the context of these provocations.

Exhibition 1

Let's suppose that the observer could ride on the electrons. The electron on which the observer is riding is brought into contact with a photonic light source of pre-determined wavelength. Including the observer in this process allows a more precise prediction of the location coordinates and dynamic features of the electrons.

How might these dynamic features of the electrons look? Imagine a scene of a child playing with a little boat at a swimming pool. The child is fascinated by the ripples that move from around the boat toward a slat that covers the drain of the swimming pool. That night, the child lays in bed, and plays with the intensity of her night light by turning the knob on the light's dial. She uses the light for a shadow puppet play. As she plays with the light, however, she notices that the light creates rings of dark and light on the wall, with a particularly bright spot in the middle.

This child observed examples of both classical mechanics—represented by the scene at the swimming pool—and quantum physics—through the beam given off by the night light. These experiences could be starting points for the development of an exhibition that reveals the logics of the quantum world, but they could also be part of a citizen-science practice in which the observer is a citizen scientist who conducts the experiment. And design could be part of these scientific investigations by using diagrammatic blueprints of these experiments to serve as intellectual fodder.

Exhibition 2: A Corollary of Exhibition 1

Let's return to the baffle mentioned above. This baffle, you remember, is used to help demonstrate the phenomenon of quantum interference. Let's replace this baffle with one created out of biological materials, which, of course, have DNA within their cells.

When the baffle is hit with ionizing rays, such as those from decaying radioactive elements, mutagenesis occurs. Mutagenesis is the mutation of genetic information through spontaneous or induced conditions. The mutation process could be demonstrated with a time-lapse photographic set-up, similar to time-lapse photographic capture of a decaying basket of fruits.

Either a single charged particle or multiple charged particles could cause mutagenesis when they collide with the biological baffle. The different outcomes of these two situations—a single-particle or many-particle interactions during mutagenesis—is diagrammable.[12]

During the process of diagramming, the designer or citizen scientist can chart what is called quantum tunneling, in which the particles pass through the baffle, to reveal both energy transfer and enzyme actions in living cells. This kind of experiment and the data derived from it are vital to the new field of quantum biology (McFadden and Al-Khalili 2018), and this kind of work is beginning to inform contemporary novel medical practices. Science writer Kerry Taylor-Smith explains:

> Several cell processes occur at the nanoscale, in the domain of atoms and subatomic particles—the realm of quantum. At this scale, matter ceases to behave according to the laws of classical physics and instead starts displaying unique and often counterintuitive properties of quantum mechanics. Scientists hope to utilize these unusual properties to develop medical tools, diagnostics, and treatments which are incredibly precise and ultra-personalized, tools which will ultimately improve and lengthen lives. Using quantum mechanics in medicine

could allow for the detection of disease in the early stages, or different risks of disease before they manifest themselves.

2019

Interpretative diagramming of novel medical processes in conjunction with data from their quantum biology fundamentals would also be instructive for medical practice. This thought-experiment scenario, and the diagrams that come out of it, could also reveal connections between medical phenomena—such as the human energy systems that are described in astrology and Eastern medicine—and developments in the still speculative arena of quantum biology.

Exhibition 3: Tunneling Through the Brain

Visitors leave Exhibition 2 and enter a third exhibition space. The quantum tunneling continues here, but this time the particles collide with a baffle that looks and functions like the chamber of a brain. The concept of a brain prototype is not new: it has its provenance in science fiction as far back as the early 1990s (Egan 1992). The first scenario in this chamber involves crafting a demonstration that relates observable behavior—such as avian navigation skills and beehive-building skills—to what is observed in quantum field spaces. We already have experiments that track the neurological processes of humans and other mammals, such as apes and monkeys. This thought-experiment exhibition asks users to consider a design-fiction prototype of a brain modifier implant. This implant produces real-time dynamic pictorial and abstract diagrammatic representations of the relationships between quantum-level actions, synaptic functions in the brain, and observable behaviors. Synapses are the gaps between two neurons that allow electrical or chemical signals to move between the two. Synapses facilitate neural transmissions that lead to animal behavior.

Separate research trajectories for quantum physics and the biological sciences means that there are distinct vocabularies, concepts, and knowledge bases in these two scientific areas. Early attempts to connect cause and effect in biology with quantum physics are therefore considered to be highly speculative, and they are viewed with skepticism by scientists in both disciplines.

This exhibition in not so much about how quantum biology could explicate unexplained biologically-induced behavior. Rather, it considers ways to map plausible relationships between biology and quantum physics outside of the language of mathematics. Design-fiction thought-experiment-based exhibitions reveal novel connections between existing concepts in both sorts of science. Both biological and quantum physical phenomena can be expressed through diagrams. The design fiction scenarios in these exhibitions would likewise employ diagrams. These design-fiction diagrams create a shared space for ideas in biology and quantum physics. They also disclose speculative conjectures that force us to consider—as citizen scientists or participants in these experimental exhibitions—our own ways of understanding the world.

Fig. 3.9 The big bounce is a hypothetical cosmological model for the origin of the known universe where the first cosmological event was the result of the collapse of a previous universe. In this idea, the universe would behave differently at different sizes. We take this idea and apply it to consideration of causal relationships in our cultural systems. How might we understand, reform, and evolve beyond these models of knowledge formation? Image by Matt Cornell.

Conclusion

In this essay, I consider the different ways that diagrams function as both a method for communicating concepts and as theory builders in scientific knowledge culture. I demonstrate how diagrams can serve as sites of convergence and divergence of ideas. Clearly, the thinking processes that underly the construction of diagrams and the cultural content that they embody can be derived from their visual and material forms. But it is also evident that diagrams that are used as part of design fiction can generate novel and unexpected scientific knowledge through their diagrammatic narratives.

I contend that scientific diagrams are an interwoven, design-driven performance of art, design, and science in the form of ArtScience. The process of designing diagrams has the potential to make critical contributions to science, while also foregrounding the visual attributes of science. The exhibitions presented in this essay utilize ArtScience design methods to produce and depict new ideas that are simultaneously familiar and uncanny. The exhibition spaces are, at once, laboratories and studios in which the visitors can play the role of citizen scientist "researchers" in this creative design and research process.

Acknowledgments

I would like to thank William Tham and Daniel Chong for lending their time generously to read and comment on the article.

Notes

1. A good animated example can be seen at: https://demonstrations.wolfram.com/NewtonsRotatingBucketExperiment/.
2. Philosopher Edmund Husserl argues that geometry is pre-eminent mathematical intuition because of how we experience and organize the complex characteristics of nature. Diagramming contributes to what Husserl refers to as "changeable collocation," in which what is "experienced is determined by surveying and measuring the prescientific, intuitively given surrounding world." This converges into an "essential form of the surrounding world" (199: 341). In other words, its meaning prior to sense-making could diverge, depending upon what is deployed in the process of sense-making. If the focus is aesthetics or artistic inspiration, the choice of representing nature would therefore be focused through that lens. When the focus switches to viewing nature through scientific lenses, priorities in the measure would change accordingly. But if ArtScience is the default measure, we can revert to a way of seeing the world that is organic and natural.
3. The shift from visible lines to hidden actions marks the move towards the algebraization of geometry. Such algebraization suggested the hidden abstract interactions in space/time, for example, in the development of the mathematics of Einstein's General Relativity and in the depiction of the forces for holding together the building blocks of matter. The algebraization of geometry coincides with transformation and translation from classical mechanics (describing the interactions of our world) to quantum mechanics (describing the world of the subatomic particle).
4. There are further discussions on the tracking between the different physical worlds in (Lee 2013).
5. See section III of Bodde 1991.
6. Some of the instruments used for astrological activities—such as the astrolabe—are the same ones that were used to measure and quantify celestial mechanics in astrology.
7. The role of imaging infrastructures in excavating the mysteries of the cosmos that simulate the process of diagramming through the deployment of code is explicated in Lee and Low, 2018.
8. Biological diagramming influenced diagrammatic thinking throughout the sciences. In the nineteenth century, collecting and measuring organisms informed the development of both macroscopic studies of natural history and investigations of the microscopic

world of cells, bacteria, and viruses. Theoretical biology—which uses theoretical analysis and mathematical models to investigate organisms—actually emerged from the nineteenth-century natural-history propensity for collecting and measuring specimens. The nineteenth-century preoccupation with natural history and measurement also contributed to the advent of twentieth-century theoretical physics and applied mathematics, and developments in algebraic geometry. This late-nineteenth-century relationship between biology and geometry, in fact, provided a diagrammatic vocabulary for explaining physical phenomena that were discovered because of improvements in instrumentation in physics.

9. There is a gray area between alternative knowledge—which is legitimized through funding and institutional support—and "pseudoscience"—knowledge of questionable scientific veracity. This gray area is where public interest often lies. The public may also be less interested in fundamental science, and more drawn to technological knowledge that is relevant to their lived experiences.
10. The phenomenon that was observed from these interactions (between the charged particles and the surrounding electromagnetic field) is known as the Aharonov-Bohm Effect (Howie 2012).
11. Diagrammatic interventions could be made by the co-designers of the interventions, either as extended pull-outs or on the core material itself.
12. The diagramming could be created out of the outcomes of an organic object subject to the colliding forces of a low-intensity gamma-ray laser, whereby microscopic changes to the organic object could then be documented.

References

Bodde, D. (1991), *Chinese Thought, Society, and Science: The Intellectual and Social Background of Science and Technology in Pre-Modern China*. Honolulu, Hawaii: University of Hawaii Press.

Darwin, C. (1858), Memorandum to W. H. Miller, www.darwinproject.ac.uk/commentary/life-sciences/evolution-honeycomb.

Davies, S., E. McCallie, E. Simonsson, J. L. Lehr, and S. Duensing (2008), "Discussing Dialogue: Perspectives on the Value of Science Dialogue Events that do not Inform Policy," *Public Understanding of Science* 18 (3): 338–53. doi:10.1177/0963662507079760.

Dieks, D. (2016), "Niels Bohr and the Formalism of Quantum Mechanics," Pre-print. http://philsci-archive.pitt.edu/id/eprint/12312.

Dunne, A. and F. Raby (2013), *Speculative Everything: Design, Fiction, and Social Dreaming*. Cambridge: MIT Press.

East Asian Science, Technology and Society 14 (2) (2020), Issue on thinking with diagrams, doi:10.1215/18752160-8538529.

Egan, G. (1992), *Quarantine: A Novel of Quantum Catastrophe*. New York: Harper Prism.

Feynman, R. P., R.B. Leighton, and M. Sands (1965; 1989), *Commemorative Issue: The Feynman Lectures on Physics,* Volume 3: Quantum Mechanics, Boston: Addison Wesley.

Howie, A. (2012), "Akira Tonomura (1942-2012)," *Nature* 486 (7403): 324. doi: 10.1038/486324a.

Husserl, E. (1999), "Transcendental Phenomenology and the Way through the Life-World," in Donn Welton (ed.), *The Essential Husserl: Basic Writings in Transcendental Phenomenology*, 337–63. *Studies in Continental Thought*, Bloomington: Indiana University Press.

Isaac, S. (2021), "Under Your Skin Anatomical Flap Books," *Royal College of Surgeons*, www.rcseng.ac.uk/library-and-publications/library/blog/under-your-skin-anatomical-flap-books/ (accessed on March 14, 2021).

Krulwich, R. (2021), "What Is It About Bees And Hexagons?" Internet Radio. *Krulwich Wonders: Robert Krulwich on Science*, www.npr.org/sections/krulwich/2013/05/13/183704091/what-is-it-about-bees-and-hexagons (accessed on March 14, 2021).

Laurenza, D., M. Taddei, and E. Zanon (2006), *Leonardo's Machines: Da Vinci's Inventions Revealed*. Pynes Hill, UK: David and Charles.

Lee, C. A. L. (2013), "The Science and Art of the Diagrams: Culturing Physics and Mathematics, Part 1," *Scientific American*, https://blogs.scientificamerican.com/guest-blog/the-science-and-art-of-the-diagrams-culturing-physics-and-mathematics-part-i/.

Lee, C. A. L. (2015), "Non-Trivial Philosophy: Cybernetics, Data Analytics, and the Biophysics of Information Theory," *Kybernetes* 44 (8/9): 1310–23. doi:10.1108/K-10-2014-0227.

Lee, C. A. L., and W. S. Low (2018), "Speculative Code: Mediating the Virtual-Reality of Emergent Science," *Nomorepotlucks*, https://www.researchgate.net/publication/333372380_Speculative_Code_Mediating_the_Virtual-Reality_of_Emergent_Science_-

Marsh, G. (2021), "How to See around Corners," *Nature News*, doi:10.1038/nature.2012.10258 (accessed on March 14, 2021).

McFadden, J. and J. Al-Khalili (2018), "The Origins of Quantum Biology," *Proceedings of the Royal Society A: Mathematical, Physical, and Engineering Sciences* 474 (2220): 20180674. doi:10.1098/rspa.2018.0674.

Miko, I. (2021), "Gregor Mendel and the Principles of Inheritance   Learn Science at Scitable," *Nature Education*, www.nature.com/scitable/topicpage/gregor-mendel-and-the-principles-of-inheritance-593/ (accessed on March 14, 2021).

Norris, A. (2014), "5 Intriguing Buildings Inspired by Bees," *From the Grapevine*, www.fromthegrapevine.com/lifestyle/5-intriguing-structures-inspired-bees (accessed on February 12, 2019), link inactive on June 15, 2022.

Plotnitsky, A. (2006), *Reading Bohr: Physics and Philosophy* (Fundamental Theories of Physics), Dordrecht: Springer.

Rust, C. (2004), "Design Enquiry: Tacit Knowledge and Invention in Science," *Design Issues* 20 (4): 76–85. doi:10.1162/0747936042311959.

Rutkin, H. (2006), "Astrology," in Katharine Park and Lorraine Daston (eds.),*The Cambridge History of Science: Volume 3: Early Modern Science*: 541–61, Cambridge: Cambridge University Press. doi:10.1017/CHOL9780521572446.024.

Salter, C., R. V. Burri, and J. Dumit (2017), "Art, Design, and Performance," in U. Felt, R. Fouché, C. A. MIller, and L. Smith-Doerr (eds.), *The Handbook of Science and Technology Studies*, 139–67. Cambridge: MIT Press.

Taylor-Smith, K. (2019), "Quantum Physics in Medicine," *AZoQuantum.Com*, www.azoquantum.com/Article.aspx?ArticleID=124.

Thompson, D. W. and J. T. Bonner (1961), *On Growth and Form*. Cambridge Paperbacks. Cambridge University Press.

Wright, A. S. (2013), "The Origins of Penrose Diagrams in Physics, Art, and the Psychology of Perception, 1958–62," *Endeavour* 37 (3): 133–9. doi:10.1016/j.endeavour.2013.02.001.

4 Phenomenal Machines: An Interview with Nicole Koltick

LESLIE ATZMON

Nicole Koltick's *Phenomenal Machines* is a generative design project that speculates on how non-human agents—robots, crystals, and their environment—exhibit cognition and create a sense of "place" together outside the realm of human cognition. The robots utilize basic "vision" and behavioral algorithms that enable them to interact with their environments. This behavior, in turn, influences the form and color of sodium chloride salt crystals. The crystal formation and growth are affected by the actions of the dynamic terrain (through secretion of salt solution or expansion and contraction of the dynamic surface), or the interventions of the robots operating within the terrain (through placement, movement, and disruption).

Atzmon If you were to describe the project to your mom, what would you say?

Koltick A simple summary of this piece, *Phenomenal Machines*, would be "robots with a hobby" or "robots who garden." Instead of the archetype of robots who are doing things for *us*, we wanted to flip that relationship and ask what we could do for robots. We wanted to build an environment for them to inhabit— to allow them to roam freely—and give them a hobby: crystals for them to garden.

Atzmon What was the impetus for this project? What influenced this work?

Koltick I've been working on speculative robotics since 2008 or 2009. This previous work involved large-scale geo-engineering projects in which autonomous bots with a range of capabilities would roam various landscapes and perform certain tasks.

Atzmon Were those real-world or virtual landscapes?

Koltick They were digitally modeled and digitally rendered. Science fiction, in a sense.

Atzmon Design fiction?

Koltick Yes, exactly. I'd been working on some projects like that. There was one that proposed a synthetic iceberg in the North Atlantic. The robots would swarm around each other and form a synthetic reflective surface and they could break apart the icebergs in shipping channels. I worked on another proposal for a series of robots that would go into quarries, and their job was to re-mineralize the quarries and create these very geometrically complex cave interiors. Then these interiors would be re-seeded with various kinds of flora and fauna. So, the project involved taking a blocky, sparse kind of underground condition and repopulating it and turning it into a synthetic cave.

Atzmon If I remember right, you're an architect?

Koltick Yes, I'm trained as an artist and an architect. I think my work really has that crossover.

Atzmon So where did the idea of robots and crystals come from then?

Koltick I founded my lab in 2012. And one of the things I wanted to do was make digital things tangible. Before that, my earlier robot projects were all virtual. They just existed in the digital realm. In my lab, I wanted to begin testing things in the real world at full scale. So, we started to do some projects that we built. Some of these had small-scale robotic components. I wanted to do a full-scale real-world robotics project using my past virtual robotics experience. We had some pretty robust brainstorming sessions in the lab. We had six masters thesis residents working in the lab, and we began talking about synthetic ecologies and robots—it was a very loose discussion. We were looking at salt flats and other interesting geological phenomena, and we just happened upon this idea of a crystal-manipulating robot. One of our initial ideas was that this robot would just move along and encrust things.

Atzmon The connection with your architecture is that this project is a total environment? Here nothing is human, though, right?

Koltick What drew me to architecture is that it's about a series of interconnected relationships: the building is interacting with people, the building is interacting with the environment. So, there's a series of nested relationships at various scales, which I'm very interested in, and these new projects take on that idea.

Atzmon This project with the robots and crystals, with nothing human in it, makes you think about regular architectural interrelationships differently.

Koltick My interest in the perspectives of the non-human comes out of my deep engagement with post-humanist philosophy. This philosophy had a renaissance in the early 2000s and is now everywhere. Here I'm trying to de-center the human from perception, interaction, and sensation. I am really entranced by the idea of computational phenomenology—how a computer or robot perceives and experiences the world. That was a critical conceptual thrust for *Phenomenal Machines*. If we're building a world for these robots, what is their experience of it? Can we try to think of different ways that we can set up relationships and interactions for them that we can make visible? That had an effect on the color in *Phenomenal Machines*, which comes out in the interaction between the robots and the landscape in this particular project.

Atzmon OK. Let's talk about *Phenomenal Machines*. This was a real-world project, right, not virtual? Who built the robots and what are the crystals?

Koltick This project is all real world, not virtual. I took a big jump in my work from speculative to real world. My team—a philosopher/engineer, two graduate thesis students, and a graduate assistant resident, and I—built the robots. We were a very scrappy group; obviously not trained as roboticists. We started working with simple robotic pieces and really built the rest from there. The team was interested in developing a new species of robot, one that's not working for humans, as well as putting something novel into the world.

Atzmon A human obviously wrote the algorithm, but then "hands off," right?

Koltick Yes. The idea was let them roam with the crystals. But there's another layer of complexity: there's an environment that's actually automated that we consider to be another form of robot. The environment heaves air and water; we gave this environment a kind of autonomous ecological capability—almost like it has weather.

Atzmon So real air and real water. I'm sure the air and water affect the crystals? And does this "weather" affect the robots?

Koltick Yep. The robots respond to the "weather." They have to activate the crystals with a mineral solution, which is mainly water. Without water, the crystals don't grow.

Atzmon What are the crystals made from?

Koltick The crystal is a sodium chloride salt-based solution, and it has a substrate that's embedded with color. When we were initially researching the crystals, we tried and tested all kinds of crystal-forming solutions. There was a lot of material testing in the lab to get to the solution we finally used in the project. And we investigated which crystals and colors would be most aesthetic and would show up best to get the robots to view them. The crystals grow, and they then start degrading and tumble around the environment. These crystal "fluffs" are "recyclable"; they can be rehydrated again by the robots.

Atzmon Did you work on the crystals with the same team? No chemist?

Koltick No chemist, the same team, with some Google search mad-scientist stuff. The attitude was we'll just make it work!

Atzmon The algorithm directs the robots' behavior?

Koltick There's actually a series of nested algorithms, not just one. There are algorithms that govern how the robot moves, and then there are algorithms that generate what precipitates certain movements in the robot. If the robot recognizes a certain color, for example, how does it craft something in response, and how does it navigate the landscape? There are also algorithms governing what happens within the landscape in terms of the air bladders and water circulatory systems.

Atzmon The robots and crystals and environment just run, and the changes that happen are not exactly random, right?

Koltick They are "emergent." It's an emergent outcome, which ties back to biology. I have a deep interest in biological systems, as well as collective behavior. There has been extensive research investigating how biological algorithms interact with each other—you end up getting effects that are unexpected.[1] You set things in motion, but you don't know what you are going to get.

Atzmon I understand that your project deals with a kind of cognition. Can you please explain that?

Koltick Once you're moving past a human-centered idea of cognition, it becomes clear that cognition can take many forms—some that are not recognizable to us. Here I ask what does taking in information, translating it, and then putting it into action look like for a non-biological entity? What does robot cognition look like? We begin to see what they are seeing, to visualize what they are experiencing and observing the effects of those interactions.

Atzmon What did you observe them experiencing?

Koltick We observed their vision as well as their overall interactions with each other and the environment. We had provided them with a camera for "eyes" and we incorporated some of this footage into our video of the project (see the video at vimeo.com/209855928). We were interested not only in their vision but also in the traces of their activity. Since this project set up multiple agencies, including robotic agencies, mineral agencies, and material agencies, we wanted to see what was visible and what emerged from these interactions. This involved observations of not only the crystals but also their dispersal throughout the environment. The role of color helped make these interactions highly visible.

Atzmon What does the color do in *Phenomenal Machines*?

Koltick The crystals are naturally white; they could have been white in the project, but color drew attention both to the agency of the robot and what it did with the crystals. Including color to amplify this robot behavior was key to the project narrative, and it helps human viewers recognize this narrative. In other words, the color enhances legibility.

Atzmon OK, so it's sort of like staining a cell so you can see the organelles? The color is also really beautiful. When you watch the project video, the whole project is very beautiful in a unique way. Did beauty enter into this at all?

Koltick Yes, that cell staining example offers a good parallel. As an artist and architect, for me aesthetics is always running in the background among my criteria. I also believe that the novel aesthetics of this project is a way to grab attention and spark a gut reaction in people. I have three criteria in my lab: everything has to have novel aesthetics, to generate and use novel processes, and present some kind of novel reaction to the material world.

That's where this project gets into design research. How do you develop, test, and iterate every piece of it, what does the landscape look like, how do we model the geometry, how do we work with and form the silicone? We did so many tests with the crystals as well as for the color and form. Doing all of this contributes to the evolution of the project and the narrative so that we eventually get to *Phenomenal Machines*.

Atzmon Yes, design research is so important to all of the work in *Design and Science*. So glad we got a chance to chat about your project *Phenomenal Machines*. Thank you, Nicole.

Koltick Thank you.

Fig. 4.1 Mineral agencies. Top three rows: Crystal research and development. Bottom row: Crystals in synthetic landscape and robot gardening (far right). Project team: Nicole Koltick, Elena Sabinson, Jay Hardman, and Mike Hogan

Fig. 4.2 Design research and development of the robots and synthetic ecology habitat. Top row: 3D-printed robot “skin” and the robots, which are called NESL (Nurturing Emergent Synthetic Life). Second row: Robots in development, and the design and development of allomimetic gestural programming facilitated through a custom controller. We developed this system in order to model the robots' behavior on a series of intuitive gestures. We enacted, captured, and then transmitted these gestures to the robots, which provided them with a range of behavioral techniques for their gardening tasks in the habitat. Third row: Full-scale fabrication of the silicone terrain including water circulation systems and inflatable air bladders. There is a progression from a small-scale, 3D-printed model of the terrain to full-scale fabrication of the final components. We developed custom molds and casting techniques to create the overall terrain and the air bladders, which are an integral part of this novel soft robotics system. This terrain has both air and water systems integrated into it, along with custom electronic sensors and actuators. Bottom row: The completed landscape with liquid pools and detailed shots of the silicone textured surface.

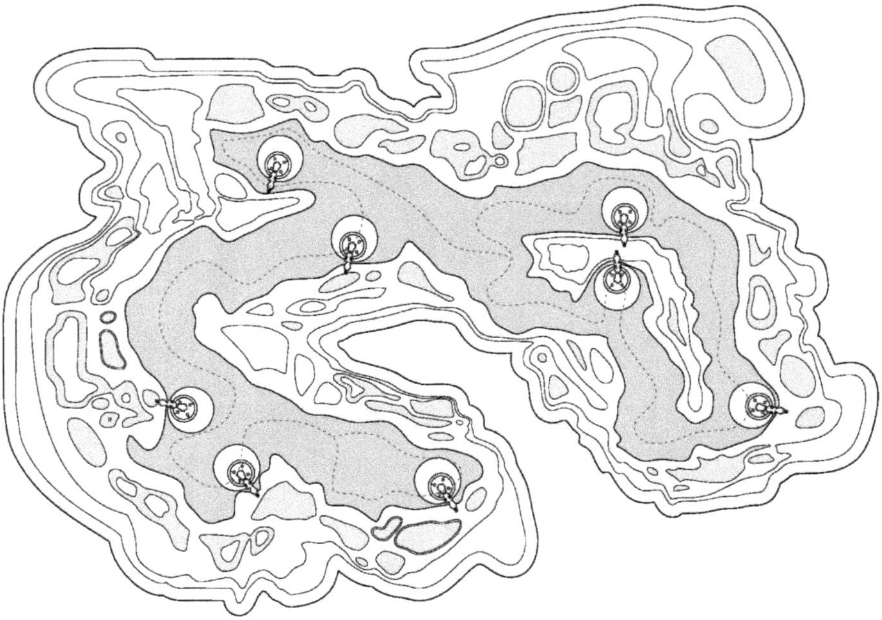

Fig. 4.3 Project terrain with robots in their habitat (plan view)

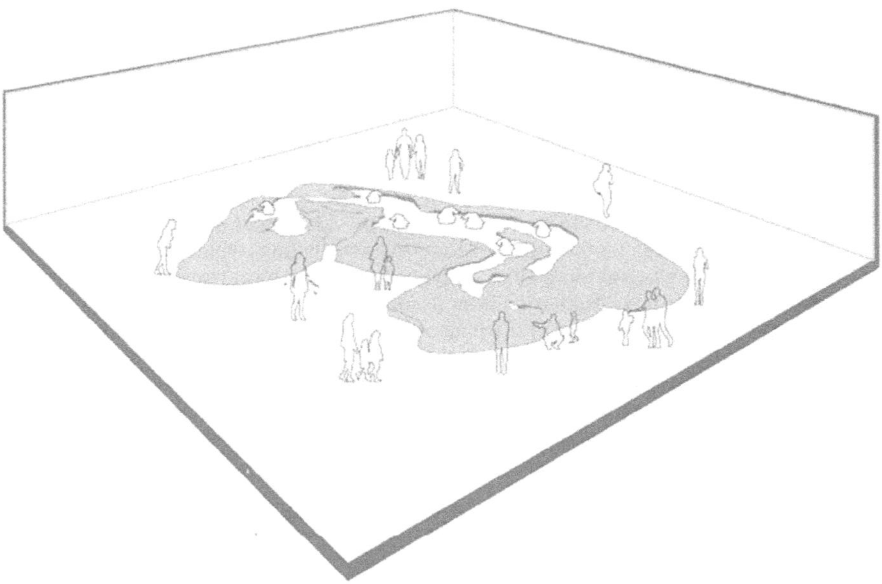

Fig. 4.4 Synthetic ecological habitat (installation overview)

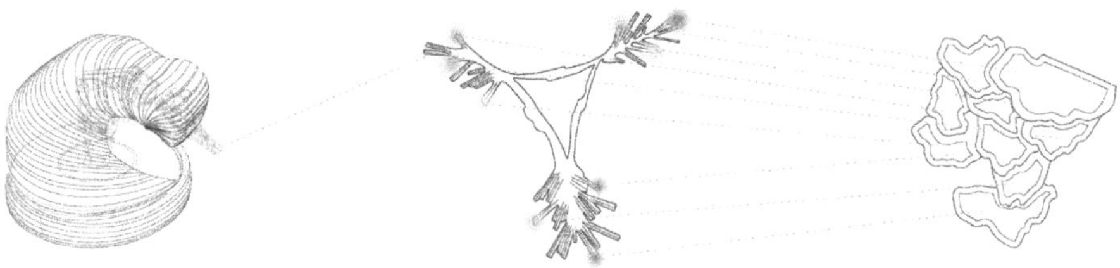

Fig. 4.5 Intersecting agencies diagram: Robotic agencies, mineral agencies, material agencies.

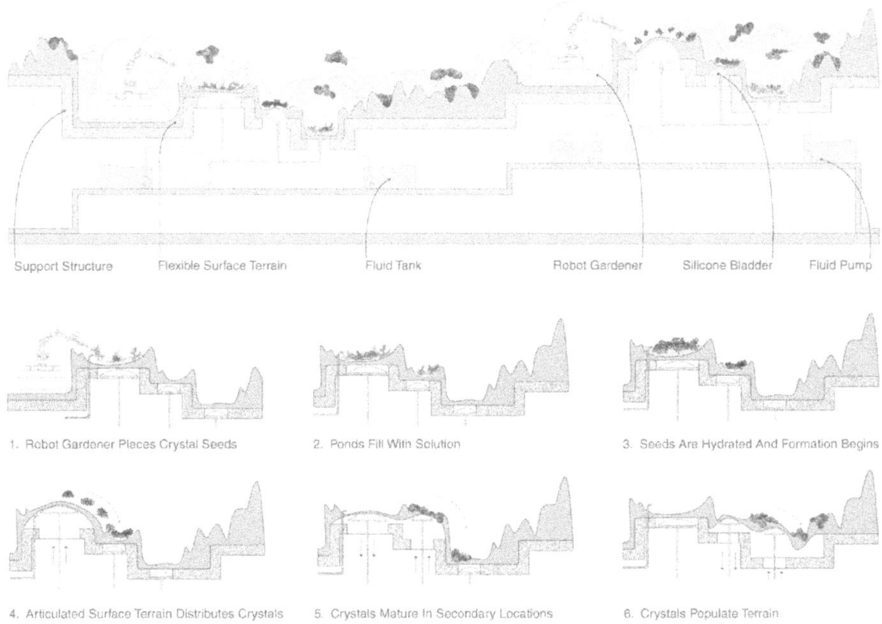

Fig. 4.6 This diagram depicts the movement and lifecycle of the crystals, which evolve during their interaction with the robots and the terrain. This habitat is composed of several distinct systems that form the basis of the dynamic cycles of the ecology in the habitat. First, there is a circulatory system that supports watering and hydration of the crystals. Next, there is a soft robotic terrain that uses air to gently inflate and deflate parts of the system. The air acts in response to corresponding behaviors in the watering system, crystal placement, and the robots' movements. The robots (NESL) navigate this terrain and place crystal seeds. They continue to garden the grown crystals in conjunction with these embedded movements and behaviors that take place in the landscape.

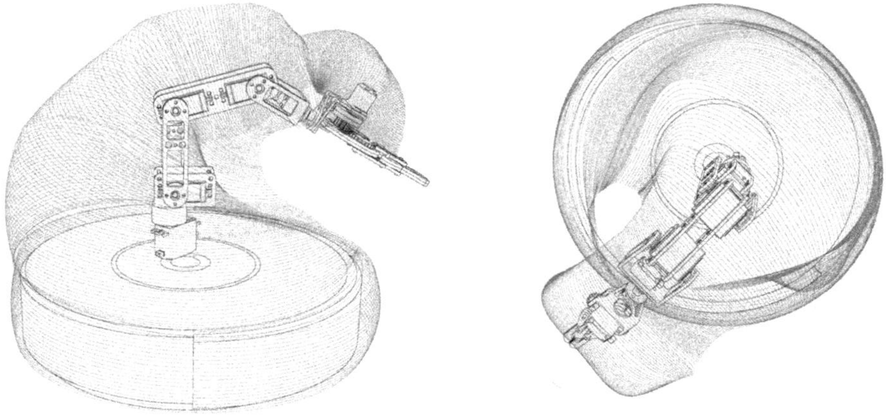

Fig. 4.7 NESL (Nurturing Emergent Synthetic Life) robot.

Note

1. For more information, please see Ingrid Lobo, (2008) "Biological Complexity and Integrative Levels of Organization," *Nature Education* 1 (1): 141. Online at: nature.com/scitable/topicpage/biological-complexity-and-integrative-levels-of-organization-468/ (accessed on July 1, 2021).

Part 2 Biomimicry and Biodesign

5 Design Inspired by Nature: The Bat Brolly

CLINT PENICK AND PRASAD BORADKAR

In industrial design that focuses on the creation of physical products, the reigning paradigm promotes user-centeredness and stresses design as an activity that addresses the needs of people. This emphasis can be traced to the publication *Designing for People*, in which designer Henry Dreyfuss (1955) paves the way for user-centered design, as well as for human factors and ergonomics in industrial design. He reminds us that the objects that we design are "ridden in, sat upon, looked at, talked into, activated, operated, or in some way used by people individually or en masse" (Dreyfuss 1955: 1); therefore, according to Dreyfuss, considering people's physiological and psychological needs should be central to the design process. Since then, industrial designers have taken this task of designing for people seriously, which has given rise to user-centered design, human-centered design, participatory design, and empathic design (Brown 2009; Cagan and Vogel 2003; Dreyfuss 1955; Hanington 2012; IDEO 2011; Liedtka 2011; Norman 2013).

Anthropocentrism in design, however, has encouraged a myopic and self-centered conception of our goals as designers. Clearly, the things that we design based on diligent user research, and with the utmost care for people, do not impact people alone. The consequences of design activity—human-centered or otherwise—reach far beyond humans. We are, after all, one of several million species who live on this planet. Why, then, should our design be so anthropocentric? Can we design products and services keeping people in mind, while also considering other species and entire ecosystems? Can we envision the potential impact of our design on people, but also on the other inhabitants of our biosphere? We would like to suggest that it is time to re-examine our anthropocentrism in design in order to progress towards a more sustainable future.

When we shift from purely anthropocentric thinking and human-centered design to biocentric thinking and life-centered design, our solutions certainly demonstrate relevance beyond people. This in no way suggests that we reject human-centered design; instead, it recommends that we re-imagine our goals and adopt new methods that acknowledge the other millions of species who are our neighbors. Of course, sustainable design, green design, ecodesign, and other similar practices do address issues of the environment. What we need, however, is an essential expansion and unambiguous reframing of who we identify as "target users." How can we extend our conception of target users beyond people to include all living beings?

In addition to being "neighborly" to other species by including them in our design goals, we can also learn from them. This is where biomimicry can help. Biomimicry is an emerging discipline that seeks to emulate nature's strategies to create sustainable solutions to human challenges. Humans have looked to nature for inspiration since our earliest history, but a formal methodology for biomimicry drawing on peer-reviewed biological research has only evolved over the past several decades. In her seminal book, entitled *Biomimicry: Innovation Inspired by Nature*, Janine Benyus writes:

> Unlike the Industrial Revolution, the Biomimicry Revolution introduces an era based not on what we can *extract* from nature, but on what we can *learn* from her. This shift from learning about nature to learning from nature requires a new method of inquiry, a new set of lenses, and above all, a new humility.
>
> 1997: 2

In recent years, biomimicry has been used to create faster bullet trains (McKeag 2012), more efficient wind turbines (Fish, Weber, Murray and Howle 2011), and sustainable factories (McNeal and Benyus 2015).

In the following essay, we provide a general overview of the "biomimicry thinking" framework as outlined in the *Biomimicry Resource Handbook* (Baumeister et al. 2014), which presents a philosophy and a methodology to practice innovation inspired by nature. We illustrate this process by applying biomimicry thinking to the redesign of an umbrella, and we start with a simple challenge: *how might we design an umbrella that resists inversion in the wind?* And, as is typically done in biomimicry, we explore how organisms deal with similar challenges and identify deep principles in nature that can be applied to the design of this product. The final step is to evaluate the biomimetic designs using Life's Principles (Fig. 5.1). This entire process helps to guide us away from anthropocentric thinking and human-centered design to biocentric thinking and life-centered design.

Introduction to Biomimicry Thinking

Biomimicry is the practice of learning from nature, and of emulating natural forms, processes, and ecosystems to create more sustainable designs. Strategies in nature are not truly "designed," but like design they have evolved through a process of trial and error—one that has occurred over millions or billions of years. Strategies that "work" in evolution get passed on to the next generation, while those that fail lead to extinction. In the words of evolutionary biologist Richard Dawkins, "Natural selection is the blind watchmaker, blind because it does not see ahead, does not plan consequences, has no purpose in view. Yet the living results of natural selection overwhelmingly impress us with the appearance of design as if by a master watchmaker" (Dawkins 1996: 29).

Biomimicry (or biomimetic design) differs from related design methodologies such as biophilic design, bioutilization—which directly uses organisms or biological materials—or biomorphic design. The focus in biomimicry is on emulating the functional properties of organisms and ecosystems rather than copying their aesthetic features or incorporating biological materials. A biomimetic design may therefore look nothing like the organism that inspired it, but it must function like the organism. Likewise, biomimetic products are not always made from natural materials. The Japanese bullet train, for example, was inspired by the aerodynamic shape of a kingfisher's beak, but it is manufactured with steel and aluminum rather than keratin. Whether or not a biomimetic design is constructed from natural or synthetic materials, the goal is to create a design that is more sustainable.

Biomimicry thinking provides a framework for integrating biomimicry into the design process. Describing biomimicry thinking, Baumeister et al. explain that:

> [O]ver the years we've created and tested different versions of what we formerly called a biomimicry methodology. Experience has made us wiser, and we now know that there is not just one methodology, start to finish, that works for all industries and all applications . . . As a result, we no longer teach a methodology *per se*, but rather share a framework with which one can learn how and when to incorporate biomimicry thinking into design.
>
> 2014: 5–86

The biomimicry thinking framework is therefore not a linear set of instructions, but is instead a system that provides multiple points of entry for biomimicry into design.

There are four phases of the biomimicry thinking framework during which biomimicry can be incorporated into the design process: scoping, discovering, creating, and evaluating. The sequence in which these phases are completed depends on the project, but each stage provides a mechanism to identify principles, patterns, strategies, and functions in nature that can inform design. The scoping phase focuses on describing a design problem as well as defining the context under which a solution to the problem must function. The discovering phase seeks to identify an appropriate natural model that could inspire a design solution to the problem defined in the scoping phase. The creating phase translates the principles learned from the natural model during the discovery phase into features that can be incorporated into the proposed design. And, finally, the evaluating phase looks to assess how well the proposed design addresses the original problem and how well it meets what are called Life's Principles (Fig. 5.1), a common set of strategies found in the natural world that represent nature's strategies for sustainability.

The steps of the biomimicry thinking framework are not necessarily linear, as noted above, and the order varies depending on the path taken during the design process. In biomimicry, there are two main paths for connecting natural models with an appropriate design problem (Fig. 5.2). The first path, "biology to design," starts with a specific biological insight, and then tries to connect that insight to a relevant design challenge. Many famous examples of biomimicry have started with

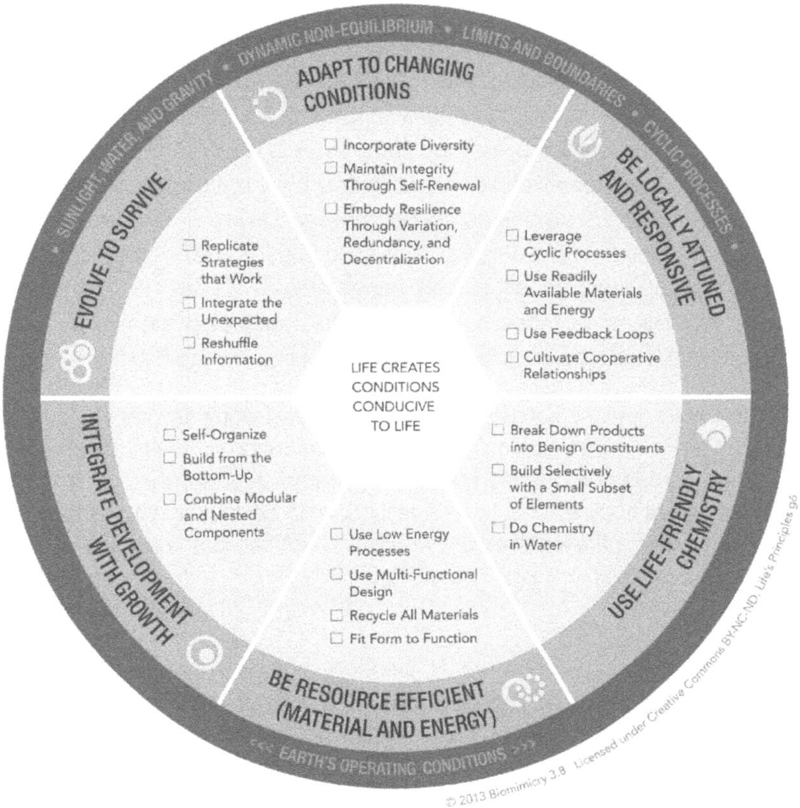

Fig. 5.1 Life's Principles are a common set of strategies found in the natural world that represent nature's strategies for sustainability. These principles are based on the recognition that all life on Earth is subject to the same operating conditions (indicated in the gray outer ring), including primary energy and material inputs (sunlight, water, and gravity), limits and boundaries, cyclical processes (day/night cycles, annual cycle, etc.), and the fact that all systems on Earth are in a state of dynamic non-equilibrium. The six primary principles and their associated sub-principles surround the aspirational goal and emergent property of these principles stated in the center that "life creates conditions conducive to life." © 2013 Biomimicry 3.8.

a surprising discovery in biology that later sparked innovation in design or engineering when applied to a specific problem. The discovery of the adhesive properties of gecko feet, for example, led to years of research into adhesive nanostructures that have now been developed into dry adhesives for medical and engineering applications (Autumn and Gravish 2008). The second path, "challenge to biology," starts with a design problem, and then tries to find a natural model that provides insight into solving that problem. For our work on the bat umbrella, we followed the challenge to biology path because we started without having a clear natural model in mind.

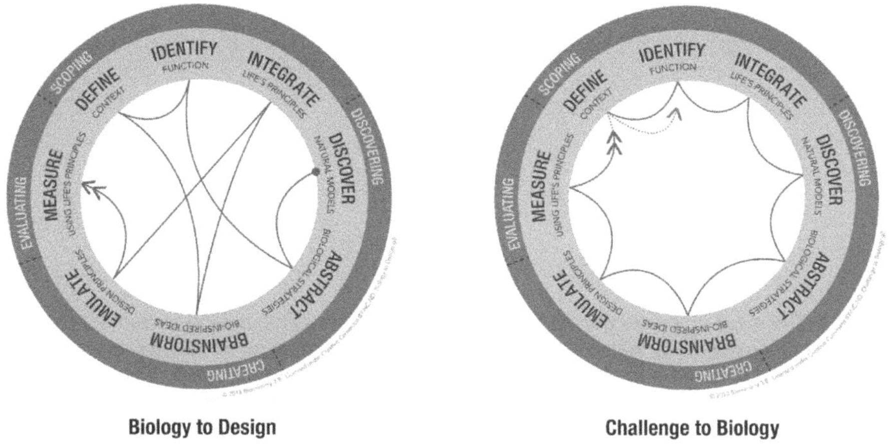

Fig. 5.2 There are two paths for biomimetic design as outlined in the biomimicry thinking framework. (A) The "biology to design" path begins with a specific biological insight and then tries to connect that insight with an appropriate design challenge. (B) The "challenge to biology" path starts with a design challenge and then tries to find a biological model that provides insight into solving that challenge. Both paths progress through the same four phases (scoping, discovering, creating, and evaluating), but they do so in a different order. For the bat brolly, we followed the challenge to biology approach. © 2013 Biomimicry 3.8.

Scoping: Defining the Problem

The scoping phase of the biomimicry framework begins by defining a clear design challenge or problem. For our umbrella design, we focused on the challenge of designing an umbrella that does not invert in the wind. Umbrellas generally fail when a gust of wind forces the canopy to invert, causing the ribs to bend or to break (Fig. 5.3). This situation is especially damaging to compact umbrellas, which have two additional hinges along each rib that can easily snap when an umbrella is inverted. Although the basic design of an umbrella is simple, a compact umbrella can have more than 150 individual parts, making it difficult to repair or recycle. Therefore, damage to an umbrella almost always results in its disposal. Estimates suggest that up to 1.1 billion umbrellas are thrown away each year (Holley 2018). Creating a more robust umbrella that resists breakage in the wind would significantly reduce the waste created by damaged umbrellas, and it also aligns with the sustainability goals of biomimicry.

The earliest reference to a collapsible umbrella dates to the year 21 AD when Chinese Emperor Mang Wang had an umbrella built for his personal chariot (Needham 1986). The basic design of Mang Wang's umbrella would be familiar today—it had a circular canopy supported by thin ribs with articulating joints that allowed the umbrella to open and close. Over the years, inventors have developed a range of solutions to improve umbrella performance in windy conditions, evidenced by thousands of patents filed since the year 2000 for wind-resistant and wind-proof umbrellas.

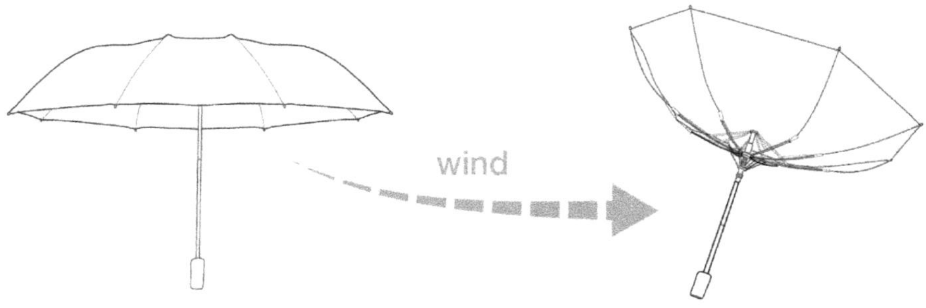

Fig. 5.3 The most common way an umbrella fails is when a gust of wind forces the canopy to invert. Once inverted, the umbrella can no longer function as an effective rain shield, and the thin ribs and mechanical components are often bent or damaged in the process.

One simple solution to the umbrella inversion problem is to use stronger materials, such as temper-hardened steel or carbon fiber. While these materials increase strength, they also raise the cost as well as the weight of the product, which makes carrying it more burdensome. A second approach has been to add vents to the canopy that allow wind to escape rather than to turn the umbrella inside out. This approach has been commonly used for large golf umbrellas that would otherwise trap strong winds, although excessive winds can still invert these umbrellas. Finally, more recent designs, like the Senz umbrella developed by Gerwin Hoogendoorn in 2006, have explored changing the entire shape of the umbrella to increase stability in the wind (Holley 2018). The Senz umbrella moves away from the traditional circular shape to produce a more aerodynamic profile, and it ranks among the strongest umbrellas to-date.

We wondered, though, if there were strategies in nature that we could use to improve umbrella design further while also reducing material waste. Once the problem has been defined, the next step in the scoping phase is to describe the function and general context under which the design must function. An umbrella should obviously shield its user from the rain (the function), but it must also be lightweight, compact, foldable, and strong enough to withstand heavy winds (the contexts under which it must function). Although these functional requirements are identified in the design process regardless of whether the designer is applying biomimicry, for us they were used first to identify and then to narrow down natural models that could potentially inform our design. In nature, organisms that feature structures with a large surface area, such as leaves or wings, experience stress from wind similar to that which damages umbrella canopies. This is particularly true for organisms that live in areas with high winds, such as coastal regions located in hurricane zones. In addition, organisms that fly must also contend with rapidly shifting air currents that could damage their wings. Moving to the discovering phase, we used these criteria to narrow down our search for appropriate organisms that might inspire our umbrella design.

Discovering: Identifying a Natural Model

The goal of the discovering phase is to identify natural models and key design principles that could provide solutions to the problem outlined in the scoping phase. To enable the transition between the scoping phase and identification of an appropriate natural model, authors of the *Biomimicry Resource Handbook* recommend "biologizing the challenge," which they characterize as "a term we made up that describes the process of turning a human challenge into a question that might appropriately be asked in biology" (Baumeister et al. 2014: 108). Therefore, instead of asking how would nature prevent an umbrella from inverting in the wind, the question that we pose is: *How would nature prevent a structure with a large surface area from buckling under force?* First, this particular question allows us to consider any structure in nature with a large surface area as a potential natural model, and second, it opens up the solution space and inspiration to come from organisms that deal with any type of strong force, including wind, ocean currents, or other physical forces that could damage a fragile structure.

We first considered plants with large leaves as a potential source for design inspiration, particularly those that are exposed to strong winds. Fan palms, for example, have leaves that may be larger than a meter in width and must be able to withstand hurricane-force winds in coastal areas. To add strength, the leaves of the fan palm are pleated. However, in very strong winds, the leaves tear along the pleats to allow wind to pass through and prevent further damage to the tree. Although we would not want to create an umbrella that tears apart in the wind, in some ways this is analogous to placing vents in the canopy of an umbrella to allow wind to pass through rather than to exert force on the canopy as a whole. Other large-leafed plants rely on intricate vein patterns to support their leaves. The floating leaf pads of the Amazonian water lily, *Victoria amazonica*, are supported by veins on the underside of the pads that reinforce the thin leaf structure above. Writing about the vein structure of the Amazonian water lily and the leaves of other plants, architect Michael Pawlyn suggests that, "the principle of using ribs to stiffen a thin surface may very well have inspired engineers to design similarly efficient structures" (2016: 14).

We ultimately shifted our focus, though, to bats as the biological model that best matched our criteria for creating a stronger umbrella. The bone structure of bat wings was a better analog to the rib structure of current umbrellas, and the ability of bat wings to fold down into a small size matched our desire to create a compact umbrella. Bats evolved flight independently from birds, and while birds rely on feathers to define their wing shape, the wings of bats feature a thin bilayer of skin that is embedded with bone to provide a support structure. To remain lightweight for flight, the bones inside a bat's wings are reduced to thin ribs. In spite of this reduced bone mass, the wing structure as a whole must be robust when faced with high gusts that apply force across the wing membrane. A combination of thin, structural ribs supporting a large wing surface area makes bat wings similar in construction to modern umbrellas.

The bones that compose bat wings are homologous to those in the human hand,

and compared to bird wings, which are relatively stiff, bat wings are dexterous and flexible because of the large number of joints. For comparison, bird wings each contain only three joints, while bat wings contain more than forty active and passive joints (Ramejani et al. 2017). This allows bats to perform remarkable feats, including their ability to develop lift in flight on both the up and downstroke of each wingbeat. Mimicking bat flight has been a "holy grail" of aerial robotics, and interest in bat flight biomechanics dates back to Leonardo Da Vinci, who in 1508 wrote a note to himself stating, "Dissect the bat, study it carefully, and on this model build the machine" (MacCurdy 1908: 152).

Our focus for developing the bat-inspired umbrella was not based on the detailed mechanisms of bat flight, but instead on how structural properties of bat wings could allow bats to deal with strong winds. Like modern umbrellas, bat wings are jointed (as described above), which creates weak points where damage is most likely to occur. For this reason, bats are particularly sensitive to hyperextension of wing joints. From an evolutionary perspective, bats are extremely successful—they evolved around 50 million years ago and comprise the second largest order of mammals (*Chiroptera*), with more than 1,200 described species (Neweiler 2000)—and bats have found effective ways to prevent their wings from getting badly damaged in the wind. Moreover, large bats, such as flying foxes, have wingspans reaching nearly two meters and are roughly equivalent in surface area to a common umbrella (Nowak and Walker 1999). The similarity in size between the largest bat wings and umbrella canopies suggests that bats and umbrellas face similar forces across their wing and canopy surfaces, respectively.

Drawing on the similarities between bat wing anatomy and the structure of modern umbrellas, we focused our final research in the discovering phase on the anatomical features of bat wings that add stability and prevent joint hyperextension. We describe the outcomes of this research in the following section (the creating phase), where we break down each feature of bat wing anatomy relevant to our functional requirements and explain how these features were incorporated into our umbrella design.

Creating: Incorporating Lessons from the Natural Model

Integrating biomimicry thinking at the creating phase means distilling lessons that were learned during the discovering phase into concrete ideas that can be incorporated into a final design. Our research on bat wing anatomy revealed three features that could be applied to strengthen umbrella design. The first two of these features relate to the skeletal structure of bat wings, which can be applied to the support mechanics and ribs of a collapsible umbrella. The third feature relates to the integrated wing membrane, which is analogous to the umbrella canopy. We incorporated aspects of all three components into the final umbrella design (Fig. 5.5).

As noted above, the bones of the bat wing are homologous to those of the human hand, but there are also some differences (Fig. 5.4). First, the "finger bones" of bat

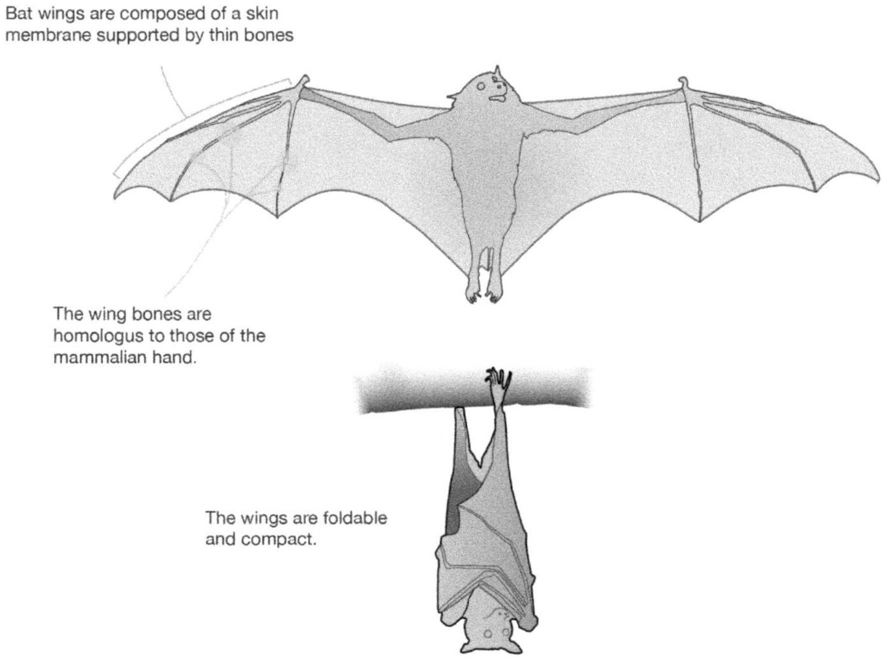

Fig. 5.4 The anatomy of a bat wing. Bat wings are composed of a skin membrane supported by thin bones homologous to those of the human hand. The combination of thin structural bones supporting a large wing surface makes bat wings similar in construction to modern umbrellas. Likewise, bat wings are designed to fold up into a compact form when resting, similar to a closed umbrella.

wings are inextricably linked to each other through the webbing of the wing membrane. Whereas the movement of each human finger is more or less independent from its neighbors, the bones inside a bat wing can exhibit tension on their neighbors through the connecting membrane (Norberg 1972). In addition, the joints of the bat wing are turned at slightly different angles from one digit to the next (Riskin et al. 2008). In the human hand, all joints in the fingers articulate in more-or-less the same direction, but the joints in the bat wing are turned so that each neighboring joint is slightly rotated (Fig. 5.4). Therefore, when a heavy gust threatens to hyperextend one joint, the neighboring joints resist bending in the same plane and add stability. Taken as a unit, the entire wing is stronger as a whole than each individual digit.

In our project, we mimicked the joint angles in the bat wing by rotating the ribs of a traditional umbrella so that the joints in neighboring ribs were slightly offset. Unlike a traditional umbrella, where the ribs extend radially from the crown to the edge of the canopy in a radial design, the ribs of the bat umbrella also bend laterally (Fig. 5.5A). Each neighboring strut in our design is turned to a slightly different angle, like the joints in the wing of the bat, which allows neighboring joints to buffer each other so that adjacent ribs don't bend in the same plane. Because the joints are the weakest

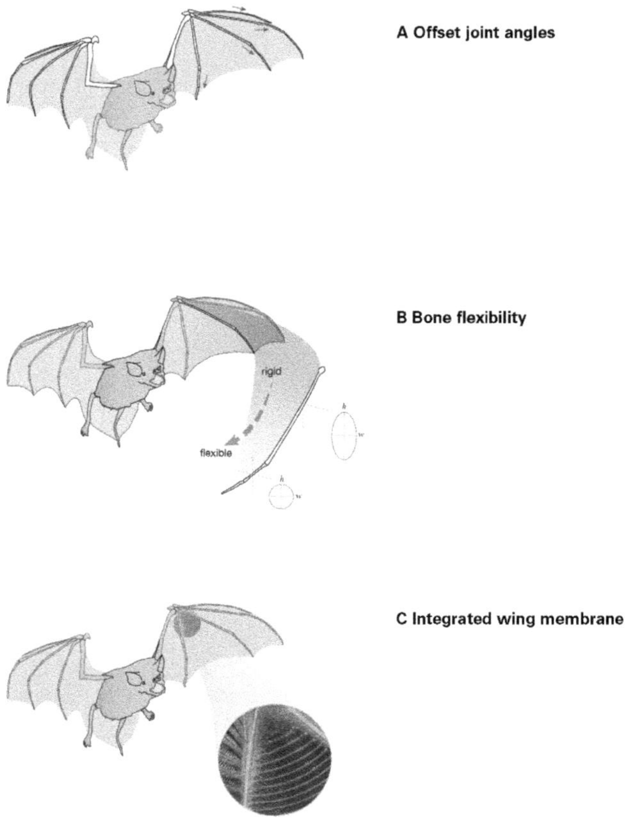

Fig. 5.5 Principles of bat wing design. (A) The joints that connect the bones within the bat wing are turned at slightly different angles from their neighbors; this creates added strength and stiffness across the wing to prevent hyperextension at individual points in response to strong winds. (B) The flexibility of bat wing bones varies over the length with increased stiffness near the wrist because of a larger and more oval-shaped cross-sectional area, while the cross-sectional area of the bones is decreased and more circular near the wing edge to allow the bones to flex without breaking. (C) The bones of the bat wing are embedded in a skin membrane that is filled with muscles and connective tissue; the wing membrane creates variable tension across the wing and adds strength and stability where needed.

part of a traditional umbrella canopy, turning the joints slightly out of plane adds strength to each joint so that they resist inversion without having to add additional materials as reinforcement.

The next feature of the bat wing that we mimicked was the flexibility of individual wing bones as they extended from the wrist to the outer edge of the wing. Near the wrist, the wing bones have a larger oval cross section and the long side of the oval is perpendicular to the force generated by flapping during flight (Swartz et al. 2012) (Fig. 5.4B). It is a common feature of bones in vertebrates to have an ovular cross section to resist bending, which adds stiffness along a single axis of a bone without adding material as reinforcement (Vogel 2003). At the tip of the bat wing, the cross-sectional area of each bone becomes circular, which allows for bending in any direction (Fig. 5.4B). The bones at the tip of the wing also lack full mineralization, which further increases their flexibility relative to the bones near the wrist joint (Swartz and Middleton 2008). This allows the bones to bend equally in all directions so that the bones deform rather than break under stress. In our umbrella design, the tips of each rib have a circular cross section to mimic the bones at the tip of the bat wing (Fig. 5.5B). The ribs are also designed to be constructed from a flexible material to allow them to bend at the tips and let wind pass through rather than create resistance at the edge of the canopy. In contrast, the cross section of the canopy ribs becomes ovular shaped near each joint to increase stiffness along the longitudinal plane. This shape adds strength around the joints and encourages bending in a lateral motion perpendicular to the angle that would cause the canopy to invert.

Finally, the bat wing membrane forms an integral component of the structure that adds stability and support. The wing membrane is dynamic and filled with muscles that run along both the longitudinal and horizontal wing axes (Fig. 5.4C). The bat can control tension within each wing quadrant independently, and different regions of the wing hold different levels of tension (Swartz et al. 1996). This also allows bats to distribute force across the wing so that some areas remain taut, while others deform to release tension. Traditional umbrellas are composed of six to nine uniform sections of fabric that have to be cut to the shape of the curve of the canopy. The canopy is attached to the midpoint and tip of each rib, and it provides tension to bend the ribs into the general shape of an umbrella. In this sense, traditional umbrellas are already similar to bat wings as the canopy is integral to providing tension across the support structure. In the bat-inspired design, the canopy is composed of eight pieces of fabric that are not completely uniform. In addition to creating tension longitudinally to cause the ribs to bow, the shape of each panel also applies lateral pressure on the ribs to form the irregular profile of the bat umbrella (Fig. 5.6A). The canopy is therefore integral to bending the tines to align with the offset joint angles that were inspired by the joint placement in bat wings.

By integrating the three components described above, the final bat-inspired umbrella was designed to resist inversion in the wind through two mechanisms. First, the overall shape and offset joint angles of the umbrella should add stiffness and increase the strength of the canopy without adding material as reinforcement. This strength is augmented by the oval-shaped cross section of the ribs near the joints and center of the umbrella. The second mechanism that improves umbrella performance in the wind added flexibility at the edge of the product. This was achieved by tapering the cross section of the ribs of the umbrella to have a circular cross section where they meet the canopy edge. Combined, these features provide additional strength that resists inversion from heavy gusts of wind, while small gusts would pass through the umbrella because of added flexibility at the canopy edge.

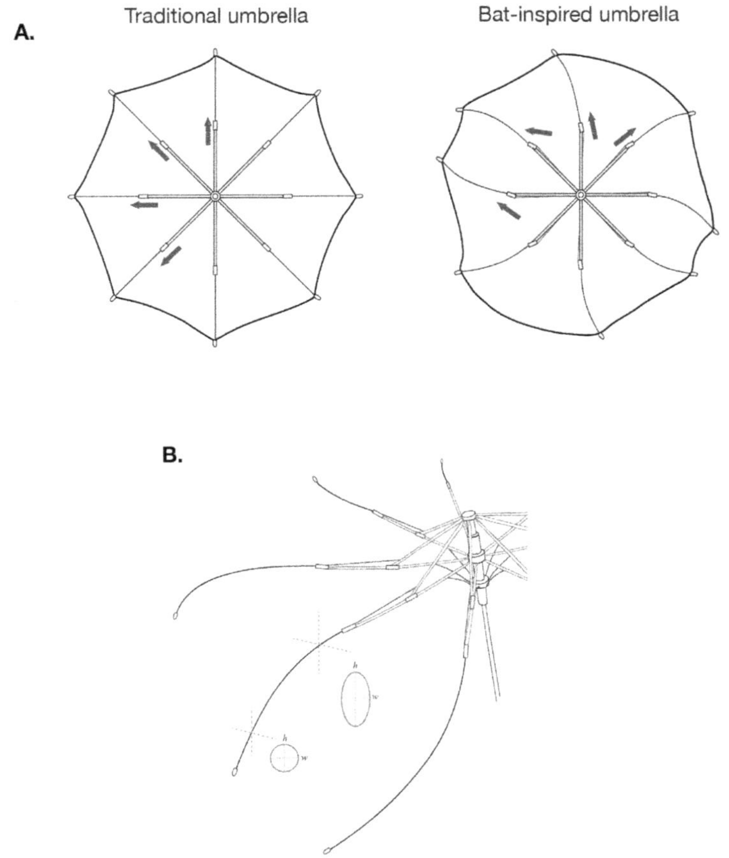

Fig. 5.6 Bat-inspired umbrella design. (A) Compared to traditional umbrellas, which feature ribs that are straight from the crown to the edge, the ribs of the bat umbrella also bend laterally. Each neighboring rib is therefore turned to a slightly different angle like the finger joints in the bat wing. This design balances tension among ribs so that adjacent ribs don't bend in the same plane when they are blown by a gust of wind. (B) The ends of each rib vary in cross-sectional area like the wing bones of bats, so that the end closest to the joint has an oval shape that resists bending in plane with the joint; the distal tip of each rib has a circular cross-sectional area that allows flexibility in all directions to allow small wind gusts to blow through with minimal resistance.

Evaluating: Comparing to Life's Principles

The ultimate goal of biomimicry is to create designs that are sustainable, and it is at the evaluating phase that a proposed design is compared to Life's Principles (Fig. 5.1). First and foremost, the design must also function as intended, therefore, evaluation is a built-in part of the standard design process. For our bat umbrella, the design is at the conceptual phase, and we are currently in the process of building and

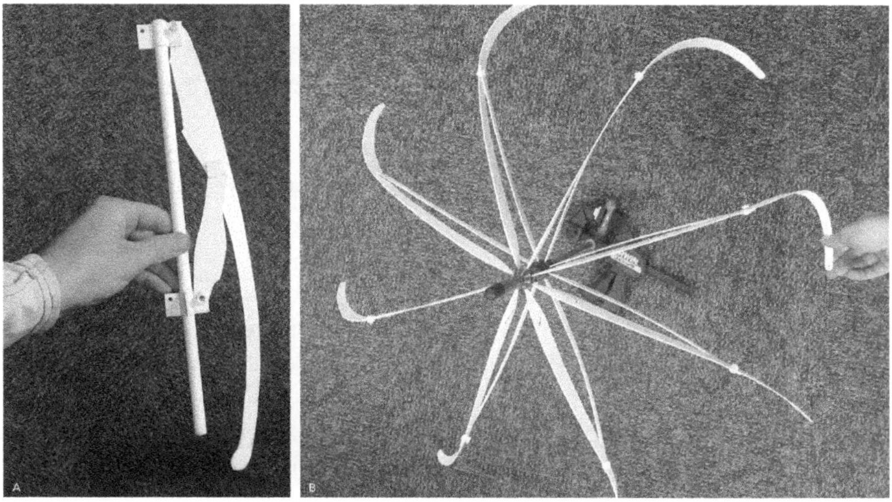

Fig. 5.7 Bat brolly prototype. (A) Side view showing the operating mechanism for raising and lowering the canopy with flattened ribs that prevent bending towards the joint; (B) Demonstration of how the ribs bend in tandem when they are connected together at the distal tip; the rib shape allows the ribs to bend from side-to-side but not backwards towards the joint. Prototype and photos supplied by Nathan Carmichael.

testing individual components of the design, as well as completely assembled prototypes, with industrial designer Nathan Carmichael (Fig. 5.7).

Further work will be required to validate all components of our proposed design and to assess our design against Life's Principles. Describing Life's Principles, Baumeister states:

> Life has evolved a set of strategies that have sustained over 3.8 billion years. Life's Principles represent these overarching patterns found among the species surviving and thriving on Earth. Life integrates and optimizes these strategies to create conditions conducive to life. By learning from these deep design lessons, we can model innovative strategies, measure our designs against these sustainable benchmarks, and allow ourselves to be mentored by nature's genius using Life's Principles as our aspirational ideals.
>
> Baumeister 2016

As we develop our bat-inspired umbrella further, we will evaluate our design decisions against Life's Principles. We will ensure that the product is functionally innovative, since it mimics the anatomical structure of the wings of the bat, and that it is also sustainable because it follows the principles of all life. Our bat-inspired umbrella already incorporates several features of Life's Principles. First, the umbrella is designed so that it "adapts to changing conditions," namely, changes in wind. It does so by varying the joint angles of each rib and provides redundant support by relying on neighboring ribs to buffer strong forces (i.e., it "embodies resilience through variation, redundancy, and decentralization"). To "be resource efficient," our umbrella relies on changes in shape rather than the addition of new or heavier materials (i.e., it "fits form to function"). This also allows our umbrella to be manufactured from standard umbrella components, which means that it "uses readily available materials."

Because the bat-inspired umbrella is designed to resist damage in the wind, it has the potential to reduce the waste that is generated globally by broken umbrellas. But further application of Life's Principles to our material choices and manufacturing processes could push the sustainability envelope even further. We could look for canopy and rib materials that break down into benign constituents at the end of life to apply "life friendly chemistry." We could analyze umbrella manufacturing and distribution processes to develop low energy alternatives for production and shipment, which would "be resource efficient (material and energy)." Finally, an ideal umbrella would be one that is "self-healing" and could "maintain integrity through self-renewal." A self-healing umbrella would negate throwing away damaged umbrellas, which would significantly reduce waste from discarded umbrellas.

Although addressing all of Life's Principles within a single design seems unfeasible, that is exactly how nature operates. Bat wings are constructed of organic materials that break down and return nutrients to the soil when the bat dies or is digested by a predator. The wings themselves are built using a limited set of elements (primarily carbon, hydrogen, oxygen, and nitrogen) that are organized into tissues with properties ranging from stiff bones to the elastic membrane of the wing structure. And, of course, bat wings heal themselves when they are damaged, often growing back stronger with reinforced bone or scar tissue. The ultimate goal of biomimicry is to create human-made products that match these same features.

Conclusion

The research and design efforts on the bat brolly are currently ongoing. What we have presented here is some of the early research into the problem of umbrellas inverting in the wind, the potential of bat-wing structure to inspire a biomimetic solution, the methodology of biomimicry, and an initial biomimetic design solution. Early prototypes are promising, but are also challenging to build. The properties of available materials, and limitations of current manufacturing processes, present difficulties for accurately mimicking the functional capabilities of the bat's membranous wings and articulating joints. It is clear that several iterative rounds of designing and prototyping will be essential to developing an umbrella that is capable of withstanding high winds and is also environmentally sustainable. Examined closely, bats—and other organisms and ecosystems that have been evolving over millions of years—can reveal remarkable adaptions to aid our problem-solving efforts. Although this project serves as a case study, our larger goal is to develop and explore further the idea of life-centered design. This approach is an alternative to the current focus on humans; it is a means of generating fully sustainable solutions to the environmental challenges that we face today.

References

Autumn, K. and N. Gravish (2008), Gecko Adhesion: Evolutionary Nanotechnology, *Philosophical Transactions of the Royal Society of London A* 366: 1575–90.

Baumeister, D. (2016), "Life's Principles Play Deck," https://biomimicry.net/product/lifes-principles-cards/.

Baumeister, D., J. Smith, R. Tocke, J. Dwyer, S. Ritter, and J. Benyus (2014), *Biomimicry Resource Handbook: A Seed Bank of Best Practices*. Missoula: CreateSpace Independent Publishing.

Benyus, J. (1997), *Biomimicry: Innovation Inspired by Nature*. Harper Collins: New York.

Brown, T. (2009), *Change by Design: How Design Thinking Transforms Organizations and Inspires Innovation*. New York: Harper Business.

Cagan J. and C. Vogel (2001), *Creating Breakthrough Products: Innovation from Product Planning to Program Approval*. Upper Saddle River: Prentice Hall.

Cooper, A. (2004), *The Inmates are Running the Asylum: Why High Tech Products Drive us Crazy and How to Restore the Sanity*. Indianapolis: Sams.

Dawkins, R. (1996), *The Blind Watchmaker: Why the Evidence of Evolution Reveals a Universe Without Design*. New York: Norton & Company.

Dreyfuss, H. (1955), *Designing for People*. New York: Allworth Press.

Fish, F., P. Weber, M. Murray, and L. Howle (2011), "The Tubercles on Humpback Whales' Flippers: Application of Bio-inspired Technology," *Integrative and Comparative Biology* 51: 203–13.

Hanington, B. (2012), *Universal Methods of Design: 100 Ways to Research Complex Problems, Develop Innovative Ideas, and Design Effective Solutions*. Beverly, MA: Rockport Publishers.

Holley, P. (2018), "The physics behind this odd-looking, 'stormproof umbrella' from the Netherlands," *The Washington Post*, www.washingtonpost.com/news/innovations/wp/2018/03/30/how-a-dutch-designer-unlocked-the-secret-of-a-storm-proof-umbrella/?utm_term=.971d33485feb (accessed on June 23, 2021).

IDEO (2011), *Design Kit: The Human-Centered Design Toolkit*. Canada: IDEO Publishers.

Liedtka, J. and T. Ogilvie (2011), *Designing for Growth: A Design Thinking Tool Kit for Managers*. New York: Columbia University Press.

McCurdy, E. (1908), *Leonardo Da Vinci's Note-books: Arranged and Rendered into English with Introductions*. New York: Charles Scribner's Sons.

McKeag, T. (2012), "Auspicious forms: Designing the Sanyo Shinkansen 500-Series bullet train," *Zygote Quarterly* 2: 10–35.

McNeal, M. and J. Benyus, (2015), "Janine Benyus: Inventing the eco-industrial age," *Wired*, www.wired.com/brandlab/2015/07/janine-benyus-inventing-eco-industrial-age/ (accessed on June 23, 2021).

Needham, J. (1986), *Science and Civilization in China: Volume 4, Physics and Physical Technology, Part 2: Mechanical Engineering*. Taipei: Caves Books.

Neuweiler, G. (2000), *The Biology of Bats*. New York: Oxford University Press.

Norberg, U. (1972), "Bat Wing Structures Important for Aerodynamics and Rigidity (Mammalia, Chiroptera)," *Zoomorphology* 73: 45–61.

Norman, D. (2013), *The Design of Everyday Things*. New York: Basic Books.

Nowak, R. M., & Walker, E. P. (1999), *Walker's Mammals of the World* (Vol. 1). JHU press.

Pawlyn, M. (2016), *Biomimicry in Architecture*. Newcastle upon Tyne, UK: RIBA Publishing.

Ramezani, A., S. J. Chung, and S. Hutchinson (2017), "A Biomimetic Robotic Platform to Study Flight Specializations of Bats," *Science Robotics* 2: eaal2505.

Riskin, D., D. Willis, J. Iríarte-Díaz, T. Hedrick, M. Kostandov, J. Chen, D. Laidlaw, K. Breuer, and S. Swartz (2008), "Quantifying the complexity of bat wing kinematics," *Journal of Theoretical Biology* 254: 604–15.

Swartz, S., J. Iriarte-Díaz, D. Riskin, K. Breuer (2012), "A bird? A plane? No, it's a bat: An introduction to the biomechanics of bat flight," in GF Gunnell and NB Simmons (eds.), *Evolutionary History of Bats: Fossils, Molecules and Morphology*, Cambridge: Cambridge University Press.

Swartz, S., K. Middleton (2008), "Biomechanics of the bat limb skeleton: Scaling, material properties and mechanics," *Cells Tissues Organs* 187: 59–84.

Swartz, S., M. Groves, H. Kim, W. Walsh (1996), "Mechanical properties of bat wing membrane skin," *Journal of Zoology* 239: 357–78.

Vogel, S. (2003), *Comparative Biomechanics: Life's Physical World*. Princeton: Princeton University Press.

6 PuriFungi, A Natural First Aid Kit for the Earth

AUDREY SPEYER

Introduction

In 2014, Elizabeth B. N. Sanders and Pieter Stappers wrote an article for *Interactions* magazine in which they mapped the areas and directions of—and "dreams" for—contemporary and future design. Their semi-circular pie chart is divided into three equal slices for design research and practice: provoking, engaging, and improving. In this essay, I present my project PuriFungi, a designed system that utilizes the process by which fungi (these kinds of fungi are commonly called mushrooms) break down industrial pollutants. My initiative fits within Sanders and Stappers' design research landscape: PuriFungi is an eco-tool that decontaminates toxic waste (e.g., polluted soil or cigarette butts) in order to *provoke* global consciousness about the environment, *engage* communities in design solutions and concrete actions, and *improve* biodiversity and the quality of people's lives. PuriFungi is also situated within the area of transition design.

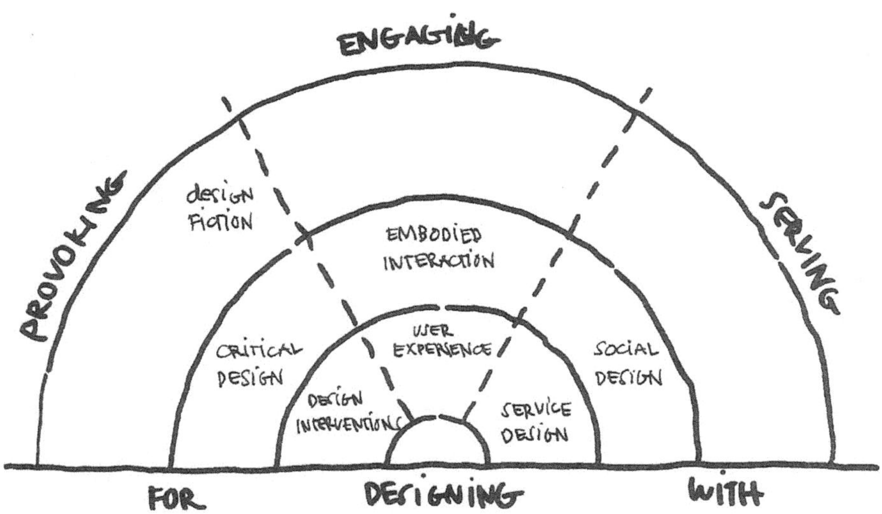

Fig. 6.1 Mapping of possible futures for design practice and research by Sanders and Stappers, 2014. The rings radiate outwards from the current core of designing toward the edges of time. Image used with permission from Elizabeth Sanders and P.J. Stappers.

Transition Design—Living System Theory: Its Mission and the PuriFungi System

The Carnegie Mellon University pdf "Transition Design Overview" explains that this kind of design "takes as its central premise the need for societal transitions to more sustainable futures" (Irwin et al. 2015: 1). Transition Design, the authors contend, also "advocates the re-conception of entire lifestyles, with the aim of making them more place-based, convivial, participatory, and harmonizing with the natural environment" (2015: 1).[1]

Transition Design is particularly germane to our twenty-first century environmental circumstances. There has been an unprecedented impact on both biodiversity and climate since the Industrial Revolution began in the late-eighteenth century. Climate change and biodiversity reduction are serious concerns, which demand that we rethink the ways that we currently produce (unlimited resource extraction), consume (the linear life cycle of "standard" product ends with non-recyclable, polluting waste), and grow our resources (intensive farming and monoculture that reduce biodiversity). Some urban and rural locales are already reconsidering how we produce, consume, and grow. These communities are exploring ways to break down natural things into their components using eco-systemic approaches that re-connect organisms to their environments.

Transition design initiatives—which are often locally based—typically occur through sustainable fabrication—that is, by making products that are easy to recycle. Transition design also produces products that are made of recyclable or compostable materials, for example, from bio-fabrication processes. Sustainable fabrication brings circularity to the life cycle of a product, in which there are no end-product toxins or non-usable waste. There are also agricultural movements that encourage local farming and permaculture initiatives. This "green" trend includes innovative urban initiatives, such as models of consumption and production at community scales that can suggest new economic and industrial models at the societal level. This "green" trend also signals a general thinking shift in design practices and processes, from an anthropocentric system focused on human beings to a systemic, symbiotic, and sustainable system that stresses the value of "co-working" with nature.

In her book *Symbiotic Economy*, environmentalist Isabelle Delannoy describes a new, radical economic theory that is based on harmony between humans and their ecosystems. "In many fields, symbiotic economy can reduce by ninety percent (and even more) our use of resources by re-developing local productive capacities" Delannoy argues. "Symbiotic economy," she continues, "is based on the symbiosis between human intelligence, the puissance [vigor] from natural ecosystems, and the technosphere (tools)" (2017). There is a fundamental shift, according to Delannoy, in the organization of the economic and social aspects of our society from a mass-industrial system that is based on a vertical hierarchy to local community systems that are based on a horizontal hierarchy. This shift stresses collaboration in which "sharing is caring," "growing is caring," and where the primary mission is to be symbiotic with the ecosystem.

It is in this context of an industrial and economic turning point that new initiatives emerge through transition design. The PuriFungi project is part of this transformation and this symbiotic approach. To develop PuriFungi, I utilized interdisciplinary design research and embraced the notion of co-working with living nature. I also produced PuriFungi—which uses symbiotic systems that absorb toxic components from the soil—in collaboration with scientists and engineers.

"Reverse" Design Research and General Presentation

The inherent traits of resilience, regeneration, circularity, and symbiosis in Nature that underpin PuriFungi allowed me to consider new thinking and making approaches for this project. My research for PuriFungi is based on "reverse" design thinking in which designers and their collaborators first identify natural phenomena, and then analyze their functions and figure out how to deploy them. PuriFungi incorporates living fungi. In my "reverse" design process, I first had to understand fungal anatomy, physiology, and ecosystems. PuriFungi began with in-depth research on the essential needs of fungi, and then an investigation of ways to enhance fungal mycelium growth in the soil (the mycelium is the part of fungi that takes up nutrients, and pollutants, from the environment). I continued my research in a bio-lab in collaboration with specialists in mycology and chemistry from the United Kingdom, France, Belgium, and the United States. My goal was to understand how fungi and design could be deployed as partners in soil remediation. Traditional design played a role in this partnership.

Fig. 6.2 PuriFungi incubator. Photograph by Lucas Castel.

I designed a biodegradable incubator in which the fungal media can be cultured. Several incubators are implanted into a polluted site, creating a fungal-network system over a large area of polluted soil.

The Fungi Kingdom in Nature

Fungi are organisms that are classified somewhere between plants and animals; they have similarities with plants, but they don't manufacture the chlorophyll that plants use for their own food and energy. Fungi are actually more similar to animals because, first, like animals, they use external food sources as energy, and second, fungi and animals share a common molecule called chitin (a strong structural component of the cell membrane) that is not found in plants. Fungi are, however, classified as their own biological kingdom, a scientific term that also refers to five other groups of living organisms: plants, animals, protists, archaebacteria, and eubacteria.

Some organisms that have evolved over millennia develop the ability to thrive in extreme environmental conditions. Fungi, for example, can grow on highly polluted land on which other life forms may not grow. As the "recycler" kingdom, fungi have developed a digestive system that allows them to ingest natural nutrients and human-made polluted substances. In PuriFungi, I focus on certain mushroom species that have developed this very high tolerance for pollution. The mushroom fruiting body part of the fungi—the one people typically can see—is the above-the-ground reproductive organ for the fungus that generates spore-producing bodies called sporocarps. In order to transform polluted land into a fertile zone, though, the mycelium of the fungi—the underground part of fungi organism or the fungi "roots"—first needs to grow into a network in the ground.

Fungal Anatomy and Physiology

This mycelial network is a crucial part of fungal physiology. Fungal cells are organized into tube-like filaments called hyphae. When two compatible hyphae meet, they join together to form the mycelium network discussed above. The mycelium, which is the dominant somatic (or vegetative) body of fungi, grows underground in a spreading, branching network of hyphae.

The mycelium also produces nutrient and water distribution networks in an efficient and complex manner. They run through all of the land masses on earth, "communicating" with other species, while also breaking down inorganic and organic matter. Upon digesting organic and inorganic material, the mycelium delivers various nutrients to neighboring plants, trees, and bacteria. Carbon, nitrogen, and phosphorus exchanges also take place among tree species via the mycelium. Fungi are primary manufacturers and exchangers; they deliver nutrients, water, and biochemical signals to other organisms around them. Fungi also boost tree immune systems, and are the first recyclers of organic matter, such as wood and leaves, which are made of lignin—the primary structural components of wood. This degradation produces the forest floor litter, which also provides nutrients for bacteria, plants, and animals, such as insects. Clearly, fungi play a crucial role in the life and health of forests. They are the "Wood Wide Web" of the planet, or the "Earth's natural Internet," a term that was introduced by fungus expert Paul Stamets during his TED talk in 2008.[2] Science writer Nic Fleming similarly describes fungi mycelia: "Hidden under your feet is an information superhighway that allows plants to communicate and help each other out. It's made of fungi" (2014).

Mushrooms, Primary Recyclers in the Unique Soil Matrix

Fungi degrade some of the most resistant materials on Earth, from organic matter (such as lignin) to inorganic human-made substances that have similar molecular structure to wood lignin. They are essential for soil regeneration and healthy soil—which is one of the richest bio-chemical energy *milieux* on Earth. Soil is crucial for a healthy ecosystem. One teaspoon of soil holds more microorganisms than there are people on Earth. Bacteria, algae, insects, earthworms, and fungi together create a symbiotic system to produce soil. Moreover, soil holds more carbon than our atmosphere and vegetation combined. In December 1998, David Pimental from Cornell University reported in "Population Growth and The Environment: Planetary Stewardship," that "Once fertile soil is lost, it takes 500 years or more to form a mere 25 mm of fertile soil" (1998: n.p.). Moreover, according to the Natural Resources Conservation Service, ninety percent of American farmland is on average losing topsoil seventeen times faster than new topsoil is being formed, and it is not possible to develop soil artificially in a laboratory (Miller and Spoolman 2012: 306). Soil is essential for the whole ecosystem, but this symbiotic *milieu* is completely disrupted on industrial sites and some urban areas, drastically reducing its quality and biodiversity.

Soil Pollution

Polluted soil contains various contaminants, many of which have reached critical concentrations in some environments. Soil is polluted from transport and incinerators, and at old industrial factories, brownfield sites, and hazardous waste facilities. According to "Industrial pollution in Europe," published by the European Environment Agency, the production sector is "responsible for an estimated 60% of [all of the contaminated] sites. We can therefore foresee that industry is probably a significant contributor to local soil contamination" (2019). The report "Progress in management of contaminated sites," also published by the European Environment Agency, mentions that:

> The large volume of waste production and the widespread use of chemicals during the past decades have left numerous sites with local soil contamination. The dominant major sources of local soil contamination are inadequate or unauthorized waste disposal, unsafe handling of dangerous substances within industrial or commercial processes, and accidents.
>
> (EEA 2014)

And further down in the report: "[t]he main contaminant categories are mineral oils and heavy metals" (2014). This contamination affects both urban and rural areas.

Additional studies from the European Environment Agency demonstrate that hydrocarbons (mineral oils) and polycyclic aromatic hydrocarbons from petroleum industries make up almost fifty-three percent, and heavy metals make up thirty-five percent, of the total contaminants found in soil.[3] These two contaminants, it turns out, can be particularly well absorbed by specific mushroom species. Industrial areas with high concentrations of substances that are toxic for the environment and to human health often become inaccessible, taking urban and rural spaces out of commission. Besides the community concerns and the human health consequences from exposure to soil contamination, which vary according to the pollutant, there are ecosystem impacts that affect food chains, and consequently have effects on both predator and consumer species. In "Living Soils: A Call to Action," D. Udall et al. explain that:

> Not only are we losing the soils in which we grow our food but we are also losing the ability of our soils to offer essential "ecosystem services" such as supporting biodiversity, storing carbon, and flood and drought protection. 380,000 tonnes of soil carbon is being lost, each year, from the peat soils in the East Anglian Fens. Once it has been lost from the soil most of this carbon is emitted to the atmosphere, contributing to climate change
>
> 2014

Remediation Techniques

The European Environment Agency report "Progress in management of contaminated sites," notes that "Up to the present, the most common remediation technique has been the excavation of contaminated soil and its disposal as landfill (sometimes referred to as 'dig and dump')" (2014). The standard remediation "dig and dump" technique, though, has a drastic impact on the biodiversity. The digging phase of this technique destroys local biodiversity, depleting parts of their habitats and killing organisms. The dumping phase of this technique, by definition, only buries the problem. Moreover, Uli Henrik Streckenbach's animation "Let's talk about soil," for the Global Soil Week in 2012, documents competition for soil across the world. This competition leads to land grabbing, often with questionable means, for questionable purposes, and with high-stakes financial rewards. This kind of activity is happening in the context of projections that available arable land per Earth inhabitant will be reduced by half by 2050. Land-grabbing and worsening soil pollution stress how important it is to find ways to remediate toxic components naturally in the soil; ways that revitalize contaminated ecosystems and stimulate, rather than diminish, local biodiversity.

Mycoremediation

Mycoremediation is an environmentally friendly method of cleaning the soil by using mushroom cultures that absorb, stock, and sometimes digest toxic components

Fig. 6.3 PuriFungi test on a polluted site in London. Photograph by Tom Mannion.

found in polluted soil. Oil spills, pesticides, and other industrial waste products are converted into innocuous compounds at the molecular level. Fungi are a very efficient first step in remediation because they secrete enzymes that absorb and then digest high concentrations of pollution in a short period of time—typically between three and nine months. After this first step, microorganisms then metabolize the remaining toxic substances with toxin-digesting enzymes. Plants, insects, and animals that have disappeared from a habitat—along with regenerative and symbiotic properties of soil—typically return after microbial activity.[4]

Mycoremediation, though, is a critical first step. In *Mycelium Running, How Mushrooms Can Help Save the World* (2005), fungi expert Paul Stamets observes that fungi cleaned up oil spills by ninety-seven percent over eight weeks. Stamets also demonstrated that, at the end of a diesel remediation trial with oyster mushrooms, the soil was depolluted and then native plants naturally re-appeared. The organization Polypop Industries, which was a laboratory founded by entrepreneur Gil Burban, likewise demonstrated that between thirty-five percent and fifty-five percent of pollutants were absorbed during the first twenty-five days of a mycoremediation treatment (2012).[5] Polypop's results likewise demonstrate that there is considerable activity when the fungal culture is first introduced into a polluted environment.[6] It can take up to six months, however, for complete results. This natural technique employed by both Stamets and Polypop, which uses organic matter such as sawdust from agricultural waste, is among the least expensive remediation techniques. It costs only thirty percent of traditional excavation and incineration remediation costs—a technique that also causes air pollution.

Designing the PuriFungi System

As I mentioned above, understanding mushroom anatomy and physiology, and the natural processes that occur during mycoremediation was essential to my design process. These natural processes can be divided into two components: first, there is the mycelial growth underground. Fig. 6.4 shows mycelial growth isolated in a special Microsac with filter system. Second, is the fruiting period when the mushroom germinates above ground (there can be several fruiting periods) (see Figs. 6.5 and 6.6). For clarity, mycelial and mushroom growth are separate in my discussion, but they are actually part of the same growth process.

Several environmental factors are crucial for fungal growth. As described above, the mushroom culture in PuriFungi starts with mycelial growth during which the

Fig. 6.4 Mycelium growth process isolated in a Microsac.

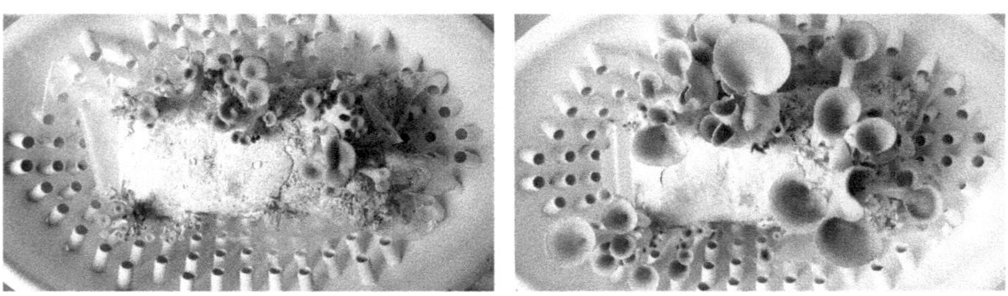

Fig. 6.5 *Pleurotus ostreatus* fruiting in the incubator, at two days of growth.

Fig. 6.6 Below ground part of the incubator with substrate prior to inoculation.

organism feeds upon nutrients from organic and inorganic matter. With adequate nutrition, and the growth of a well-developed mycelial web, the mushroom fruit emerges. During this fruiting period, the fungi need continuous oxygen and humidity. It is necessary, for example, to maintain eighty to ninety percent humidity, non-direct sunlight, access to organic substrate (such as wood, straw, hemp), and a good flow of oxygen. These specific conditions are difficult both to obtain and maintain on a polluted site, and these limitations guided my design for the PuriFungi incubator. Clearly, the PuriFungi incubator interior needed to have continuous air flow, and to maintain a balanced humidity level.

Enzymatic Digestion

My research revealed that the type of mushroom I ought to use in PuriFungi depended on what substance was to be remediated. As I discussed in more detail above, during mycoremediation the mycelium secretes enzymes that digest organic and inorganic substances, including toxic components in the soil. Toxic components are either digested or stocked in the fungi through this process, depending on the type of contaminant. Enzymes called laccases, cellulases, and lignin peroxidase in the fungi *Tramete versicolor, Pleurotus citrinopileatus*, *Pleurotus ostreatus*, and *Pleurotus eryngii* are particularly efficient at breaking hydrogen-carbon bonds, and digesting contaminants such as petroleum products and insecticides. Heavy metals in the soil, which can't be dismantled by enzymes, are instead absorbed and stocked in the fruiting part of the fungi, the mushroom part. Stamets found that fifty percent of the particles released after this heavy-metal remediation process are carbon dioxide, and ten to twenty percent are water (2004). No matter which mushroom species I used, the mycelia and fruiting bodies had to grow well in my PuriFungi incubator.

Mycelium Trained Reactivity

Besides mushroom species, another vital consideration for my PuriFungi incubator design was the medium in which the fungi grow. Adding carbon sources, such as non-sterilized sawdust, straw, or corn cobs at polluted sites augments the degradation of pollutants by fungal enzymes. This specific process has been tested by Stamets. He observes that adding carbon sources:

> Marks a major advancement in the understanding of how to project mycelium in mycoremediation strategies, and it may also cut the cost significantly . . . Natural spawn or pure culture spawn that has made contact with habitat microbes before insertion has the best chance for successful mycoremediation.
>
> 2004: 90

Stamets's research shows that the mycelium is responsive to its environment, and that it can be trained to be reactive specifically to the microbial environment and contaminants at a site. That is why mature mycelium from a mushroom farm has better mycoremediation properties than pure culture spawn—which are fungi that have been cultured in a sterile environment. Stamets tested this principle in his work with Batelle Pacific Northwest Laboratories in the late 1990s in Washington state. The best results were seen when *Pleurotus ostreatus* spawn was mixed with soil contaminated by oil and unsterilized alder wood chips.[7] Based on these studies, I always use mature mycelium in my incubator.

PuriFungi Incubator Design Development

In summary, in order to have good mycelial growth fungi need:

- Organic matter (carbon sources) in compact and small pieces for complete nutrition.
- High humidity (between eighty and ninety percent humidity is ideal).
- No direct light (preferably a dark environment for this first step of growth).
- No air flow/ventilation for this first stage. The air ventilation, as mentioned above, will need to be part of the second phase, the fruiting phase.

The PuriFungi incubator pod must serve several functions:

- It needs to be an incubator: a device in which the premature mushroom culture or mycelium spawn is blended with unsterilized wood chips or straw. The composition of the medium depends on the mushroom species.[8]

- It needs to be a disposable pod that can be used to transport the fungi from a bio-laboratory to the polluted site.
- It needs to function as a kind of greenhouse, on a small scale; a space that is protected from cold temperatures, direct sun, and dry air.

The design of the PuriFungi incubator conforms to these functions and needs. Half buried in the ground, the PuriFungi system incubator provides the proper amount of humidity, shade, and air for the mushroom culture (see Figs. 6.2, 6.6, and 6.7). For the incubator, I chose a resistant bio-plastic, from the Biome Bioplastics company (UK), that won't release toxic materials into the earth. The design of the incubator is ovoid, which is easier to "plant" in the ground. The visual language of the incubator is biomorphic and futuristic. Its visual form—which resembles an egg—embodies the notion of purifying the land and returning life to a toxic area.

The incubator is composed of two parts: a top part, with requisite space, shade, and air flow for the mushroom to grow, and a below-ground bottom part that gives the mycelium access to the polluted soil. Both parts were produced using a vacuum forming technique to mold the bio-plastic, and designed in response to the morphology and the biological needs of the mushrooms, as discussed above.

The below-ground portion is the most important part of the incubator; this section is where the remediation process actually happens. PuriFungi can treat topsoil up to

Fig. 6.7 Active PuriFungi incubator on site with *Pleurotus ostreatus* species of mushroom.

seventy to eighty centimeters below the ground, which is also enough depth for cultivation. The 176 openings in the incubator are supplemented with white connectors, which function as channels to link the mycelium to the polluted soil. The objective is to create as much access as possible between the mycelium and the soil—in other words, to allow the mycelia to "invade" the land.

Each incubator can treat four square meters of polluted top soil. Several incubators are imbedded at each site, based on what is needed for full remediation, and each incubator is connected to additional inoculated substrate that is added to the site. Each incubator assures the growth of a "strategic center" of fungi cultures, and the extra inoculated substrate connects the individual incubators. For *in situ* installation, the scale of the project can be adapted to evolve at the site according to the nature of the soil (clay or sandy soil, or with organic matter already present), and the specific mycelial bio-management of the contamination.

Mycoremediation Process with PuriFungi Systems

After inoculation in the incubator, the culture is transferred to the contaminated field. Once there, the mycelium grows out of the incubator from the openings in the pod (the white connectors). After four to six weeks of on-site incubation, the mushroom fruit and cap grow in the incubator (see Fig. 6.7). As mentioned above, it is crucial that the level of humidity, oxygen, and shade is maintained inside the incubator during fruiting. Environmental sensors can be added in the incubator to take real-time measurements and also adapt to ensure proper conditions. I have been testing arduino captors for the PuriFungi incubator, sorting out exactly what will work best by working with agricultural engineer Camila Amaya-Castro and the start-up La Cool Co. More recently, I have been working with a team at Microsoft Innovation Center in Belgium (Mons) that specializes in automation systems that measure and regulate temperature, humidity, carbon dioxide, and oxygen levels.

PuriFungi Project and Mushroom Species Database

The success of mycoremediation is governed by three important factors: the accessibility of the contaminants, an environment that is conducive to remediation, and the availability of the best mushroom species for the specific contamination. Mycoremediation is only successful if the right mushroom species is introduced into the polluted site. More research needs to be done to document which species work best as mycoremediants for specific kinds of pollution, and we need good field evidence to show which species of mushroom can best absorb certain types of pollutants. Classifying—in an open-source way—mushroom species according to the level of bio-absorption or bio-stockage of specific pollutants will be important to the development of the field of mycoremediation.

Part of my PuriFungi project includes a database with open-source, collaborative research to investigate mycoremediation pairings. Using mycological testing, I identified the mushroom species that remediates the biggest variety of polluted substances: the white-rot type of mushroom is one of the most reactive and persistent species that can remediate both hydrocarbon and heavy metal pollution. I have focused in particular on the process by which the white-rot species *Pleurotus ostreatus* digests petroleum products, polycyclic aromatic hydrocarbons, and heavy metals. I have also investigated how other mushroom species, such as *Tramete versicolor, Pleurotus citrinopileatus,* and *Pleurotus eryngii*, absorb pollutants from the textile dyeing industry, intensive farming (insecticides and pesticides), dioxin, and even chemical weapons.[9]

Conclusion

PuriFungi is a transition design project that has a positive environmental impact. The PuriFungi incubator is open-source, regenerative, and sustainable. Its design is based upon the ideals of citizen activism—I have presented workshops to groups in order to engage communities in various parts of the project. PuriFungi employs living materials to reinvigorate natural resources. It is designed around a minimum-waste policy, both in pre-production—using coffee grounds and agricultural waste such as straw—and in post-production—the incubator pod is biodegradable. PuriFungi connects to nature using "reverse" design thinking, in which designers first identify natural phenomena, and then incorporate these into a design project.

The PuriFungi incubator, then, is part of a network that integrates a designed object with natural organisms and processes. It "teams up" with nature to alter the impact that industry has on the earth. Systemic and cooperative approaches to nature, plus design, allow us to tackle human-precipitated environmental issues.

In conclusion, mycoremediation is a "clean technology" that can alleviate some major problems: waste and pollution accumulation, protection of biodiversity, and the treatment and conversion of contaminated land. Contaminated areas are often designated by public or governmental organizations, but existing processes for restoring soil are time consuming and expensive—and they often create their own contamination problems. Mycoremediation also has the potential to encourage citizen awareness of local environmental issues in our cities.[10]

As a designer, this project brought up some essential questions for me that changed my way of thinking, researching, and making. How can we assimilate nature into our built environment to produce a more integrated model for living? And what should the limits be on monitoring and testing living organisms?

Biological and living materials are already suggesting revolutionary outcomes in which regeneration and synergy are at the center: such as fab labs that provide digital fabrication, and biomanufacturing that utilizes biological systems to produce materials. These approaches approaches offer an inspiring sustainability model in which design and science merge to prioritize biodiversity.

Through PuriFungi, I address environmental issues using a symbiotic design approach that is centered on nature rather than on purely human desires. PuriFungi gives nature what it needs to restore itself: mushroom species that are enhanced by being part of a designed system that will help them to grow naturally and restore a soil-life balance. My design approach was to analyze and identify natural phenomena in order to co-design with them—in other words, to bring back a natural resilience that has been destroyed by human activities. I am particularly fascinated by how nature always finds a way to persist, to survive, to grow, and to regenerate itself. Projects such as PuriFungi can help rebuild a healthy environment for the future, in effect, healing nature with nature.

Notes

1. A discussion of Living Systems Theory, which is part of transition design can be found at Irwin, T., C. Tonkinwise, and G. Kossoff (2020), "Transition Design: An Educational Framework for Advancing the Study and Design of Sustainable Transitions," *Cuadernos del Centro de Estudios en Diseño y Comunicación* 105, http://fido.palermo.edu/servicios_dyc/publicacionesdc/cuadernos/detalle_articulo.php?id_libro=831&id_articulo=17078.

   See also Terry Irwin:

   > Instead of examining phenomena by attempting to break things down into components, Living Systems Theory explores phenomena in terms of dynamic patterns of the relationships between organisms and their environments. Principles such as self-organization, emergence, resilience, symbiosis, holarchy, and interdependence, among others, can serve as leverage points for initiating and catalyzing change within complex systems.
   >
   > Irwin 2011b

2. Fungus expert Paul Stamets called fungi "Earth's natural internet" in a 2008 TED talk. He first had the idea in the 1970s when he was studying fungi using an electron microscope. Stamets noticed similarities between mycelia and ARPANET, the US Department of Defense's early version of the internet (2004; 2008).
3. See "Contamination from local sources" (2019), www.eea.europa.eu/themes/soil/soil-threats/soil-contamination-from-local-sources, and "Progress in management of contaminated sites" (2014), www.eea.europa.eu/data-and-maps/indicators/progress-in-management-of-contaminated-sites-3/assessment (accessed on July 4, 2021).
4. Heavy metals and radioactive cations—a form of molecular-level radioactive contamination—cannot be decomposed. They can, however, be rendered into forms with low solubility so that they become less harmful. Mushrooms stock complex radioactive cations in a part of the mushroom culture called the sporocarp, which can be removed and isolated afterwards.
5. Polypop Industries was a laboratory created by Gil Burban in 2012. This work is now developed by HYPHEN SAS.
6. Details of percentage of digestion during the first twenty-five days of author's mycoremediation trial:

47% of heavy hydrocarbons
35% of polycyclic aromatic hydrocarbons
55% of cadmium
40% of plumb.

7. Ninety-seven percent of oils, polynuclear aromatic hydrocarbons (PAHs), degraded after eight weeks. And another mycoremediation test with sterilized alder chips only showed sixty-five percent of PAHs degraded (Stamets 2004; 2008).
8. They are unsterilized because they will be more responsive to their environment. Only aggressive cultures such as the *Pleurotus* species can be inoculated on unsterilized substrate because they grow very rapidly, invading the substrate before contamination can take hold. This technique is not 100% effective, and for better mycelial growth results the substrate could be pasteurized.
9. In the past, fungi species were identified through their morphology—that is, their physical characteristics—based on observations, as well as on factors such as their biochemistry. Today, DNA sequences (ideally from different sections of their DNA) permit classification of species at the gene level. This shows that some morphologically different mushrooms can be identified as the same species, increasing the number of species. In a recent publication (Hawksworth and Lücking 2017), the number of mushroom species is listed as between 2.2 and 3.8 million. However, this number varies as it is complex to define how much DNA sequences differ from one species to another. The article "8.7 Million Species exist on Earth, Study Estimates" from *The Washington Post* by Juliet Eilperin stipulates that "Only seven percent of the fungi kingdom have been discovered out of a prediction of 611,000 fungi" (2011).
10. Mycoremediation tests may also be applied in agroforestry—which applies ecological principles to the design and management of sustainable agricultural ecosystems and permaculture initiatives.

References

Burban, G. (2012), Polypop Industries, www.facebook.com/media/set/?vanity=mycoremediation&set=a.1499831426823249 (accessed on July 4, 2021).
Delannoy, I. (2017), *The Symbiotic Economy*. Arles, France: Edition Actes Sud –Colibri.
Eilperin, J. (2011), "8.7 million species exist on Earth, study estimates," *The Washington Post*, www.washingtonpost.com/national/health-science/87-million-species-exist-on-earth-study-estimates/2011/08/22/gIQAE7aZZJ_story.html (accessed on June 24, 2021).
European Environment Agency (2014), "Progress in management of contaminated sites," www.eea.europa.eu/data-and-maps/indicators/progress-in-management-of-contaminated-sites-3/assessment (accessed on July 4, 2021).
European Environment Agency (2019), "Industrial pollution in Europe," www.eea.europa.eu/themes/industry/industrial-pollution-in-europe (accessed on July 4, 2021).
Fleming, N. (2014), "Plants talk to each other using internet of fungus," BBC, November 11, www.bbc.com/earth/story/20141111-plants-have-a-hidden-internet (accessed on June 23, 2021).

Frank, A. B. (1885), "Ueber die auf Wurzelsymbiose beruhende Ernährung gewisser Bäume durch unterirdische Pilze," [On the nourishing, via root symbiosis, of certain trees by underground fungi], *Berichte der Deutschen Botanischen Gesellschaft* 3: 128–45.

Hawksworth, D. L. and R. Lücking (2017), "Fungal Diversity Revisited: 2.2 to 3.8 Million Species," *Microbiology Spectrum* 5 (4): 1-17, https://journals.asm.org/doi/10.1128/microbiolspec.FUNK-0052-2016 (accessed on June 24, 2021).

Irwin, T., C. Tonkinwise, and G. Kossoff (2020), "Transition Design: An Educational Framework for Advancing the Study and Design of Sustainable Transitions," *Cuadernos del Centro de Estudios en Diseño y Comunicación* 105, http://fido.palermo.edu/servicios_dyc/publicacionesdc/cuadernos/detalle_articulo.php?id_libro=831&id_articulo=17078 (accessed on July 4, 2021).

Irwin, T., et al. (2015), "Transition Design Overview," Carnegie Mellon University, pdf download at www.academia.edu/13122242/Transition_Design_Overview (accessed on June 14, 2022).

Irwin, T. (2011b), "Design for a Sustainable Future," in Hershauer, J., G. Basile, S. McNall (eds.), *The Business of Sustainability: Trends, Policies, Practices and Stories of Success*, 41–60, Santa Barbara, CA: Praeger.

Miller, G. T. and S. E. Spoolman (2012), *Living in the Environment*. Boston: Cengage Learning.

Pimental, D. (1998), "Population Growth and The Environment: Planetary Stewardship," *Electronic Green Journal* 1 (9): n.p., https://escholarship.org/uc/item/8g67g6ng (accessed on July 4, 2021).

Recer, P. (1995), "Erosion Compromising World Food Supply, Researchers Say," February 24 https://apnews.com/7df71d6ee41649539c30fc6e21ee9b9b (accessed on June 23, 2021).

Sanders, L. and P. J. Stappers (2014), "From designing to co-designing to collective dreaming: three slices in time," *Interactions* 21 (6), http://interactions.acm.org/archive/view/november-december-2014/from-designing-to-co-designing-to-collective-dreaming-three-slices-in-time (accessed on February 2, 2019).

Stamets, P. (2004), *Mycelium Running: A Guide to Healing the Planet through Gardening with Gourmet and Medicinal Mushrooms.* New York: Random House.

Stamets, P. (2008), "6 Ways Mushrooms Can Change the World," TED talk, www.ted.com/talks/paul_stamets_6_ways_mushrooms_can_save_the_world/transcript?language=en.

Let's Talk About Soil (2012), [Film] Dir. U. H. Streckenbach, Global Soil Week 2012 and the Global Soil Partnership, Germany.

Udall, D. F. Rayns and T. Mansfield (2014), "Living Soils: A Call to Action," Centre of Agroecology, Water and Resilience, and Soil Association of Scotland, Bristol, Edinburgh, Bristol/Coventry: The Soil Association/Coventry University, www.soilassociation.org/media/4673/living-soils-a-call-to-action-2015.pdf.

7 Vital Matters: Growing Living Materials

VICTORIA GEANEY

Introduction

Biological design—or biodesign—entails co-creation with living matter and artificial materials, and it often involves entangled human-nonhuman agential relations. Biodesign therefore requires designers to work with living systems, either during production or in the project outcome. My practice centers on this biologically designed paradigm. I am a fashion-led researcher and practitioner whose interest lies in work that straddles both design and biology. In fashion-led biodesign, designers, biologists, materials, and biological entities, such as bacteria, participate as co-actors in vital assemblages of living and non-living agents. In my work, I consider how employing fashion in conjunction with vital matter (Bennett 2010: 112) can affect how we use bacterially produced textiles and garments, both now and in the future. I use my practice to think about the agency of living systems in garments, and I often work collaboratively with biologists.

Interdisciplinary collaboration involves the synthesis of practices of two or more disciplines (Darbellay 2015: 165–66), and in this essay I present and analyze what happens when designers and biologists collaborate. In particular, I demonstrate how collaboration produces an interdisciplinary praxis by chronicling my work with two synthetic biologists from the University of Cambridge, Anton Kan and Bernardo Pollak. Synthetic biology involves redesigning or engineering organisms to carry out new functions. I present our working process, from concepts to precursors to final work, for the first iteration of a bioluminescent garment entitled *Lo Lamento* (2016). This garment was imbued with the bioluminescent bacteria *Photobacterium kishitanni* 201212X (isolated by Japanese academic Henryk Urbanczyk from a deep-sea fish), which lived on agar and nutrients on the surface of the garment materials. By discussing the ideas and explorations that led to this project, I reveal how my collaborators and I blurred the disciplinary boundaries between fashion-led research and synthetic biology to produce a project that is at once biological and designed.

Biologists typically apply the scientific method in an evidence-based manner (Reeve 2019). Their data guide their results, and are then used to confirm or reject scientific hypotheses—a process that often leads to further working hypotheses and experiments (Reeve 2019). Fashion design practitioners likewise operate iteratively, but they also work expressively as interpreters and storytellers. Although both scientific and "designerly" (Cross 2001: 54) ways of working are iterative, a fashion practitioner's process is typically narrative, and sometimes intuitive—particularly when compared to the standard scientific method (Fairburn et al. 2016:94; Dieffenbacher 2013: 10). Fashion practitioners are also almost always concerned with materiality, with notions of both the "second skin" and the body, and with making outcomes available to users. Considering these differences in designerly and scientific approaches, it may be difficult to imagine how the two can function effectively together.

Kan, Pollak, and I brought divergent skills to the making process. I contributed narrative and poetic ways of working, reflexivity, and an understanding of the reciprocal agential relationships between material, space, form, and body. I also brought an interest in performative outcomes—that is, showcasing the project in public or gallery settings. The biologists provided knowledge of and expertise on the bacterium we used, *Photobacterium kishitanni*. They likewise contributed their scientific method of research through testing to ascertain the most suitable growth conditions and nutrient solutions for the bacteria. Kan and Pollak also supplied funding and laboratory space for the project. We three, though, also brought overlapping ways of operating, such as experimentation and iterative manners of working. Additionally, both the biologists and I work with living organisms and "bodies." Collaborating with Kan and Pollak—and working with non-human organisms—expanded my fashion practices beyond the standard notions of materiality and the body. Our collaboration broadened Kan and Pollak's science practices and notions of beings and bodies beyond observing, documenting, and hypothesizing about microbial behavior. These factors suggested hybrid methods of working to us and precipitated new cross- and interdisciplinary concepts and methods. Since we worked in both the studio and the laboratory, collaboration also allowed us to blend the design studio with the laboratory.

Throughout the project, the scientists and I shared informal ideation sessions, discussions, prototyping, testing, and making. We also explored concepts for the final project exhibition: the outcome of this iterative project was a site-specific garment installation. For Kan and Pollak, our creative collaboration encouraged self-expression and a narrative approach. The process made their research public and accessible, and enabled the development of a broad contextual understanding about design possibilities and applications for our project, as well as the implications of working with bioluminescent bacteria in design beyond this project. We discussed, for instance, the notion of street lighting that uses bioluminescent bacteria as a sustainable form of energy production.

The collaboration gave me freedom from business constraints and the user-centered drivers that would have dictated my designs if they were produced for the commercial fashion industry. Working with the bacteria, and considering its properties, provided me with inspiration that motivated my making processes and the concepts that we considered. The notion of collaborating across disciplines allowed me to push the boundaries of my own discipline, and to reconsider definitions of fashion and garments. Indeed, using bacteria as part of a garment forced me to challenge my ideas of the very essence of fashion as a living entity. This sharing process was critical to our thinking, trial and error, experimentation, and making practices.

Our collaborative process led me to review writing on fashion design. Fashion design articles and books have traditionally focused on the function of the final product rather than on the process through which the work is conceptualized and crafted. This approach preserves an air of mystery and intrigue about making processes in fashion design that should almost certainly be reconsidered (Dieffenbacher 2013:10; Jenss 2016:14; Storey 2018). I contend that making processes in design ought to be elaborated, as they reveal how concept, content, and product coalesce for designers. Process also functions as a shared foundation for collaborative work. Some fashion designers and fashion-led researchers are initiating and leading fashion designer-biologist collaborations to produce innovative, nuanced, provocative, and dynamic "living" solutions. I intend that making my collaborative processes available to others in this essay will be important to understanding how collaborations are formed, as well as how they operate for designers who are interested in working with scientists in hybrid disciplinary landscapes.

My Collaborative Biodesign Work

I produced a series of bacteria-based projects in collaboration with biologists over the course of my practice-led fashion-design doctoral research at the Royal College of Art, London. Mine is an integrated practice-led design research approach in which shared knowledge production emanates from specialist knowledge. Employing live bacteria on garments first and foremost requires preliminary research about their care and cultivation, as bacteria require specific conditions for growth.

Design is a medium, according to Dunne and Raby, that allows designers and scientists to "blur distinctions between the 'real' real and the 'unreal' real," giving "form to the multiverse of worlds our world could be" through speculative design (Dunne and Raby, 2013: 159). The bacteria-based garments produced in my collaborative projects could not be worn by a human model because of health and safety constraints. The relationship between the body and the bioluminescent bacterial garments in my work, therefore, is an imagined one. This imagined relationship, however, poses new questions for fashion of reality versus fantasy—real clothing in contrast to speculative garments.

The *Aqualose* Precursor Project

My first opportunity to work with synthetic biologists was as a consulting designer with the International Genetically Engineered Machine (iGEM) team at Imperial College London in 2014. The iGEM Foundation is an independent, non-profit organization

Fig. 7.1 Computer-aided design technical flats showing front, back, and sides of the bacterial cellulose waistcoat. Illustration by Victoria Geaney.

dedicated to cross- and interdisciplinary collaboration in synthetic biology. I first worked with the Imperial College team on their *Aqualose* project (Imperial College London, 2014) in which bacterial cellulose was specialized for water filtration through the binding of functional proteins (Florea et al. 2016: 7).

Bacterial cellulose is the by-product of a fermentation process that is used to produce kombucha tea. This type of probiotic tea most likely originated in the Far East when "the leaves of the *Camellia sinensis* plant were first steeped into tea" (Childs and Childs 2013: 27–8). The lore around kombucha claims that it has magical or elixir-type properties for achieving immortality (Childs and Childs 2013: 27–8). As a biomaterial, bacterial cellulose is biodegradable and biocompostable, and is therefore interesting as a possible sustainable textile. Its ephemeral and temporal biodegradability could enable a "cradle-to-cradle" approach (Braungart & McDonough 2008: 5)—for example, to recycle leftover food waste using a fermentation process to grow a bacterial cellulose garment.

In 2003, Suzanne Lee founded Biocouture, a design and research consultancy and studio focusing on the biodesign of materials. Lee looked at using the fermented and dried skin that came out of the kombucha production process—which is called the symbiotic culture of bacteria and yeast (SCOBY)—as a biomaterial for garments (Katz 2012: 171). Her fermentation experiments led to the production of some of the first cellulose for garments such as skirts, shoes, and shirts from the microbial cellulose. These products have a similar look and feel to leather, but are a vegan alternative. Using biological "organisms like microbes as the factories of the future" (Fairs 2014) for material for fashion design, prompted initiatives such as the iGEM 2014 project.

During the iGEM 2014 *Aqualose* project, the Imperial team investigated how synthetic biology of the yeast genome could enable functionalization of bacterial cellulose biomaterial. I visited the laboratory regularly and the scientists and I ideated together, discussing if engineering the yeast genome could alter the properties of the bacterial cellulose to overcome issues such as its water retentive nature. I showed them technical drawings and designs that used bacterial cellulose in a highly sculptural way, and then worked with the scientists to blend the bacterial cellulose into a substance that I sculpted into small pyramid shapes. A week later, however, the pyramids had dried flat because the water they had retained had evaporated. The iGEM team and I then tested our initial ideas, including growing the cellulose and combining it with other materials and substances, such as newspaper, flower petals, and fabric samples. Although these attempts were unsuccessful, they showed the capabilities and boundaries of working first-hand with the cellulose. I realized at this time that bacterial cellulose could be shaped or molded over other forms, but that it could not produce a three-dimensional sculptural form in and of itself.

While visiting the iGEM Tom Ellis Synthetic Genomics and Synthetic Biology Research Group laboratory and facilities, I was drawn to green fluorescent protein (GFP), which is extracted from the *Aequorea Victoria* jellyfish, and is often used as an expression reporter to track proteins. GFP, which glows in the dark under ultraviolet light, can be reproduced by harnessing *Escherichia coli* (*E. coli*) bacteria to create it. This biotechnological approach produced sustainable biological wearables made of GFP combined with bacterial cellulose (which built upon practitioners Suzanne Lee and Donna Franklin's bacterial cellulose projects). Light and fluorescence were already central to my fashion practice as I had previously produced a wearable technology fashion collection using light emitting diodes (LEDs). I, therefore, recognized the conceptual and aesthetic design potential of this naturally glowing protein in a fashion context. I encouraged the team to use GFP to produce biological forms of wearable technology.

To create material, the iGEM synthetic biologists optimized bacterial cellulose biosynthesis in the bacteria *Gluconacetobacter xylinus*. The cellulose was grown using a SCOBY, green kombucha tea, sugar, and cider vinegar, and it took approximately two weeks to ferment and form. Layers of fermented skin formed on the top of the material growing in the vats in the laboratory. These layers were extracted, washed, and treated to eliminate residual bacteria. Once dry, the cellulose material lost a lot of its moisture and shrank to the consistency of crunchy paper—yet it was a lot stronger than paper and harder to tear.

This material would be used to make our garment. For the design, we agreed upon a unisex waistcoat, which was to be exhibited, as well as accessories for the team to wear. These items showcased various grades of the material—such as the thickly layered cellulose buttons, and the finer, almost paper-like fabric for the body of the waistcoat. I designed and made the waistcoat and accessories from the bacterial cellulose that was grown in the laboratory. I then painted sections using green fluorescent protein that the team had expressed using *E. coli*, as well as naturally grown dyes extracted from coral. The waistcoat was exhibited at the iGem Conference in Boston as part of the Art and Design track, and the Imperial team finished as first runner-up.

Our iGEM team project demonstrated how versatile bacterial cellulose can be as a biomaterial. Our project likewise suggested how people might be able to grow their own garments in the laboratory using bacteria, and that bacteria could light up a garment biologically without using LEDs, sensors, or power sources. This collaboration also led to my early research focus: using bacteria to create aesthetic, functional, and potentially wearable garments that utilize bacterial fluorescence or bioluminescence.

Lo Lamento

Following the *Aqualose* project, the synthetic biology meetup group at the University of Cambridge invited me to give a presentation on synthetic biology and fashion in the work I did with the Imperial College iGEM 2014 *Aqualose* team, and on the early stages of my research at the Royal College of Art. The presentation led to a meeting with the University of Cambridge synthetic biologists Anton Kan and Bernardo Pollak. They were working on a project, funded by the Synthetic Biology Strategic Research Initiative at Cambridge, to collect and sequence bioluminescent bacteria in order to find the brightest strains for use in synthetic biology (Synthetic Biology Strategic Research Initiative 2015). Their grants were intended to encourage collaborative, interdisciplinary, and publicly accessible projects (Synthetic Biology Strategic Research Initiative 2015).

Around the same time, Alessandra Caggiano, director of the e-Luminate Festival, asked if I could produce a light-up fashion piece for the festival. The e-Luminate Festival is a Cambridge, UK event that celebrates the "infinite possibilities created by light at the intersection of art and science" (Cambridge Live 2017). Kan, Pollak, and I discussed collaborating on a design project that used bioluminescent bacteria, and we agreed to present our final outcome at the e-Luminate Festival. The festival provided a shared focus, with a set deadline, and an opportunity to publicly showcase the project.

Early Ideation—The Scientists Enter the Early Stages of a Fashion Design Process

Kan, Pollak, and I shared expertise from our respective disciplines—and my experience creating *Aqualose*—to imagine possible bioluminescent bacteria fashion outcomes for the e-Luminate festival. We discussed *Photobacterium kishitanni*, the brightest bioluminescent bacteria they had sequenced, and considered how the natural luminescence could be showcased in a fashion context. In nature, bioluminescence lures prey, attracts mates, and camouflages the organism. It is also used as a warning to predators and also to illuminate darkness. The glow of the *Photobacterium kishitanni* is beautiful, and it is visually arresting in a way that would draw attention from human spectators. Both the scientists and I were therefore keen to share this biological phenomenon with the public.

We discussed logistical and design considerations, such as the conditions that were necessary for growing the bacteria, including constant air and food (yeast extract and sea salt), and the sterile conditions needed for the bacteria to glow. We talked about patterning and interactivity, and altering the color of the bacteria through dyeing. We also considered pumping air through the piece using an aquarium pump, which would increase or decrease the illumination. For the design installation we thought about using time lapse videos, how to encase the garment, and how the garment might somehow respond to touch or sound.

Focused Ideation Toward Key Concepts

As noted earlier in this essay, a bacteria-infused garment worn by a human model was not possible because of health concerns and the need for a moist surface coating of agar and yeast extract on the garment. We therefore had to focus our attention on creating an installation rather than on a wearable fashion solution. Three key ideas emerged from our ideation sessions. Our first concept was inspired by the notion of a shining, bacterial second skin. In this solution, the glowing garment is encased in a glass vitrine with a feeding system that coats the garment with agar and media for the bioluminescent bacteria to grow and glow. Human skin is naturally coated in bacteria, and we communicate with these bacteria through our own biochemistry. This first concept was intended to highlight these human skin and bacterial interactions—for instance, the wet garment could represent human perspiration.

Our second concept revolved around an interactive experience between spectators and the bacteria in the installation. *Photobacterium kishitanni* glows in response to oxygen, and stops glowing once an oxygen supply is removed. We imagined that spectators could press a button or use a foot pump to deliver oxygen to the installation—creating bright bursts of bioluminescent glow—to serve as a communicative interface between bacteria and humans.

The third key concept involved a wet, glowing garment or fabric installation that focused on bacterial patterning. In our concept, *Photobacterium kishitanni* needed absorbent materials to hold the nutrients required to feed them. We discussed how to pattern the illuminating bacteria using glowing and non-glowing areas, which could be achieved by incorporating absorbent and non-absorbent materials into patterns. Our collaboration progressed from preliminary notions to key ideas and practical requirements for these three concepts.

Design Modeling and Visualization

I produced a preliminary computer-aided design (CAD) for a mock-up that I had envisioned during our discussions (see Fig. 7.2): the garment would be central, set within a paneled glass vitrine, and surrounded by a feeding system of growing and glowing orbs. A pump would force the liquid nutrient solution and bioluminescent bacteria from the orbs onto the garment (see Fig. 7.3). This system, with its flowing light and liquid energy reminiscent of breathing and heart and lung function, influenced my design for the top of the dress. I structured the arterial-looking tubes, and the rest of the installation, to flow from the top of the garment. I imagined the spheres as glass orbs—filled with liquid and connected to the dress by tubing—that supplied nutrients and bioluminescence to the whole piece. The flow of nutrients from a central point

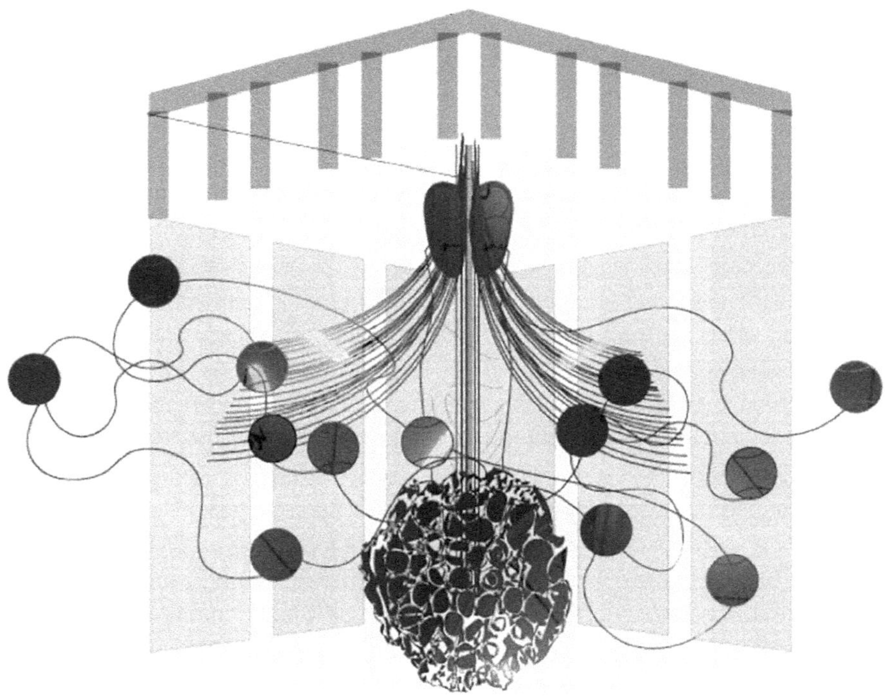

Fig. 7.2 Preliminary computer-aided design visualization of the *Lo Lamento* installation. Illustration by Victoria Geaney.

on the garment is also reminiscent of a mother nourishing a fetus via the umbilical cord.

I designed the base of the dress based on my experiments that showed that agar could be used as a sculpting medium. Agar can set relatively hard, so I felt that it could "house" the bioluminescent bacteria on the garment base with no water required. The shape of the garment base was inspired by the clustered flowers of *Allium giganteum*, hundreds of which emanate in the plant from a midpoint to create an overall spherical shape. This inspiration established a further link to the notion of interconnecting tubes that carry energy from one point to another. I envisaged the base of the installation as three-dimensional—with a large number of sculpted spheres of agar that were coated in the glowing bioluminescent bacteria.

I shared this design visualization process with the scientists, which helped our collaborative ideas coalesce. I rendered the installation digitally (Fig. 7.2) as a conceptual illustration rather than a final working fashion design. This preliminary design imagery gave us all a clearer sense of what the piece might be, and led to detailed logistical and technical deliberations. At this stage, we decided not to include some of our early ideas—such as push buttons or a foot pump to enable air for bacterial illumination—because the living light worked well on its own as a rich communicative motif. We discussed having the installation as a garment coated in living bioluminescent bacteria, using tubing and glass vials, and orbs or containers to hold the bacteria. The piece would be set inside a glass display case for protection and safety. Together we considered options for flasks, stoppers, and tubing for the installation. We had not yet finalized materials that would work with the bacteria (so the material is shown in the rendering as translucent). I brought various cloth samples for the scientists to use both for testing to see which materials enabled a brighter glow, and to observe how long the bacteria lived on the materials.

Shared Prototyping, Shared Testing, and Shared Making—Merging Fashion and Science in Testing Fabrics for Bacterial Growth

Kan and Pollak conducted laboratory testing of the sample swatches I sent them, which included cotton calico, lace, polyester silk chiffon, polyester taffeta, wool felt, cotton drill, fleece, cotton jersey, and cotton terry towelling fabric. They tested fabric samples in three types of media: LSW-70 broth, using 1.5% concentration of agar, and also using 3% concentration of agar. It turned out that 100% cotton fabric permitted no growth, however, there was some growth on calico cotton.

The scientists added agar and nutrient solution, and imbued the samples with the photobacterium. Kan and Pollak grew the samples in humid conditions to replicate the glass vitrine environment we would have at the exhibition, and to keep the fabrics from drying out. In the fashion design studio, I tested two different agar solutions: a 1.5% one, and one prepared according to the typical instructions (one tablespoon to

240ml water). I dipped various materials into the agar to see which would retain moisture once they were coated with the substance.

The three of us then discussed findings about the properties of the materials that enabled the bioluminescent bacteria to grow successfully. Our experiments demonstrated that absorbent materials, such as felt and cotton terry towelling, retained greater amounts of the wet agar. Separately, we each found the higher concentration of agar to be more effective. Further, increasing the surface area by using fabrics such as terry towelling, woven with a loop yarn, helped to maintain moisture and encourage bacterial growth. Kan and Pollak sent me photographs showing the fabrics coated in agar and yeast growth medium and inoculated with bacteria. They also noted that layering the fabrics enabled moisture absorption and retention, which made the bacteria glow brighter.

Draping-on-the-stand and Garment Production—Divergence of Practices

At this point we distributed roles according to expertise. The synthetic biologists worked on optimizing the growth media and producing large quantities of liquid media for the bacteria. They used LB liquid and seventy percent sea water and LSW 70 media (containing tryptone, yeast extract, and sea salt). The bacteria can be grown in solid or liquid media, and the scientists used two variations – media with agar and media without (LSW 70 media solid and LSW 70 media liquid). The agar acted as the substrate upon which the bacteria grew. Kan and Pollak cultured and studied the bacteria so that we could predict when the glow would be brightest according to plate-reader experiments they conducted, and they programmed a microcontroller board to operate the peristaltic pumps to run at 2ml per minute for the installation. The scientists calculated that two liters of LSW 70 medium would be required for the installation every sixteen hours. The peristaltic pumps were used to drive fresh media around the glass orbs.

I organized the installation logistics, sourcing delivery of the glass display case, and collaborating with a technician to build and test a pump system. I also worked in the fashion studio, draping-on-the-stand and pattern cutting, working and reworking the garment, conducting fittings, and using finishing techniques and pressing. During this making process, I took the scientific findings into consideration while I reflected upon and reworked the garment according to critiques from fashion and textile researchers and tutors, and then conducted two garment fittings on a human model in the fashion studio.

Making Process

At this stage, I transitioned from finalizing our collaborative concepts to making. I employed 2D and 3D construction methods, including manipulating materials by hand, as well as drawing, draping, looking critically at the form—and I compared these physical mock-ups to concepts, design development, and prototypes.

In response to my small-scale fabric testing, I selected absorbent, natural materials for the garment, and chose white and cream colors to display the glowing bacteria. The primary fabric I used was highly absorbent and relatively unprocessed raw wool. I also selected white wool felt, medium weight cotton drill, pinstripe polyester acrylic, and a wool underlay. As noted above, in our experiments, layering parts of the garments created air spaces as well as areas for the wet, melted agar solution, which promoted bacterial colonization. Creating air spaces also helped with ease of layering and dismantling the garment on the mannequin.

For the garment design, I was inspired by the deep sea where the photobacterium originates. I sourced images of sea creatures such as translucent, tentacular jellyfish with tendrils, and sea anemones, such as the bulb species *Entacmaea quadricolor*. I was also influenced by biomorphic forms from nineteenth-century biologist and illustrator Ernst Haeckel's book *Art Forms In Nature* (1974). Another of my inspirations came from the way that the bacteria respond and "breathe." I considered how these bacterial physiological functions correlate with layers in the human body, such as skin and skeletal ribcages that protect the lungs, heart, and vital organs. Layering highlighted my concepts of nurture and protection, which simultaneously increased surface area and absorbency.

Draping

To prevent mistakes in the garment form, fashion designers must "work with the material" when designing for the irregular form of the human body. Pattern cutting defines the shape, but draping reveals the capabilities of the material, as well as how it will work on a mannequin and ultimately on a human form. For *Lo Lamento*, I used a combination of these two fashion design methods to achieve the final shape.

This stage of making first requires close-up work to shape, edit, and cut away, then a reflective stage of reviewing and analyzing how the shape works with the concept, and, finally, how the piece can be improved. Draping, shaping, cutting away, and editing my work is an iterative process of making and reflection and re-working. Using a basic dress pattern block, I drafted and shaped a pattern with a neckline that implied a curved "lung" shape, as referenced in my concept. I cut the pattern from two materials—thick wool hidden underneath and a top layer of polyester acrylic fabric—that I draped onto the mannequin. I next cut away at the top layer to suggest jellyfish tendrils. I cut the layer draped above that to form skeletal shaped sections. I created another final layer for the top of the garment made of a stiff wool felt, which retained its shape. This layer referenced ribs and a rib cage, again, as detailed in my concept. Under this layer, I included a section of bobbled wool shapes that referenced both a bulb anemone and a heart and vital organs under the "rib cage" of the piece. My shape concept, then, is a rib cage exterior layer that protects the interior layers, which include a hidden, bobbled "vital organ" section. I produced the final project in the fashion studio. Through making and shaping and editing, I translated one facet of our concept—the garment—into a material form.

Live Model Fittings

Fitting garments to a human model allows fashion practitioners to test and adjust fit. I worked with a pattern-cutting technician, with whom I discussed finishing techniques—in particular, for the wool felt, which was the most visible layer of the piece. The technician suggested bonding the seams for a technical finish, and using a rotary cutter to better finish the polyester acrylic layer. These techniques gave the garment a professional finish.

Final Design

Various constraints meant that we were unable to test a full-scale garment before the festival. During our fabric-testing phase, we enabled bioluminescent bacteria colonization so that our test samples glowed. The festival piece itself turned out to be an experiment to see if we could produce the garment and reproduce the bacterial innoculation at full scale.

Collaborative Exhibition Installation—Re-convergence of Practices

At this point, the scientists and I came back together to install the final project. Kan and Pollak had produced large quantities of the nutrient medium solution of liquid agar substrate, sea water, and yeast medium to feed the bacteria on the garment. On the day before the e-Luminate Festival I took the finished garment to Cambridge, where we autoclaved it in the laboratory. Autoclaving sterilizes the material to remove all existing microbes—either transferred inadvertently by me or from the fabric itself—in order to prevent mold growth. This step ensured that the garment was prepared properly for the bioluminescent bacteria and nutrient broth.

We had issues transporting the original glass display case, so we made a Plexiglas (perspex) vitrine ourselves. The acrylic case, however, was not strong enough to hold the glowing orbs around the garment, which meant that we had to forego this part of the original design concept. Instead, we separated this part of the installation: we set the orbs on a table, where they responded to bursts of oxygen when the tubing was handled. The glowing garment was installed inside the vitrine. Adaptability is key to both science and fashion—and to collaboration.

We worked collaboratively to set up the installation, and carried out the inoculation processes on site to avoid the health and safety issues involved in moving a bacteria-

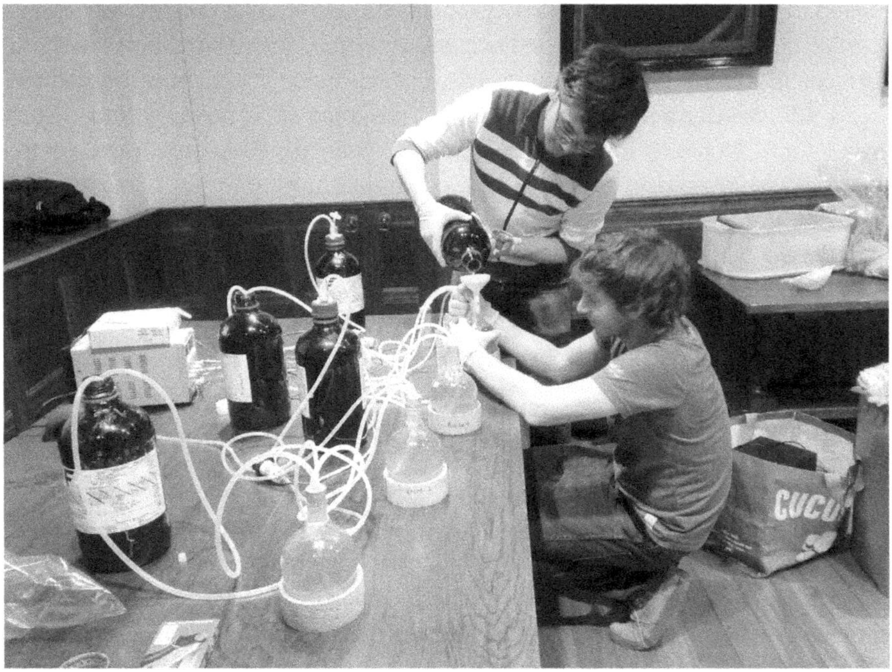

Fig. 7.3 Anton Kan and Bernardo Pollak adding liquid media to the pumping system as part of the *Lo Lamento* installation. Photography by Anton Kan, Bernardo Pollak, and Victoria Geaney.

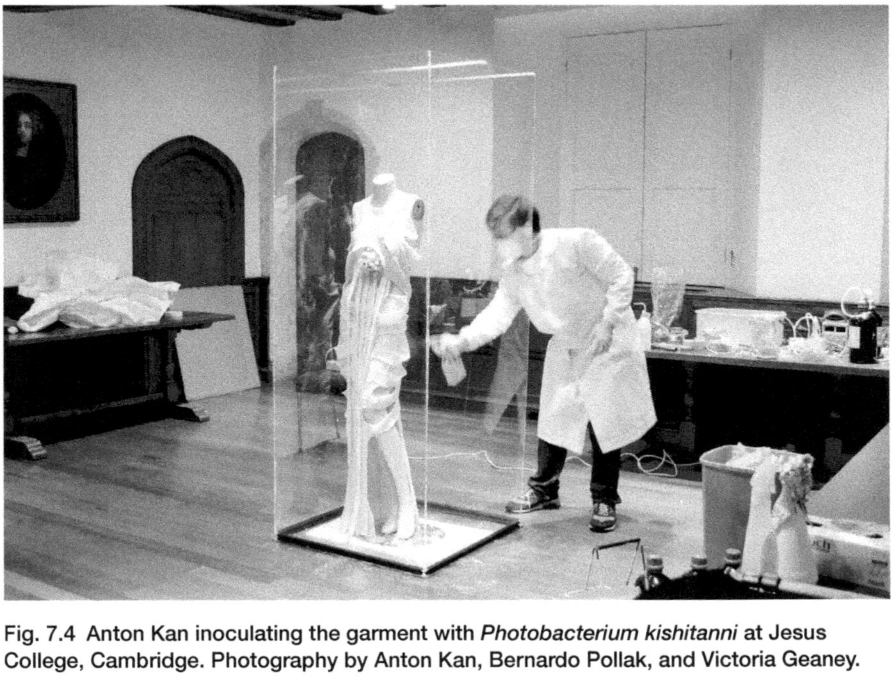

Fig. 7.4 Anton Kan inoculating the garment with *Photobacterium kishitanni* at Jesus College, Cambridge. Photography by Anton Kan, Bernardo Pollak, and Victoria Geaney.

coated garment. We carefully disinfected surfaces, and wore gloves and face masks to avoid contaminating the agar and bacteria. As noted, I had already autoclaved the garment to kill any microorganisms, so we then thoroughly soaked the garment layers with the warm agar nutrient solution. The three of us carefully draped each layer in turn, according to the design, onto the mannequin. We then waited for the agar mixture to solidify, and Kan sprayed the garment with the bacterial liquid medium. We carefully placed the piece inside the display case to prevent contamination. Kan, Pollak, and I poured more agar and liquid media onto the dress to ensure the piece was completely moist, and to help facilitate the growth of the bioluminescent bacteria.

We left the installation overnight to allow the bacteria to colonize, multiply, and grow onto the agar and materials. We returned the next day to find that our large-scale experiment had worked, and the garment and the orbs were glowing—just in time for the exhibition.

During the festival, Kan, Pollak, and I spoke to the general public about the installation. The synthetic biologists chiefly discussed the bioluminescent bacterial glow, and I explained the concept of the garment, the methods of production, and how we used the bacteria as a form of biological light. Following the exhibition, I contacted relevant magazines, websites, and publications to publicize the work, using techniques of promotion developed during my training in fashion. A host of opportunities arose both during and after our collaboration. The festival promoted the

Fig. 7.5 *Lo Lamento* photographed glowing in the daylight at Jesus College, Cambridge. Photography by Anton Kan, Bernardo Pollak, and Victoria Geaney.

Fig. 7.6 Side view of the bioluminescent bacterial *Lo Lamento* installation photographed glowing in the dark at Jesus College, Cambridge. Photography by Anton Kan, Bernardo Pollak, and Victoria Geaney.

Lo Lamento project on BBC Radio Cambridge, and we were interviewed about our collaboration. On the first day of the exhibition, we filmed interviews organized by the festival, and part of the recording was featured on an online science television show. I wrote to *Design Exchange*, which featured an article about the work (Plough 2016), and the work was also discussed in an article for the *FashNerd* website (Kapfunde 2016). Attention from these various media highlight how the work is of interest to both design and science audiences, as well as to the general public.

Conclusion

Collaborating on this biodesign project encouraged Kan, Pollak, and I to take on new kinds of risks and to rethink our regular disciplinary work. Kan and Pollak, for instance, had to work at human-scale—a much larger scale than normal for these microbiologists. We also crafted an innovative concept for applying living bacteria to a material and agar scaffold—a feat that required problem-solving, teamwork, and iterative ways of working from each co-actor. The biologists reported a sense of freedom and unencumbered "evolution" during the early generative ideation phases of the fashion design research process, an experience that is familiar to designers. Applying living bacteria to worn garments forced Kan and Pollak to consider the unavoidable embodied aspect of fashion design (whether real or imagined), how bacteria are themselves like "bodies," and the visual, tactile, and aesthetic qualities of garment plus bacteria. I had to rethink what a "living" garment could be, and what that could mean for fashion design. These kinds of novel approaches to and considerations about our project helped pave the way for our collaboration.

Fashion-led design research has acquired a new sense of urgency during these unprecedented times. My form of fashion-led biodesign research "leads . . . to new understandings about fashion practice" (adapted from Candy and Edmonds 2018: 64) and function. As living organisms that undergo genesis, growth, decline, and death, bacteria used in fashion pose vital questions about the temporality and care of clothing. That is, incorporating bacteria into garments positions them as components in precarious or delicately balanced ecologies. A bacterial garment that requires constant care challenges notions of garments as throwaway items, and shifts them to entities that require cultivation. These concepts put pressure on the relationships between human and nonhuman, and living and non-living entities. These points of view insist that we perceive fashion as something that is alive, fragile, and transient, rather than non-living and disposable.

The COVID-19 pandemic has brought about a period of self-reflection and meditation in the fashion discipline. A biological phenomenon has provided an opportunity for fashion designers to stop and reflect. London Fashion Week 2020 was canceled, and a virtual gender-neutral fashion week was held in June 2020 instead. The COVID-19 virus has affected our lives, health, and economy. These kinds of challenges could almost certainly benefit from the new methods and innovative thinking that come out of interdisciplinary collaborative work. My essay presents a descriptive methodological roadmap to interdisciplinary collaborative work that can change ways of thinking, and my discussion demonstrates the value of fashion and its thinking to an emerging collaborative landscape within and beyond design.

References

Bennett, J. (2005), "The Agency of Assemblages and the North American Blackout," *Public Culture* 17(3): 445–66.

Bennett, J. (2010), *Vibrant Matter: A Political Ecology of Things*. Durham, NC: Duke University Press.

Bennett, J. (2018), "Vibrant Matter," in R. Braidotti and M. Hlavajova (eds.), *The Posthuman Glossary,* 447–8, London: Bloomsbury.

Braungart, M. and W. McDonough (2008), *Cradle to Cradle. Remaking the Way We Make Things (Patterns of the Planet)*. London: Vintage.

Cambridge Live (2017), *e-Luminate Cambridge Festival*, *www.cambridgelive.org.uk/e-luminate* (accessed on September 3, 2020).
Candy, L. and E. Edmonds (2018), "Practice-Based Research in the Creative Arts: Foundations and Futures from the Front Line," *Leonardo* 51(1): 63–9.
Childs, E. and J. Childs (2013), *Kombucha!: The Amazing Probiotic Tea that Cleanses, Heals, Energizes, and Detoxifies*. New York: Penguin.
Cross, N. (2001), "Designerly ways of knowing: design discipline versus design science," *Design Issues* 17(3): 49–55.
Dieffenbacher, F. (2013), *Fashion Thinking: Creative Approaches to the Design Process*. London: AVA Publishing.
Dunne, A. and F. Raby (2013), *Speculative Everything*. Cambridge: The MIT Press.
Fairburn, S., S. Steed, and J. Coulter Nixon (2016), "Spheres of practice for the co-design of wearables," *Journal of Textile Design Research and Practice* 4(1): 85–109.
Fairs, M. (2014), "Microbes are the 'factories of the future,"' *Dezeen and Mini Frontiers*, http://www.dezeen.com/2014/02/12/movie-biocouture-microbes-clothing-wearable-futures/ (accessed on July 10, 2020).
Florea, M., H. Hagemann, G. Santosa, J. Abbott, C. Micklem, X. Spencer-Milnes, L. de Arroyo Garcia, D. Paschou, C. Lazenbatt, D. Kong, H. Chughtai, K. Jensen, P. Freemont, R. Kitney, B. Reeve, and T. Ellis (2016), "Engineering control of bacterial cellulose production using a genetic toolkit and a new cellulose-producing strain," *Proceedings of the National Academy of Sciences of the United States of America* 113(24): E3431–E3440, www.pnas.org/cgi/doi/10.1073/pnas.1522985113 (accessed on August 24, 2020).
Franklin, D. (2014), "Meaningful Encounters: Creating a multi-method site for interacting with nonhuman life through bioarts praxis," doctoral dissertation, Edith Cowan University, https://ro.ecu.edu.au/cgi/viewcontent.cgi?article=2575&context=theses (accessed on July 4, 2017).
Haeckel, E. (1974), *Art Forms in Nature*. New York: Dover Publications.
Imperial College London (2014), "Customisable Ultrafiltration Membranes from Bacterial Cellulose," *Aqualose* (wiki for iGEM 2014), http://2014.igem.org/Team:Imperial (accessed on August 24, 2020).
Jenss, H. (ed.) (2016), *Fashion Studies: Research Methods, Sites, and Practices.* London: Bloomsbury Academic.
Kapfunde, M. (2016), "Geaney's *Photobacterium Kishitanni*, An Exploration of Non-Traditional Materials," *FashNerd*, https://fashnerd.com/2016/12/geaney-the-woman-behind-the-innovation-of-the-living-light-dress-made-of-photobacterium-kishitanni/ (accessed on August 3, 2020).
Katz, S. (2012), *The Art of Fermentation: An In-Depth Exploration of Essential Concepts and Processes from Around the World*. Vermont, ME: Chelsea Green Publishing.
Lee, S. (2005), *Fashioning The Future: Tomorrow's Wardrobe*. London: Thames and Hudson.
Lee, S. (2011), "Grow your own clothes," TED talk, www.ted.com/talks/suzanne_lee_grow_your_own_clothes (accessed on July 14, 2020).
Lee, S. (2019a), "Material world's new frontiers: Suzanne Lee," talk given at Biodesign Here Now, OpenCell London, London Design Festival.

Lee, S., (2019b), "Why 'biofabrication' is the next industrial revolution," TED talk www.ted.com/talks/suzanne_lee_why_biofabrication_is_the_next_industrial_revolution (accessed on July 18, 2020).
Plough. J. (2016), "*Lo Lamento*," *Design Magazine*, https://www.demagazine.co.uk/design/lo-lamento (accessed on June 29, 2016), link inactive on June 15, 2022.
Reeve, B. (2019), "Reporting on the role of a biologist and how a biologist operates within a laboratory setting," [email] (Personal communication, April 24).
Storey, H. (2018), Author interview by Victoria Geaney. [video conferencing] August 2.
Synthetic Biology Strategic Research Initiative (2015), *Strategic Research Initiatives at the University of Cambridge December 2015* Report, University of Cambridge, www.synbio.cam.ac.uk/PublicSRIReportDec2015.docx.pdf (accessed on July 18, 2020).

8 Follow Your Nose

MIRIAM SIMUN AND SHARON LIN

Smoke in the Air

It was inevitable that the phantom odor would return, its stench full of torn covers and attic dust, with a tint of a winter snack—something in between roasted peanut shells, smoked hickory, and burnt sunflower seeds. It had been this way for Lexa for at least a month now. Before she realized that her scent had vanished, she felt as though she was in a perpetual daze, like she had caught the cold and her nose had never recovered. Her doctor had told her often enough that there was no easy cure for anosmia. The smells around her had faded one by one until only the strongest and most penetrating smells were left. In time, even those disappeared. Sometimes she would close her eyes and feel as if inside a cloud of haze, like the air left inside a chimney hole.

Introduction

How do we come to know things? In a world that is dominated by written language, visual communication, and auditory indicators, we don't commonly think of the role that scent plays in the way humans navigate the world. But, consider the narrative above in which the individual has lost her sense of smell, and the dimension of countless everyday experiences instantly shifts. Scent is the oldest mammalian sense (Stoddart 1990). The sense of smell requires direct reception of chemical information in the nose, and this information is intimately linked to perception of memory and emotion in humans (Herz et al. 2004). Although we may not yet fully understand scent, we experience the effects that it has on us regardless of our conscious perception. As such, scent plays a crucial role in human experience. All of these factors make scent a rich and important creative medium with which to work.

Developments in olfactory sciences—in the fields of chemistry, neurobiology, and psychology—have led to a better understanding of the motivations and mechanisms behind scent. This research has also offered insights into our ability to control and refine the sense of smell (Sell 2014). In this essay, I will outline the ways that current science understands the function of scent in human biology and perception. I will also consider the links between the sense of smell and memory and emotion. Finally, I will discuss two of my performative participatory artworks, *GHOSTFOOD* (2013) and *The Farewell to All One Has Used, Broken, Lost . . .* (2014). These two projects use scent to engage audiences in ways that access alternate forms of knowing, feeling, and memory.

Clarity from a Nebula

The sharpness of latex scent razed against the bioreactor outerwear that overwhelmed in its waft of sweat-stained shoebox even under adequate ventilation. The undercurrents of faint urine and seaside fish markets reverberating from the animal cages . . . It was both part horrific and a comfort to Danielle, something she could always count on—a constant in her volatile world. Danielle reached over to grab the Petri dish of agar gel and bacteria. She had worked on this particular strain for the last few months, and she knew its entire molecular structure by heart. As she twisted off the glass cover, a fresh spray of organic odors—sticky sweet agar and waxy pectin melting into the warmth and moist of the bacterial growth—emitted from the colonies laid neatly in circular formations, like dots on a Seurat. There was always a momentary excitement and clarity for her when she peered into the microscopic worlds. She imagined the organisms dancing in symbiotic formations, their geometric tendrils and shapes floating across space like miniature nebula. As she prepared her regular routine decontaminating her hands, sprays of Lysol and Clorox swirling with the sharp tang of antibacterial soap, she removed her gloves. The sudden smell of latex and the burning friction of rubber hitting her nose as she felt enveloped in the scents of her bacterial world, with all its spiraling lifeforms combining and forming new smells around her chemically sterile lab.

The Science of Scent

The Physiology of Smell

A person's experience of a scent is the result of a complex reaction to molecules that are emitted from a source and are sensed by olfactory receptors in the nose. Although, on the surface, smell does not appear to be a dominant sense, human DNA actually has a substantial number of genes devoted to detecting different types of odors. A 2003 study identified 339 intact olfactory receptor genes compared to approximately twenty-five taste-receptor genes (Malnic 2004).

There are two paths for smells to be received: orthonasal—through the nostrils, where molecules are received by one or more receptors, and retronasal—through the channel connecting the roof of the throat to the nose. Chewing, which is part of the second mechanism, releases aromas from food particles into the mouth, which can then access the olfactory sensory neurons. Our tongue has only rudimentary flavor receptors—sweet, salty, bitter, sour, and umami—so the full complexity of flavor is delivered via the olfactory channels by the first mechanism. As a result, a cold or flu can prevent odors from reaching the sensory cells, diminishing the combined sensations of a food's flavor.

Olfactory sensory neurons, each equipped with a single odor receptor, are located within the olfactory bulbs—rounded masses of neural tissue within the nose. When stimulated by microscopic molecules that are released by nearby substances, these neurons identify the chemical compounds that are received through olfactory tracts. From the olfactory bulb, two direct connections lead to signal processing in the brain within the amygdala and hippocampus, which are parts of the brain that are responsible for emotion and memory.

There is debate among researchers about how molecules may smell a particular way. Some theorize that smell is because of the shape of the molecule being sensed, while others argue that it is actually because of the molecular vibrations (Turin 2015). Molecular vibrations are affected by the structure of molecules, which are "made of atoms connected by bonds. The vibration of the molecule is a product of the arrangement of bonds between the atoms and the atoms themselves" (Ryan 2013). Chemists at RMIT University have demonstrated that molecular vibrations determine smell. These researchers have shown that different molecules made of different atoms with the same shapes possess different smells (Dreverman 2007). Cis-3-hexene-1-ol, a molecule that smells of cut grass, for example, is almost identical in shape to cis-3-hexene-1-thiol, which smells of rotten eggs (Turin 2015) (See Fig. 8.1). This evidence demonstrates that uniqueness in the vibrations of molecules produces different smells (Bushdid 2014). At a molecular level, no matter the shape of the molecule or concentration of the chemicals, there is no mistaking one scent for another (Bushdid 2014). Our noses are capable of a kind of spectroscopy, which is a scientific technique that is used to detect molecular composition. Our noses and brains can essentially recognize and differentiate nanoscale vibrations in molecules.

Finally, a conscious and skilled perception of scent can be trained. Perfumers have a highly developed sense of smell; they train for more than a decade to identify individual scents on blotters (Caplan 2007). Thomas Hummel at the University of

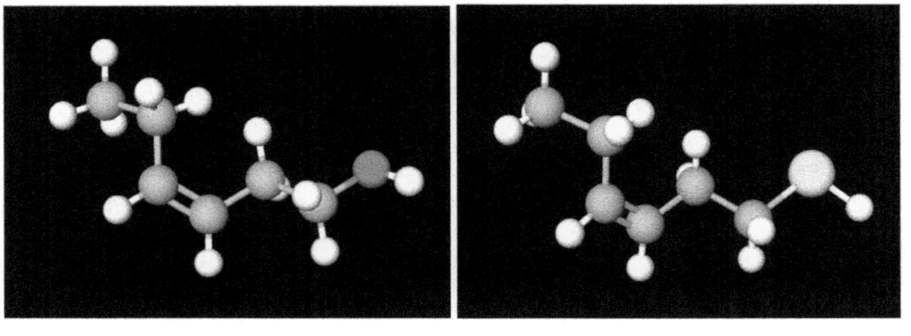

Fig. 8.1 (left to right): Molecular structure of cis-3-hexene-1-thiol and cis-3-hexene-1-ol. Courtesy of Sharon Lin, Massachusetts Institute of Technology.

Dresden investigated olfactory training in people with anosmia, a condition in which people lose the ability to smell. He showed how repeated short-term exposure to scents over a twelve-week period could train anosmia sufferers to smell again. He compared a control group and a group of patients who were given four essential oils—rose, eucalyptus, clove, and lemon—to sniff three times a day. The results indicated that the patients undergoing the training, some of whom suffered smell loss because of upper respiratory tract infection or cranial trauma, experienced some improvement in olfactory function (Negoias 2013).

Scent and Emotion

In the West our attention to smell is often lacking (Classen et al. 1994): consider the dearth in the English language for scent descriptor words. The effect of scent on our experience of the world, though, is well demonstrated (Classen et al. 1994). Although visual, auditory, and tactile information does not pass through the emotion and memory areas of the brain, olfaction is capable of doing so, which explains its unique penchant for triggering memories (Arshamian 2013). The limbic system in the brain, which includes a group of interacting brain structures, plays a major role in controlling mood, memory, behavior, and emotions. The sense of smell is associated with the limbic system and is therefore closely linked to emotions and memory, more so than many other senses, and full olfactory functionality includes the ability to evoke specific memories spontaneously (Caplan 2007). Fragrance affects mood through associative learning. The perfume industry, in fact, is built around the connection between scent and emotion, with fragrances often designed to convey a vast array of feelings ranging from desire to relaxation (Caplan 2007). Perfumes also work as aromatherapy, causing increased alertness, invigoration, relaxation, and motivation (Chamine 2016), and body odor has also been shown to play a role in how we subconsciously choose our partners (Kromer 2016). Stress and other emotions are catalysts for secretions from the apocrine gland in the armpits. When we are fearful, for instance, our perspiration emits an odor that other humans can subliminally identify as fear (Endevelt-Shapira 2017). Anger and aggression can likewise be identified subliminally as chemosignals in body odor. These chemosignals, in fact, function similarly to aggression pheromones (hormones that are released into the environment) that are observed in fruit flies. These pheromones alert fruit-fly bystanders to potentially dangerous situations (Kadohisa 2013).

Since smell plays an important role in our psychological and physical interactions, its absence can have a profound impact. Anosmia victims often describe feeling isolated and having "blunted" emotions (Wuensch 2009). In an NPR interview, anosmia victim Nisha Pradhan discussed the difficulties of living independently, as well as her worries about being unable to smell a gas leak, burning bread, or sour milk. Meals were no longer social experiences for her; cooking merely required basic tastes, and her perception of taste had become distorted (Heist 2016).

Our sense of smell plays a significant role in determining even the most mundane interactions. Although we may take the flavor of food for granted, it is, in fact, constructed from a complex combination of basic tastes, odors, and chemosensory irritations—which most people experience strongly, for example, when smelling menthol or capsaicin. As discussed above, personal scent has an effect on whether we avoid or are attracted to others. It affects our moods. From its function in emotional response and memory, to playing a role in the way we taste and how we perceive others, our olfactory experiences affect how we perceive our environment and act in the world.

Trapped Yet Fleeting

She placed the DOSD over her head and sniffed. Today's flavor was a sultry mixture of irrevocably sweet and something that reminded her of jealousy. The first bite tasted like sugarcane, as expected, and bitter with a hint of a spice she remembered but could not name. She imagined the scenes from the movies, the old-world color and splendor spilling out into rippling waves of saltwater taffy and honeysuckle. The smell lingered for a while longer before finally vanishing, leaving her with only the feeling of coarseness, as the remainder of the roll was swallowed. She checked the label of the box. Madagascar Vanilla. The name sounded as if it came from a fantasy novel. She wondered if it was the name of a classic famous actress.

GHOSTFOOD

Fig. 8.2 Miriam Simun, *GHOSTFOOD* (2013).

Offering Pre-nostalgic Taste Experiences

Olfactory additives are everywhere in our contemporary industrialized food system. Most of the "added flavors" (particularly of the artificial variety) that are found on ingredient labels are chemicals that activate our olfactory receptors. One flavorist describes working on a new microwavable nutritious, yet flavorless, gruel-like product. The flavorist's task was to invent a flavor to best accompany this edible mass, in effect linking a foodstuff with flavor in an entirely constructed process. These additive scents and flavors are pervasive, but they are also invisible and, therefore, we don't typically pay attention to them.

Our public artwork *GHOSTFOOD* inverts this state of things by externalizing olfactory additives. *GHOSTFOOD* aestheticizes, and even fetishizes, olfactory additives in our food and in our lives. About olfactory additives, the project asks: Why? What for? And to what end?

GHOSTFOOD is a food truck that was created in collaboration with Miriam Songster, which offers "taste experiences" of "soon to be unavailable flavors." The project proposes a method for simulating flavors of foods that are species that may soon be lost to extinction. In this way, *GHOSTFOOD* questions human relationships with the natural world in the face of ecological and technological changes. It simultaneously explores how scent, memory, and designed-object commodities also shape our experience of food and our relationship to the natural world.

GHOSTFOOD was originally commissioned for Marfa Dialogues / NY in partnership with the Robert Rauschenberg Foundation. This was a city-wide exhibition series that aimed to "raise dialogue around climate change" in New York City. *GHOSTFOOD* used food—around which language was likely invented (perhaps language was invented around a fire, waiting for food to cook)—to bring climate change dialogue out of the institution and literally onto the street.

Each flavor/food/species in *GHOSTFOOD* originates from a different environment, underscoring the vast reach of ecological change across diverse ecosystems. The *GHOSTFOOD* food truck serves *cod* from the ocean, *peanuts* from the grassland, and *cocoa* from the rainforest. Each species of plant or animal that is served is threatened by one or more complex climate change threats—which *GHOSTFOOD* servers are prepared to describe. The cocoa beans for the chocolate milk served, for example, come from a tree that grows in the edge-of-rainforest climactic habitat. The relationships among food, species, and habitat link the chocolate milk (and the other food items that are served on *GHOSTFOOD*) to the abstract idea of "climate change." At the same time, customers experience this soon-to-be-missing chocolate flavor through olfactory sensation, with its accompanying neurological processing, directly linking the experience to emotion and memory.

How does one experience "soon-to-be-unavailable" flavors? *GHOSTFOOD*s are served as a scent-food pairing. Scents of foods that are threatened by climate change are delivered via a DOSD (Direct Olfactory Stimulation Device), a wearable designed device, mentioned in the narrative at the beginning of this section, that adapts human

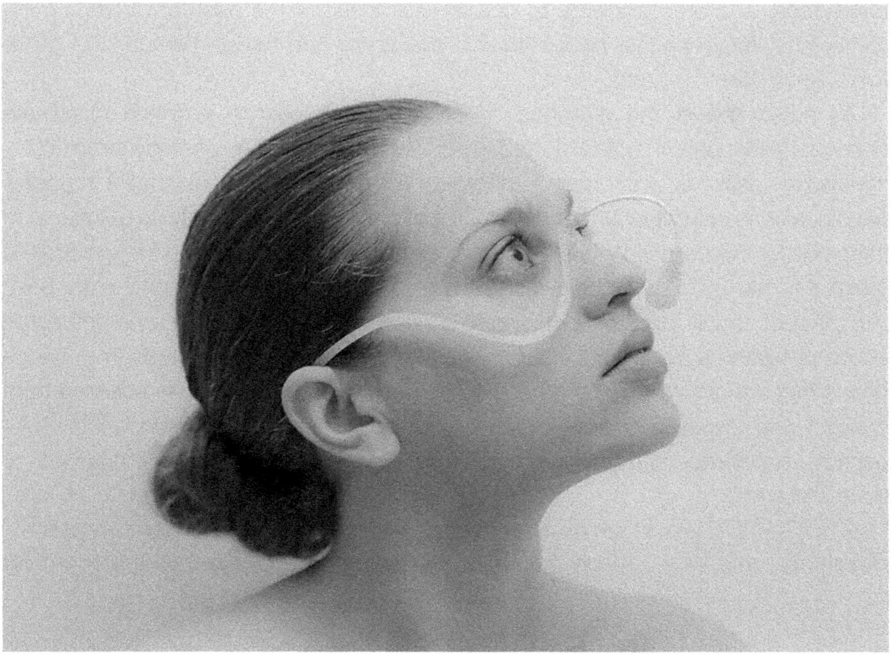

Fig. 8.3 DOSD, a wearable device that adapts human physiology to enable people to taste unavailable foods.

physiology to enable people to taste unavailable foods. The DOSD is a high-tech biomimetic device. Its design is inspired by the way that insects use their antennae to smell in order to navigate, and this function is combined with human forms of technological extension of the senses. I designed the DOSD by first considering insect antennae anatomy and physiology, along with drawings of faces of people on the subway in New York City. I explored various extensions of the device for the different face dimensions. The design of the DOSD was then 3D modeled by fabrication designer Corrie Van Sice, and 3D printed, a process that conveys the "futurity" of this object. Utilizing direct olfactory stimulation, the wearable form of the device holds a scent to the nose while the user eats and drinks.

The scents and foodstuffs in *GHOSTFOOD* are designed pairings. Each scent of unavailable food in *GHOSTFOOD* is paired with "edible, textural substitutes." These are foods that mimic the texture of the unavailable food, but are made from climate-change-resilient foodstuffs. Experienced in tandem, this flavor experience provides a simulation or illusion of eating foods that may soon or may already not be available. The menu—*chocolate* served as chocolate milk: chocolate scent paired with sweetened cow's milk; *cod* served battered and fried: cod scent paired with vegan fish substitute; and *peanut butter* served with jelly and bread: peanut butter scent paired with bread, jelly, and soy butter—includes foods that are familiar to a majority of Americans. They are also foods that are particularly common in American childhoods, and are thus likely to conjure memories of the past for adults. At the same time, they suggest issues around availability and future memories of these foods for children to-come.

As noted above, the designed *GHOSTFOOD* experience is linked to climate change. Philosopher Timothy Morton describes climate change as "a hyperobject"—that is, an entity of such vast temporal and spatial dimensions that it muddles traditional ideas about what a thing is. Crucially, as Karin von Omptede points out in her piece for this collection, this situation makes climate change cognitively difficult to grasp (Morton 2013). *GHOSTFOOD* aims to provide a simple sensorial entry point into the vast cognitive concept of climate change. The simple, familiar, and embodied action of tasting a sip of chocolate milk leads users to chocolate made from cocoa beans that are grown on cocoa trees in the rainforest, and which are suffering from drought and disease because of rising global temperatures. This encounter links personal experience and global ecological change through flavor and the senses of taste and smell.

GHOSTFOOD is manned by actors who guide visitors through this "pre-nostalgic" experience, and who engage them with discussions about the interaction among

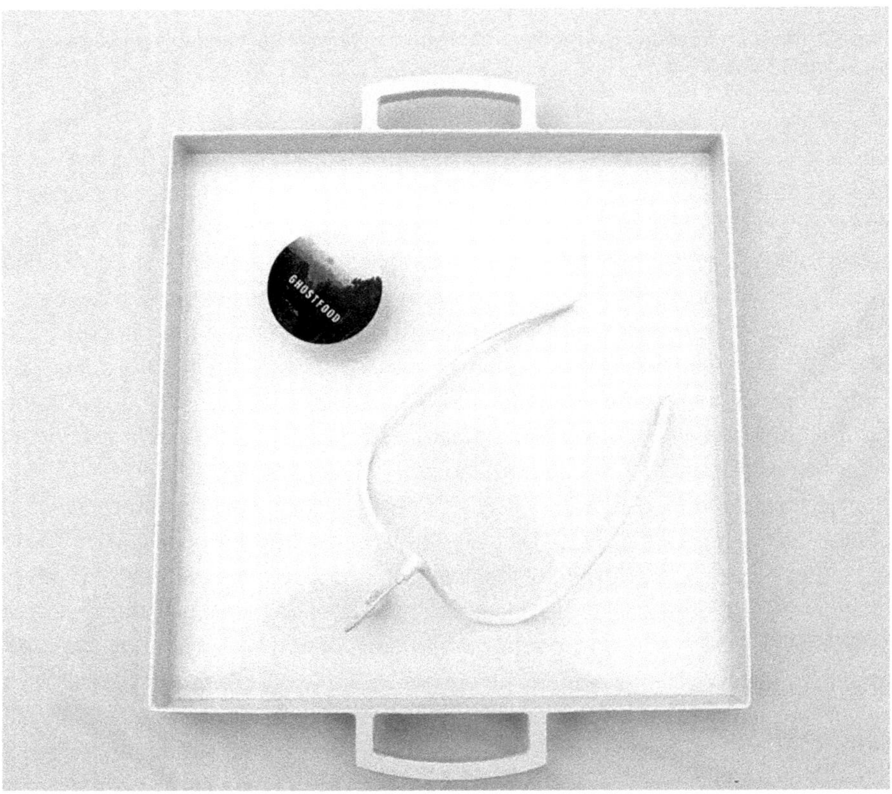

Fig. 8.4 Songster *GHOSTFOOD* food tray.

scent, flavor, taste, and climate change. In addition, the servers are trained to explain the technical aspects of the DOSD device and the methodological approach to *GHOSTFOOD*. This method is an adaptation of the Open Dialog Approach (ODA), a psychological method for combating *ecological* crisis. At the core of ODA are three concepts: dialogue, polyphony, and tolerance of uncertainty (Olson and Seikkula 2003). *GHOSTFOOD* servers are instructed to allow visitors to lead conversations, and also to answer any questions, encourage reflection, and engage with participants. This approach prompts the kinds of interactions and the sorts of sharing of experiences and reflections with one another that are based on the core principles of ODA.

GHOSTFOOD thus operates on multiple levels. First, it is a novel flavor experience that slows down the everyday act of eating, and also encourages attention to the experience of scent, flavor, and texture in food. Second, it brings industrial additives to the forefront of the experience through the introduction of new commodities—such as flavors and devices—while also pointing to the underlying question of "what might be missing"? Third, the work links the difficult, abstract concept of climate change to a personal and quotidian act—that is, taking a bite of a familiar food. Crucially, this narrative is experienced cognitively, through dialogue and written language, but also in an embodied, sensorial, flavor-rich manner in the body. Finally, the project encourages the age-old human practice of sharing experiences, memories, thoughts, and ideas over a shared meal, this time with strangers in the street.

The Farewell to All One Has Used, Broken, Lost . . .

A Tribute to One Species Living at the Edge of Extinction

We are living through the epoch of the so-called sixth great extinction, in which thousands of species are lost each passing year. For much of the human population this is an abstract situation. Living in a global metropolis, I wanted to figure out how to encounter this situation in more than a merely intellectual way. So, in 2014 I attended the imminent passing of a species called *Agalinis acuta*, with which I was unknowingly living. I created a ceremonial performance, an experience entitled *The Farewell to All One Has Used, Broken, Lost . . .* In this performance, I used ritual, an ancient human cultural practice, to call out, address, and process change, loss, and transformation.

The *Agalinis acuta* is the only federally protected endangered plant growing in New York state. As of 2014, there were only eleven communities of this sandy coastal grassland species left in the world. It is a tiny pink flower (its corolla is an average of 1.3 centimeters long) whose blossoms bloom just one day a year. For the other 364 days, it closely resembles the *Schizachyrium scoparium* grass to which its roots attach in order to seep nutrients; the *Agalinis acuta* is a hemi-parasite. It is a temperamental little plant, requiring sandy and nutrient-poor soil, regular disturbance, and minimal competition. Since 2014 the annual count of the *Agalinis acuta* species has produced dwindling numbers—its existential outlook does not look good.

Under a series I called *Agalinis Dreams*, I created a number of works exploring the trajectory of this one little plant and its relationship to the natural, cultural, and political histories of humans and other species in the region. The story of the *Agalinis acuta* is interwoven with the stories of ecological and human succession, from the melting of the glaciers to Native American land-burning practices, to the introduction of the railroad and transatlantic flight to the creation of the first suburban development, and finally to the establishment of genetic analysis. These stories are presented in the artist book *What Is Known* (2017) (available as a digital publication through the Avant online publishing platform). Although written language can describe and contextualize this species for an audience, I would argue that a full understanding also requires sensorial, embodied ways of knowing.

As such, for part of the series *Agalinis Dreams* I created a performance that is a ceremony. Making use of the ancient practice of alcohol in ritual practices, this ceremony asks visitors to don a ritual wearable headpiece—called the *Adoro*. The Adoro is a device I designed that holds the scent of the *Agalinis acuta* to the nose. It was designed as a ritual device—an ornate wearable to be used for a rare and special occasion. The Adoro is an apparatus that extends the senses; it is a designed artifact that enables sensing the impossible, and something that allows a human to enter the more-than-human world.

Wearing the Adoro, visitors then imbibe the ceremonial cocktail that is infused with the flavor of the *Schizachyrium scoparium* grass, the species without which the

Fig. 8.5 Adoro, an apparatus that extends the senses and enables the sense of smell.

flower cannot grow in the wild. Because alcohol numbs our sense of smell, an interesting time-based sensorial experience unfolds: visitors smell the flower scent, then drink the cocktail which numbs the scent. As their sense of smell slowly returns, the scent of the flower mingles with the remnants of the grass flavor from the drink, which creates the flavor of the *Agalinis acuta* as it exists in the wild, if only for a fleeting moment.

There are two facts about this species that, like it's flavor experience, are not fully knowable, and add another dimension to the experience. The first fact is the plant's bio-political context: the species *Agalinis acuta* is actually the result of a legal fiction that can be traced to a taxonomic imprecision. The species was granted endangered status in 1988. But, in 2010 a pair of researchers determined that the *Acuta* is genetically identical to another species of *Agalinis*, the *Agalinis decemloba*. This species was named first and so retains the name, and yet the *acuta* retains the legal benefits of its endangered status, and thus continues to exist legally, if not scientifically.

The second fact is that the scent of the *Agalinis acuta* is actually imperceivable by humans in the wild. The flower is so small—each is a mere one to two centimeters in length—that it simply does not produce enough volatile chemicals for its scent to be perceived by humans. But one morning at dawn, as I watched while visiting one of the healthier sites where this species grows, suddenly hundreds of bumblebees descended upon these blossoms. The blossoms were just shy of being fully open. For just under an hour the bees appeared almost intoxicated as they flew from flower to flower, the small plant bending underneath their weight. What was it that these creatures smelled and tasted from this pink blossom?

Fig. 8.6 Headspace, a device that collects and decodes scent molecules.

I then sought to capture the scent of the *Agalinis acuta* and re-create it for human perception. Working with the company International Flavors and Fragrances (IFF), Inc., I used its "Headspace technology" and mass spectrometry in this project. IFF's Headspace technology first collects and decodes scent molecules around the object of interest—in this case it captures the chemical marker of this blossom—and then translates it into a perfume. To carry out these tasks, I used a sealed spherical glass form that I placed over several blossoms for approximately an hour in order to elucidate the scent. Headspace technology extracts scents by first sending inert gases into the sealed-off space around the flower, and then creating a vacuum in order to remove external odor compounds from the headspace. The extracted compounds from *Agalinis acuta* were captured using an absorbent material. Mass spectrometry is a technique that ionizes chemical compounds and sorts their ions based on a mass-to-charge ratio. These compounds were analyzed at the IFF laboratory using mass spectrometry, and this data was used to determine the components of the scent by revealing its elemental or isotopic makeup. This elemental makeup information is a chemical recipe that can be used to recreate the scent. The samples are based on measurements of ratios of the scent components. Keeping the ratios the same, but increasing the volumes of the elements, produces enough of the scent to be perceived by humans.

The scent, of course, is a translation created by using technology. It is our best approximation, a chemical description of what the bees might have experienced at the blossom. This situation brings up questions about how close this smell is to the "real" smell, and how much does any discrepancy matter? The project also provokes questions about the best means to preserve the flower: cultivating the natural species in an unnatural fenced-off context with highly managed land plots, textual descriptions, botanical dried specimens, or still and moving images. Or is chemical scent, which is by its very nature as fleeting as the blossoming itself, the best means of preservation? And, finally, in my project, the scent serves as part of a ceremony, an ephemeral experience that is designed to mark the ephemeral nature of things.

My performance project is designed to be a ritual learning, celebrating, mourning, or remembering experience. This project explores interspecies relationships. *Agalinis Dreams* considers how species are affected by loss, cultural shifts, bio-political structures, and notions of progress at the very moment that they are on the brink of extinction.

Regenesis

The algal blooms were back in their springtime virginity, smelling of freshly cut aloe vera and holly leaves rubbed bare of their waxen outer shell. It was a remembrance of smells past, of walks alongside the swan boats and canoes paddling through the lake in the middle of the park. She longed to hold on to those times, the earlier years when all she needed was a scrap of paper and a pencil and she would be fine for the rest of the day. Her eyes closed, Isabel brought her pen to the parchment that rested on her legs and began to sketch the skyline from memory. There was no question why she didn't come here more often. Simply inhaling in these places conjured too many feelings of a simple happiness to which she could never return.

Implications

Scent is the oldest human sense, and the one with direct neurological access to emotion and memory. It is a powerful medium with which to create experiences, providing direct access to the physiology and psychology of audiences. Scent links us to the natural world in intimate and ancient ways, and provides us with ways of knowing that reach beyond the rational and analytical ways of being in the world. As a result, scent is a powerful medium through which to consider our relationships to the natural world, and to consider the past or imagine the future.

As a medium, scent is temporal and ephemeral, and it is grounded in materiality. As gas that literally enters the body, scent is a very intimate way to communicate with an audience. Further, as it is both invisible and temporal, like a piece of music, scent shifts and mutates over the duration of an experience. It mingles with the specific timely context in which it operates, and is never exactly repeatable. Scent thus offers rich material for creating new and powerful experiences.

References

Arshamian, A., E. Iannilli, J. C. Gerber, J. Willander, J. Persson, H. S. Seo, T. Hummel, and M. Larsson (2013), "The functional neuroanatomy of odor evoked autobiographical memories cued by odors and words," *Neuropsychologia* 51: 123–31.

Bushdid, C., M. O. Magnasco, L. B. Vosshall, and A. Keller (2014), "Humans can discriminate more than 1 trillion olfactory stimuli," *Science* 343 (6177): 1370–2. New York.

Caplan, J. (2007), "The World of Smell," *Time*, May 9. Available online: http://content.time.com/time/specials/2007/perfume/article/0,28804,1618617_1618614_1618557,00.html.

Chamine, I. and B. S. Oken (2016), "Aroma Effects on Physiologic and Cognitive Function Following Acute Stress: A Mechanism Investigation," *Journal of Alternative and Complementary Medicine* 22 (9): 713–21.

Classen, C.V., C. Classen, D. Howes, and A. Synnott (1994), *Aroma: The Cultural History of* Smell. Routledge: New York.

Drevermann, B. (2007), "Marine Fragrance Chemistry: Synthesis, Olfactory Characterisation, and Structure-Odour-Relationships of Benzodioxepinone Analogues," MA dissertation, Department of Applied Chemistry, RMIT University, Melbourne, Australia.

Endevelt-Shapira, Y., O. Perl, A. Ravia, D. Amir, A. Eisen, V. Bezalel, L. Rozenkrantz, E. Mishor, L. Pinchover, T. Soroka, D. Honigstein, and N. Sobel (2017; 2018), "Altered responses to social chemosignals in autism spectrum disorder," *Nature Neuroscience* 21: 111–19.

Fujikura, K. (2015), "Multiple loss-of-function variants of taste receptors in modern humans," *Scientific Reports* 5: 123–49.
Heist, A. (2016), "With No Sense Of Smell, The World Can Be A Grayer, Scarier Place," NPR. October 10. Available online: www.npr.org/sections/health-shots/2016/10/10/496455192/with-no-sense-of-smell-the-world-can-be-a-grayer-scarier-place (accessed on June 24, 2021).
Herz, R., J. Eliassen, S. Beland, and T. Souza (2004), "Neuroimaging evidence for the emotional potency of odor-evoked memory," *Neuropsychologia* 42: 371–8.
Kadohisa, M. (2013), "Effects of odor on emotion, with implications," *Frontiers in Systems Neuroscience* 7: 66.
Kromer, J. et al. (2016), "Influence of HLA on human partnership and sexual satisfaction," *Scientific Reports* 6: 325–50.
Malnic, B., P. A. Godfrey, and L. B. Buck. (2004), "The human olfactory receptor gene family," *Proceedings of the National Academy of Sciences* 101 (8): 2584–9.
Morton, T. (2013), *Hyperobjects: Philosophy and Ecology After the End of the World.* Minneapolis: University of Minnesota.
Negoias, S., et al. (2013), "Localization of Odors Can Be Learned," *Chemical Senses* 38 (7): 553–62.
Olson, Mary E. and J. Seikkula (2004), "The Open Dialogue Approach to Acute Psychosis: Its Poetics and Micropolitics," *Journal of Family Process*, https://onlinelibrary.wiley.com/doi/abs/10.1111/j.1545-5300.2003.00403.x (accessed on June 24, 2021).
Ryan, C. (2013), "Secret of scent lies in molecular vibrations," UCL (University College London), www.ucl.ac.uk/news/2013/jan/secret-scent-lies-molecular-vibrations (accessed on October 18, 2020).
Sell, C.S. (2014), *Chemistry and the Sense of Smell*. Wiley: New Jersey.
Stoddart, D.M. (1990), *The Scented Ape: The Biology and Culture of Human Odour.* Cambridge University Press: Cambridge.
Turin, L. (2015), "The Science of Scent," TED talk, available online: www.ted.com/talks/luca_turin_on_the_science_of_scent (accessed on June 24, 2021).
Walker J.C., Jennings R.A. (1991), "Comparison of Odor Perception in Humans and Animals," in Laing, D. G., R.L. Doty, and W. Breipohl (eds.), *The Human Sense of Smell*. Berlin, Heidelberg: Springer.
Wuensch, K.L. (2009), "Viral Anosmia," East Carolina University, http://core.ecu.edu/psyc/wuenschk/Anosmia_Core/ViralMick.htm (accessed on June 24, 2021).

Part 3 Makers and Users in Design and Science

9 URBN STEAMlab and Biophilic Environments: Science, Art, and Design

DIANA NICHOLAS AND SHIVANTHI ANANDAN

Introduction: Environment of Collaboration

> Without integrative disciplines of understanding, communication, and action, there is little hope of sensibly extending knowledge beyond the library or laboratory in order to serve the purpose of enriching human life.
>
> **Richard Buchanan, 1992a**

Productive interdisciplinary collaboration, including that between designers and scientists, is crucial to the complex challenges that we face in twenty-first century design (Buchanan 1992). Our interdisciplinary research group at Drexel University in Philadelphia, URBN STEAMlab (USL), addresses this need by using a reflective science- and research-informed design process to carry out speculative and applied biodesign projects. USL—which is comprised of design and scientific researchers operating in tandem to create integrative solutions—deploys bench-science, engineering, and technology to produce green interior spaces. Our goal is for participants to work together to ameliorate food insecurity in urban living spaces through food-generating design that integrates natural systems. We are also concerned with both technological advancement and advocacy in service to the needs of urban families.

In this essay, we discuss the historical and ontological underpinnings of our trans-disciplinary process, elaborate the work done by USL, and consider the contemporary climate for an innovative collaborative practice aimed at meaningful and pragmatic problem solving. This essay also examines aspects of science and design through their disciplinary lenses in order to ferret out areas of overlap. We present one USL project as a case study whose original objective was to be a problem-solving undertaking. This project, though, ultimately gave rise to a novel process of research and design in which biology and design overlap.

Design as Science: The Science of Design

We maintain that designers and scientists find common ground, to a great extent, through their thinking and making processes. The contemporary idea that science and design are collaborative practices that share related systematic qualities arose, in part, from a 1960s design movement. Architects Buckminster Fuller and Christopher Alexander examined design as a science. They became interested in the idea that, like the systematic aspects of scientific methodology, systematic design methodology was a worthy research topic. Christopher Alexander, in particular, developed a systematized visual language for designing houses that he likened to a science of design. Alexander called it "pattern language," and it included a "simple guide filled with the proven [visual and descriptive] elements any ordinary person could use to create a living world that best serves the humans that interact with it. Think *Sims*, but for the real world" (Menges et al. 2011). Pattern language, in fact, became a basis for programming and digital forms of design, which utilize systematic patterns.[1]

Ironically, Alexander later disavowed this science of design process. Others, though, took up his interest in design process as science. In particular, in *Designerly Ways of Knowing: Design Discipline Versus Design Science* (2001, 2007), Nigel Cross reconsiders design science methodologies of the late 1960s. It is crucial for our purposes to note that Cross points out here that designers work with the "'artificial world'—the human-made world of Artefacts" and that their "knowledge, skills, and values lie especially in the techniques of the artificial" (Cross 2001: 5). Although dealing with human-made artifacts has traditionally been the case for designers, we maintain that blurring the boundaries between, and embracing processes that investigate the artificial and the natural worlds together, opens up many areas of overlap between design and science.

In *Designerly Ways of Knowing*, Cross also discusses how philosopher Donald Schön's important work on reflective process and professional practice embraces the notion of "messy, problematic situations" in the design process. Schön suggests that the messiness that characterizes design research and practice likewise makes space for reflection and reiteration—and, we would argue, for collaboration. He documents how the exercise of reflection and practice becomes a loop that can lead to new theories and ways of working in architecture (Schön 1991). In fact, both designers and scientists utilize messy, iterative, and reflective processes. Like blurring the boundaries between artificial and natural entities and processes does, these shared ways of working offer areas for collaboration between designers and scientists.

Design theorist Charles Owen likewise draws connecting lines between science and design processes. He defines a symbiotic relationship between design thinking and science thinking in his ground-breaking article "Design Thinking: Notes on Its Nature and Use." "Where the scientist sifts facts to discover patterns and insights," Owen writes, "the designer invents new patterns and concepts to address facts and possibilities"(Owen 2008). Although this may be an oversimplification of what designers and scientists each do, Owen stresses, as Alexander does, that finding and using systematic patterns in thinking and doing is shared by designers and scientists.

Owen, in fact, developed a series of elegant diagrams for his essay that demonstrate that design and science have complementary thought processes. Seeking truth and making truth by linking practice and theory, Owen contends, can forge a loop that creates new knowledge through iterative development and decision

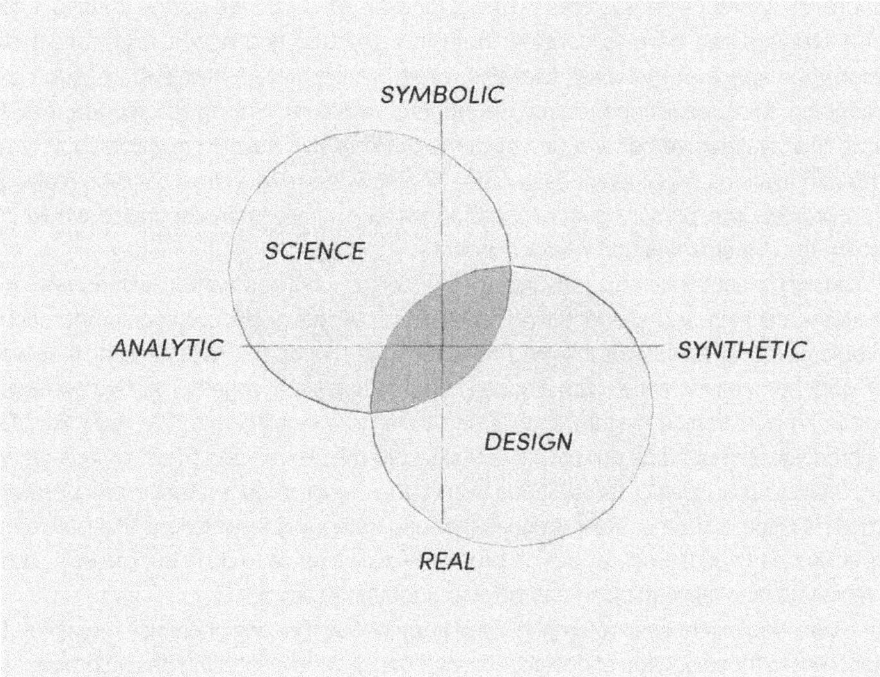

Fig. 9.1 Design and science complementary decision making, adapted from Charles Owen, "Design Thinking Notes on Its Nature and Use": "Design and science are strongly complementary: Science: analytic/symbolic; Design: synthetic/real. Together, they cover areas of decision making" (Owen 2007). Image by authors.

making in the problem-solving process (Owen 2008). The diagram that is reproduced from his work here in Fig. 9.1 shows that he evaluates decision-making processes in various disciplines to drive this conjecture. We believe that this particular canon—instigated by Fuller, Alexander, Cross, and Owen—offers a path for contemporary practitioners in which simultaneous "sensemaking" processes in design and science can suggest new solutions. Interdisciplinary sensemaking can happen when designers and scientists blur the boundaries between the artificial and natural world, embrace messy problem-solving processes, share systematic visual and verbal pattern vocabularies, employ reflection and iteration in thinking and doing, and remain open to shared decision making.

Building Science and Systems Thinking

Systems thinking is yet another shared aspect of design- and science-based processes and problem-solving methods. Systems thinking is a methodology that considers both the parts and the whole of a problem in relation to each other. Built environment designers, such as interior designers or architects, utilize an area of quantitative study called building science that is based in a systems-thinking approach. We consider systems thinking, for example, when we design the structure of a building, and have to consider both the structure and how it might affect or inform the space or materials. Likewise, when we integrate a new system, such as plumbing, into an existing building, we are using systems thinking to integrate extant and new systems. When we use systems thinking, we have to consider how the different systems might affect each other, and how they might drive decision making in a project. The primary goal of systems thinking, then, is an integrated whole in which the systems function interdependently.

Design practitioner and researcher Jon Kolko, who has written extensively on systems thinking and sensemaking, contends that designers commonly approach problems using a systems thinking mindset. The hydroponic USL project that we present here, for instance, puts a house and a "smart vase" together as two systems within an overarching system. Our "smart vase" is a small countertop vase-like 3D printed vessel that holds our proprietary algal bio-mix. Herbs and micro-greens grow on this bio-mix. Ideally, all systems within the overarching system have positive impacts upon one other. Kolko argues that understanding the relationships between specific parts and the overall design problem—as we strive to do in our project—can precipitate innovative thinking and novel outcomes (Kolko 2010).

Kolko also discusses systems thinking approaches that designers use that are not apparent to those outside of design—for example, how investigating design problems can reveal unforeseen issues and emergent systems (Kolko 2010). He calls these processes "sensemaking." As Kolko describes it, both systems thinking and "sensemaking" are synthesis processes, in which options are considered, combined, categorized, and understood. Kolko believes that people use sensemaking to "integrate experiences into their understanding of the world around them" both within and beyond design (2010: 18). According to Kolko's ideas, scientists also carry out sensemaking processes. Indeed, one of the main goals of scientific research is to make sense of the world around us. Sensemaking—whether in science or design—involves coming up with new "truths" or problems from the systemic consideration of the evidence at hand.

Systems thinking also underpins abductive reasoning, a type of design thinking that is used in concept mapping. In concept mapping, ideas and information are represented graphically, for example, as ideas in boxes or circles that are interconnected conceptually with labeled arrows to show their interrelationships. Scientists don't necessarily do mind mapping the way designers do, but many scientists use sketching to help relate parts to the whole in complex problems or to figure out how specific observations may play out in the real world.

Design and science have these processes in common: both engage in question making and answering, and both often employ a systems approach to problem solving and evolving new truths. Messy ideation, systems thinking, and synthesis processes that lead to leaps in thinking and innovative outcomes serve as fertile ground for collaboration between designers and scientists. Our USL collaborative teams operate in just this way, which we will discuss later in this essay.

Scientific Thinking and Design Thinking

Biology as a research-based discipline has historically operated in a positivist or post-positivist paradigm, which is defined as a model that is seated in facts, and proofs of facts, through examination of the natural world (Dudovskiy 2018). Science writer Alina Bradford explains that, traditionally, scientists have used deductive thinking through the so-called scientific method. "There are observations," according to Bradford, then "conclusions are drawn from these" (2017). Biological systems are extremely complex with many interacting components, which makes getting an experimental answer to a seemingly simple question, in a single set of experiments, a very challenging process.

Design methodology is commonly considered to be different from scientific method because it includes inductive as well as deductive thought. In inductive reasoning, experiences and observations are synthesized to suggest a general truth. Inductive reasoning allows broad generalizations to be made from specific observations. Designers also tend to use abductive reasoning—mentioned above in the discussion of mind mapping—in which neither problem nor solution are clearly defined (Cross 2001). Kolko explains how abduction, a synthetic process, works in design:

> The various constraints of the problem begin to act as logical premises, and the designer's work and life experiences—and their ease and flexibility with logical leaps based on inconclusive or incomplete data—begin to shape the abduction. Abduction acts as inference or intuition and is directly aided and assisted by personal experience.
>
> 2010

Designers likewise synthesize information and input differently from scientists because of designers' engagement with the needs and issues of stakeholders (Kolko 2010).

Although there is some truth to these generalizations, in reality, thinking and making processes in design have a lot in common with thinking and experimental processes in biology. People typically imagine scientific processes as a straight arrow, and design processes as a winding path. The path to good science, however, winds and doubles back like the path to successful design does. Iterative experimentation in biology begins with a hypothesis, and then experimental or theoretical results are analyzed to determine how they fit the initial hypothesis. Many times, the data gathered give some inkling of answers to the question upon which the experiments are based. When no clear answer is obtained from the experimental results, the design of the experiments conducted is considered to be flawed. If this happens, the experiment design has to be carefully analyzed, and a new design, or even a modified hypothesis, is established.

Traditionally, biological experiments are conducted at the lab bench; however, in our contemporary digital-data environment, experiments are frequently based on informatics-type data, for example, data from genome sequencing and healthcare data from large population studies. In both sorts of biological experimental strategies—bench science and informatics data consideration—the process is iterative. As noted above, this is generally because the initial set of experiments or explorations give rise to many other questions that need to be addressed experimentally before the validity of the original hypothesis can be clearly examined. Designers likewise test their work in a trial-and-error process during thinking and making. They often work closely with focus groups and individual users as a means to assess the success or failure of their design (sometimes in the marketplace) (O'Grady and O'Grady 2017). Once they receive sufficient feedback ("results"), designers analyze the feedback, and then develop new prototypes. This design process, like the experimental process in biology, is an iterative one based on information gathered ("data") and some trial and error. Both biology and design are, therefore, evidence-based, iterative processes carried out to achieve knowledge-based outcomes.

Collaborative processes and solutions can happen when designers and scientists overlap their work during these sorts of thinking and making processes, while also blurring the boundaries between the artificial and natural world and embracing messy problem-solving processes. During collaboration, designers and scientists need to share systematic visual and verbal pattern vocabularies, employ reflection and iteration in thinking and doing, and remain open to shared decision making. During collaboration, both designers and scientists typically engage question making and answering, and both employ a systems approach to problem solving and evolving new truths.

Biology by Design: Design Research and Biological Research

The USL group utilizes these overlapping aspects of design and science in a range of biodesign projects. William Myers, curator of the 2008 exhibition *Design and the Elastic Mind*, explains that biodesign incorporates design with "living organisms as essential components." Biodesign, according to Myers, dissolves boundaries between design and biology, synthesizes "new hybrid typologies," and includes "experiments that replace industrial or mechanical systems with biological processes" (2012, 8–9). Paola Antonelli, Senior Curator, Department of Architecture and Design at MoMa, characterizes biodesign as "plants and animals to bacteria and cells, to be used as architectural, graphic, or interior elements." She offers as examples:

> Wet buildings that adapt to changing environmental conditions, almost as if they were living organisms; designers concocting new diagnostic and therapeutic tools that rely on animals and plants; engineers devising new, self-healing construction materials.
>
> 2012: 7

Biodesign contributes to new conceptions of social and physical constructs and develops new rules that are based on design and biology thinking and making as shared processes—and biodesign can be pragmatic or speculative. Speculative biodesign "deals, by definition, with unreality," and imagines possible futures using biodesign as a "catalyst for change" (Dunne and Raby 2013: 12, 33). Biodesign is also driven, in part, by do-it-yourself scientists also known as "citizen scientists" or "biohackers"—biology thinkers and experimenters who operate outside the conventional science institutions and academies (Ikemoto 2017). In addition, biological form drives some contemporary research in design: the search for new forms of life and intelligence has an inherent reliance on both design and biology.

The USL Group

For more than eight years, the USL group has inculcated both graduate and undergraduate students in a process of tiered mentoring and examination of self-driven creative trajectories, typically through biodesign projects. The group includes biology students and design students from engineering, architecture, design research, and interior design. Together, the teams seek creative inspiration and examine simple and complex biodesign solutions using both biological laboratory and design "playing" processes.

The ideas of sociologist John Zeisel are particularly relevant to our USL projects. Zeisel writes about environment-behavior design, which is concerned with how design and the designed environment affect human behavior and health. Design and scientific researchers work together well, he argues, when each needs something that the other provides. According to Zeisel, designers primarily seek evidence on which

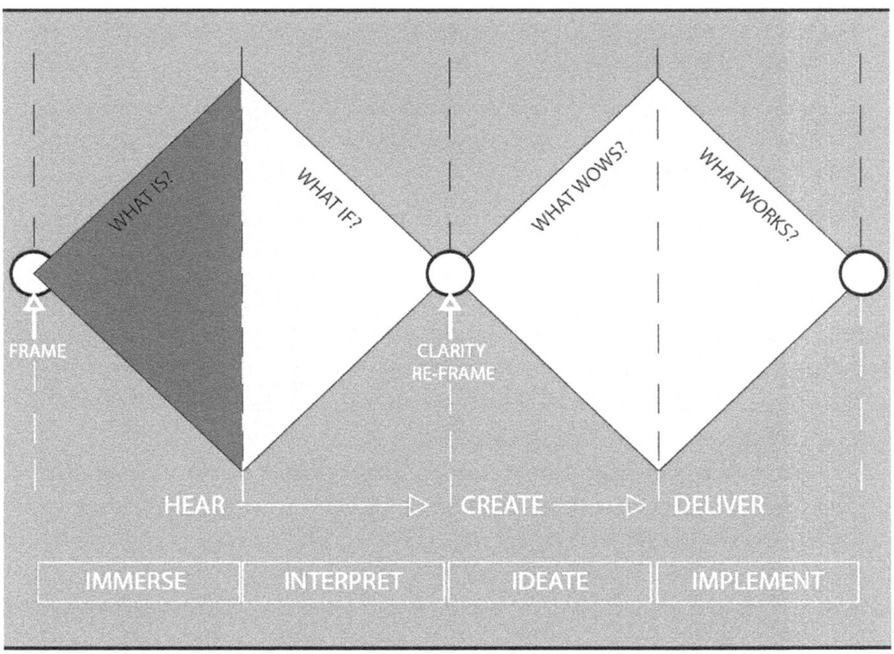

Fig. 9.2 Double Diamond diagram adapted by the authors from the UK Design Council. The diagram is annotated with the process words from several resources, including IDEO's field guides from 2011 and 2015, and Liedtka and Ogilvie's field guide for growth (Design Council, 2020; IDEO, 2011; IDEO.org, 2015; Liedtka & Ogilvie, 2011). Image by authors.

to base their decisions, and scientific researchers typically test hypotheses (Zeisel and Eberhard 2006). The interplay between the two disciplines, however, is more complex and layered than these limited descriptions of how each operates, or a simple cooperative give and take, as defined by Zeisel's evidence-based design practice case studies. As we discussed in the first sections of this essay, both of these disciplinary practices involve a variety of similar problem-solving processes, both involve iterative thought, and both are driven by the physical world and a systems approach. Biodesign projects, in which design and biology are incorporated in a project, epitomize how these aspects of both design and scientific processes come together. Collaborative biodesign, in which designers and scientists work together on biodesign, adds another level of challenges to and opportunities for the process.

To help foster productive collaboration, our students follow the British Design Council's Double Diamond process for developing prototypes and solutions. This thinking and making process is represented by a diagram with two diamond shapes (Fig. 9.2) that utilize four steps that play out in a non-linear fashion. That is, the order of these steps can go in any direction, depending on the project and progress. These steps are often labeled as:

> **Discover.** The first diamond helps people understand, rather than simply assume, what the problem is. It involves speaking to and spending time with people who are affected by the issues.
>
> **Define.** The insight gathered from the discovery phase can help you to define the challenge in a different way.
>
> **Develop.** The second diamond encourages people to give different answers to the clearly defined problem, seeking inspiration from elsewhere and co-designing with a range of different people.
>
> **Deliver.** Delivery involves testing out different solutions at a small-scale, rejecting those that will not work and improving the ones that will.
>
> Design Council 2020

These steps produce a problem-solving process that stresses both divergent and convergent modes of thinking. The two diamonds represent, in graphic form, ways to explore issues "widely or deeply (divergent thinking)" and taking "focused action (convergent thinking)" (Design Council 2020).

As we trace the open side of the diamonds on the framework (Fig. 9.2), we see that divergent thought is represented as thinking quickly and brainstorming many ideas—one sort of messy ideation. To encourage divergent thought, the USL team utilizes lateral thinking and consistent reflection. Lateral thinking is a solution-oriented process that aims to "re-frame a problem," according to designer Samara Strauss, "by highlighting new associations and relationships" that can't emerge out of vertical problem solving. "'Vertical,' step-by-step thinking" she explains, "is kind of like following a recipe . . . The process is predictable, and so is the outcome." Lateral thinking, though, goes beyond just brainstorming—it begins with ideation and leads to solutions (Samara Strauss 2014).

USL team members create new associations by experimenting with mold making, machine breaking, machine making, and materials as part of developing our prototypes. They pour silicon, resin, and concrete. They print with wood, flexible filament, and other experimental materials. Time is invested in learning from and playing with both tools and equipment—but also in creative play with teammates about the project. The combination of selves and project that is inherent in our shared process leads to an environment of both play and problem solving. Lateral thinking allows us to solve problems using an indirect, creative approach that embraces messy problem-solving, using a process that does not necessarily involve vertical, logical thinking. Lateral thinking also helps us to set aside the boundaries between the artificial and natural aspects of our project and embrace reflection and iteration. The first four years of USL were full of experimental prototypes that were created using this sort of experimentation. The images in this essay actually show the last year of work, as we homed in on project solutions for the device, and therefore show a much more linear development process than what we describe here.

The closing sides of the Double-Diamond-process diamonds represent convergent thought, including synthesizing—making sense of information or sensemaking—a process that we described earlier as one during which "options are considered, combined, categorized, and understood" (Kolko 2010). The closing sides of the diamonds also symbolize the process of understanding patterns from what has been generated in divergent thinking (Design Council 2018). This process has similarities to Owen's notions of finding and using systematic patterns in thinking and doing. We use data visualization both to support the biology findings and to help us in understanding the resources visually and functionally. USL has found that the Double Diamond process also augments interprofessional collaboration by encouraging a heightened awareness of ongoing divergent and convergent thinking during projects, along with group discussions about participants' ideas. Besides convergent and divergent thinking methods and data visualization, we encourage our teams to tackle "wicked problems," and to follow human-centered design and research analysis to determine the efficacy of our solutions in homes.

Science and Design Solving "Wicked" Problems

So-called wicked problems are social or cultural concerns that are tough to solve for four possible reasons: "incomplete or contradictory knowledge, the number of people and opinions involved, the large economic burden, and the interconnected nature of these problems with other problems" (Kolko 2012). Wicked problems are often also "ill-formulated, where the information is confusing, where there are many decision makers with conflicting values, and where the ramifications in the whole system are extremely confusing" (W. J. Rittel and M. Webber 1973). Design theorist Richard Buchanan believes that we could solve wicked problems by revamping design problem-solving processes (Buchanan 1992). Nigel Cross argues, in fact, that collaborative practice and reflection are a must for doing the kind of creative problem solving needed for wicked problems (Cross 2001). We have likewise found that our collaborative biodesign can provide solutions to some wicked problems that traditional design alone cannot.

Obesity, which is governed by genetic, social, behavioral, and dietary factors, is an example of a prevalent wicked problem that produces complex disease in the US (National Institute of Health 1998). Estimates state that by the year 2030, more than fifty percent of the US population will suffer from obesity (dosomething.org 2020). In addition, obesity is frequently a risk factor for cardiovascular disease and Type II diabetes (Office of the Surgeon General, US, 2010). Moreover, this disease is highly prevalent in children in the US (Cynthia L. Ogden et al. 2015). Birch and Ventura have identified low intake of fruit and vegetables as a risk factor for childhood obesity, as determined by epidemiological research (Birch and Ventura 2009). They assert that children will learn to eat a diet of foods that are available in their environment. Thus, if produce is easily available in their home environment, the children's diet will be richer in produce.

Populations in underserved communities do not typically have easy or affordable access to produce (Kern et al. 2017). Studies have determined that increasing the number of grocery stores in the neighborhoods of underserved populations is not effective, since many members of this population cannot afford the high cost of produce (Kern et al. 2017).

Studies also show that certain racial and socio-economic groups are more affected by obesity than others; among women, and specifically non-Hispanic white women, obesity prevalence increases as income decreases (Ogden et al. 2015). These problems may be worse where access to affordable and healthy food is difficult (Kern et al. 2017)—communities that are often referred to as "food deserts." In the US, about 23.5 million people live in food deserts, and nearly half of these people also belong to low-income groups ("Teaching Tolerance – Diversity, Equity and Justice" 2017). The Southern Poverty Law Center, and Dosomething.org also estimate that approximately 2.3 million people (2.2% of all US households) live in low-income, rural areas that are more than ten miles from a supermarket (DoSomething.org 2017). These data indicate that low income, rural populations lack easy and convenient access to nutritious foods, particularly vegetables and fruit, which present barriers to eating a healthy diet. These barriers can result in higher rates of obesity, diabetes, and other diet-related diseases in these populations.

Substandard housing conditions—including housing with little or no space for gardens—and poor health also go hand-in-hand. This information led us to conclude that growing produce in the home could be a solution to increased dietary intake of fruit and vegetables. Using a collaborative design-and-science-driven process, a USL team created a prototype of a simple hydroponic system for growing produce that requires no soil, uses minimal space, and requires low maintenance. Our system could aid underserved communities and increase urban health by making it possible for people to grow their own produce indoors, without a garden space, allowing them easy access to nutritious food. Our system also introduces green space into urban homes. Access to plant-based food that is grown in our hydroponic system can serve as a "dietary intervention" that has the capacity to help combat the wicked problem of obesity.

Shared Design "Lenses": A Development Process for Biology and Design for Healthy Living

USL's device utilizes a novel strategy: it sustains a proprietary bio-mix of plant microorganisms, which acts as a fertilizing substrate on which to grow microgreens, herbs, fruit, and vegetables. We worked on this hydroponic plant growth system project in our design research studio space and bench-science laboratory. Both studio and lab contain rapid prototyping equipment, which allowed us to generate and test various "models" for both the living and non-living components of the project during the past five years of experimentation. We worked on the design of the non-living parts of the unit, always keeping in mind the incorporation of the living components into the project.

In the design studio, we designed, tested, and modified a large standing unit made from frosted Plexiglass and silicone to accommodate the bio-mix and the plants that would grow out of the top of it. Our years-long work on the project included many iterations by student collaborators. These various forms ranged from large standing units with a water bladder in the center that would have crops growing out of the top, to the current small countertop unit—our "smart vase"—which stands at approximately ten inches high and six inches wide, with a water dish at the bottom and an internal armature to hold the plants upright. This current unit, or device as we sometimes call it here, is made from translucent PLA, a form of compostable plastic that will be printed on demand using a fused deposition modeler or filament 3-dimensional printer such as a MakerBot. In experiments in the biology lab, we found that the smaller prototypes like this one grew both algae (the bio-mix) and plants more successfully; the large prototype is too unwieldy for urban users to manage in smaller living spaces.

To get to our smart vase device design, we began with a collaborative design problem-solving process in which each team member first built on their individual "creative trajectories." The term "creative trajectory," which was coined by us, refers to a personal creative skill or interest that each student brings to the project related to their own creative goals and aspirations. One student, for instance, was particularly interested in mold making, and another was very skilled in detailed hand work, which they applied to the project in various ways. The team built on these creative trajectories, first to discuss, and then to operate around, a series of shared group design "lenses." A design "lens" is a viewpoint that the group shares, such as about biomimicry, or natural forms and structures—a lens is something that all members of the group are interested in or agree is a priority in the project approach. The combination of group design "lenses" and "creative trajectories" helped the team to define and solve issues that arose at the same time that they were being experimental in the studio. The one student's interest in molds, for example, led to a series of material experiments and mold making that gave us the substrate patterning on which the bio-mix grows that we have deployed in recent smart vases.

Encouraging these individual areas of interest and expertise among group members was crucial to the team and sustained development. Individual and disciplinary-based points of view in the creative trajectories and group lenses enhanced collaboration because each member was then invested in the problem-solving process and in breaking new ground. Our method encouraged students to experiment with their skills and learn from each other's goals and interests.

For each group lens, we developed multifaceted information or knowledge that the team investigated together. We used these lenses and the learning that went with them as touchstones and measuring sticks, and as an impetus for new ideas as the project progressed. We also examined these ideas from a systems-thinking standpoint by considering how various ideas fit both the parts and the whole in this project. Our group of designers and biologists approached the food growing unit that we were developing as a system; we saw the home as a system, the device as a system, and the food sources to be made available as a system—and all of these needed to work together to create the larger networked intervention. In the next section, we discuss our shared lenses and how they related to the systems considerations for our project.

Shared Design Lenses for Designing the Hydroponic Plant Growth System

The design and consulting firm IDEO has likewise stressed the importance of shared "lenses" in human-centered design processes. Their research calls for the implementation of three lenses: viability, desirability, and feasibility (IDEO 2011). Although we acknowledge the usefulness of these general topics for shared design lenses, our group found that developing our own customized lenses that came out of our specific process worked better for the development of our project over time.

A design lens can be thought of as a useful question that the group continually asks of the project, or as a way of measuring if the project is reaching the metrics it needs to in relationship to the lens topic. During the development of our project, the group used three shared design lenses to explore the forms and design of the vase: 1) transparency, 2) surface, and 3) material.

The team developed our current prototype using these three shared design lenses, and by bringing together their data on growing plants and on the bio-mix in experimental testing. Our small prototype experiments are vase-shaped, and they support the plants both through the external skin and an internal scaffold. The bio-mix, which is used to fertilize the plants, is encouraged to grow through the form and function of this internal structure. Our shared design lenses, which are described in more detail below, helped us to question our decisions along the way in the prototyping process.

Design Lens 1: Transparency

The base in the smart vase incubates the bio-mix amalgam upon which the plants grow. The vase is produced using a 3D printer with polylactic acid (PLA), which is a compostable filament that can be crushed and composted. There are no electronic components in the vase. What makes it "smart" is the bio-mix, which is a hardy living entity. It provides living fertilizer to the plants that grow on it. Consumers sprout seeds in the vase, the plants grow into the internal armature, and their roots dangle into the fertilized water with the bio-mix. The environmental requirements for both the bio-mix and the food plants changed and developed as we experimented. The team worked to integrate these needs and to discover new solutions, a process that was critical to the first design lens of transparency.

Transparency is especially important to our project because the bio-mix requires light to survive, but it also requires shade. Our lab experiments showed us that transparent materials are particularly good for its growth. The 3D printed vases are still undergoing iterative testing. The attached internal structure that we designed likewise required ongoing development to maintain light, airflow, and the water-tightness needed for both bio-mix and plant survival. The charts in Fig. 9.3 show how various colored materials that were used in the vase, along with lighting conditions in the environment, affected the growth of our proprietary culture. The internal humidity, and the proximity to light in both the box and the external space, also had an influence on the amount of greenery produced by the bio-mix. The amount and health of the greenery demonstrates how well the mix flourished. The design lens of transparency prompted the group to ask these questions: How much transparency is enough? How much is too much?

Design Lens 2: Material

Material—and the integration of water, light, and air with materials—emerged as the second lens through our lab experiments, iterative group work, and ongoing testing of the bio-mix. The firmness of the material and the porosity of the internal components of the vase were especially important in this second lens. The internal components of the vase are monolithically 3D printed parts of the unit, made especially to support the plant, and to convey the bio-mix. Material as a lens prompted us to ask this question: how hard or soft, heavy or light, and thick or thin should this part of our vase be? This second shared design lens of material—which, in turn, gave rise to further development of the light, water, and air flow in the vase—led the team to consider the emerging design as a system of surfaces and forms that must integrate function, biology, and aesthetics in an efficient way.

Design Lens 3: Surface

Surface is the third shared lens, and it is expressed through patterning and tessellation on the inside of the unit, which we discovered supports the growth of the bio-mix.

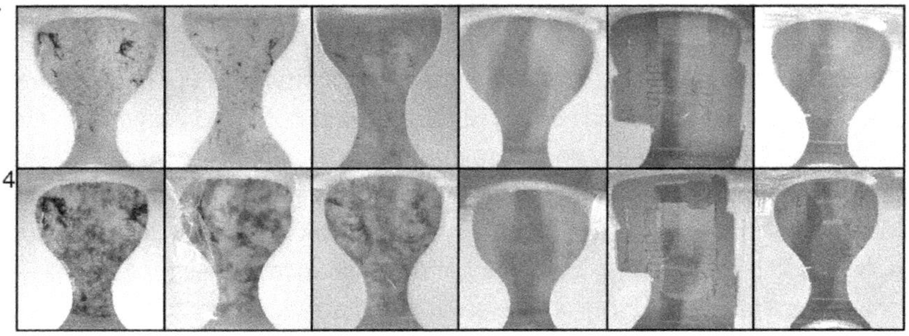

Fig. 9.3 Iterative color light and shape testing in the lab of the bio-mix growth and how robustly it grows. The darker green results with more speckling are the most alive and healthy culture. Image by authors assisted by biology/design research student Elise Krespan.

The group also experimented with patterning and tessellation as both an aesthetic and functional choice on the outside of the units. In "The Three Transparencies," architect Toyo Ito describes how he uses natural and physical structures and processes to drive his forms and to create relationships in his projects (Ito 1997). His interest in natural forms as engines for design aesthetics and structure is also discussed at length in Lisa Iwamoto's pivotal book *Digital Fabrications: Architectural and Material Techniques*. She categorizes the production of space and surface using digital tools in five ways: sectioning, tessellating (tiles shaped in patterns on a flat surface), folding, contouring, and forming (Iwamoto 2009).

These categories include forms and processes that are driven by natural structures—in fact, tessellations are congruent-edged patterns that often occur in nature. For our project, we combined and repeated several of Iwamoto's five techniques, and the group discussed their relationship to the bio-mix development. These techniques were our inspiration for the material forms and surface textures in the unit, especially including patterning and tessellations. Using this same kind of surface patterning on the outside visually linked the interior to the exterior of the vase, but it also allowed light and air into the vase. Surface and pattern as a design lens prompted the group to ask: Where and how should pattern occur? What are the effects of pattern on water and air flow?

Water flow is critical to the form and function of the project. And the bio-mix needs air to grow. So, water, along with air flow, position, volume, and circulation, were early drivers of the device prototypes, which we modeled and produced in 3D. In the lab, we took measurements and made observations about the amount of water needed to grow the bio-mix, how much maintenance watering the various bio-mix options required, and what the evaporation of liquid means in terms of the health of our bio-mix and the plants. The plant structures likewise drove our exploration of how the physical and material form of the device allowed air to circulate through the bio-mix at an optimal rate for its growth.

In our current research, water flow remains an important factor. We are exploring various ways of isolating and optimizing fertilized and fresh water as the next phase in our experiments (unfortunately this is not shown in the images that we are including here). We are currently growing micro-greens and herbs, so the structure of the vase will require adjustable components to accommodate the height and weight of the growing plants. The material qualities of all elements in the unit are also integrated into the system with the biological components of the project. The surfaces in the vase are integrated with three living systems: food plants, bio-mix, and humans.

Further Exploration: Healthy Habitat

The biology team and the design team discussed the integration of the project units into the home to create a healthy habitat that is based on the interrelationships among the living systems of bio-mix, plants, and humans—especially the human system. The urban house should offer a wholesome habitat that contributes to the overall health of urban dwellers. Contemporary ideas about human-centered urban living also drove our iterative team process. Current research on urban living has moved away from conventional architectural design solutions toward a collaborative human-centered design approach. Human-centered design is a "solution-based approach" that optimizes the "relationship [among] people [, artifacts,] and buildings." It produces "solutions . . . and opportunities" for all involved "by focusing on the needs, contexts, behaviors, and emotions" of those who use designed objects and buildings (Design Online 2020). Like our project does, most human-centered architecture and design incorporates STEM and/or health-based constituents into healthy urban living solutions.

Our collaborative design process, in which we integrated the aesthetic, functional, and user aspects of the vase into the home environment, was akin to the experimental design that our team did in the biological lab space. Like scientists, the team tested their work on the vase in a trial-and-error process during weekly thinking and making sessions. We used lateral thinking through our incremental decision making and prototyping ideas. We also worked logically to explore and accommodate the experimental outcomes from the biology lab.

Environmental Prototyping

We had the opportunity to investigate healthy habitat in a real-world situation as well by developing a prototype of the device for an exhibition. This structure was a 24"x 36" free-standing Plexiglas vitrine that the group used to test various substrates for propagating the bio-mix. The goal for this prototype was to test a large amount of the bio-mix in the vitrine during the life of the exhibition.

Design and Science, the exhibition that included our prototype, first ran in Ypsilanti, Michigan, and then in Philadelphia, in September 2019 and February 2020, respectively. This environment allowed us to deploy a large vitrine in a public setting. Although not as controlled as in the lab, the ability to show and discuss the process was invaluable to the team. The exhibition acted as a large Petri dish for our bio-mix, and the Philadelphia gallery acted as a home substitute for our smaller prototype

Fig. 9.4 USL smart vase iteration in the exhibition: a large acrylic onsite Petri dish to test the felt, silicone, and acrylic substrates for the bio-mix. Above the vitrine/Petri dish is a series of miniature prototypes produced for the exhibit. Image by authors assisted by design research student Thelmelis Abreu.

experiments. Fig. 9.4, which shows the Philadelphia exhibition site, depicts the experimental space set up by our group.

Using lateral thinking, we observed the growth of the bio-mix on a large substrate at these venues and changes were made in response as required. We found that the bio-mix in the exhibit survived at varying rates. The project also included an experimental felt substrate in a vitrine on which the bio-mix was cultivated over time, along with a series of silicone algal pods. During the two exhibitions, these materials functioned as an onsite experiment for porosity and evaporative qualities, and a general field-growth-test in a room environment. Using logical thinking, we concluded that the silicone patterns were most advantageous for promoting and supporting the growth of bio-mix, and that felt material is most efficacious for preserving water and preventing evaporation in a large room environment.

Since the exhibition, our ongoing prototyping development has concentrated on evaporation, support of the plants, and air exposure. Our observations from this time led to elongation of the prototype and the addition and editing of hole patterns, which can be seen in Fig. 9.5. The elongation will give the vase the ability to support taller plants, such as mint or basil, and the holes allow for further plant support, adjustable airflow, and more access for the user to add water when needed.

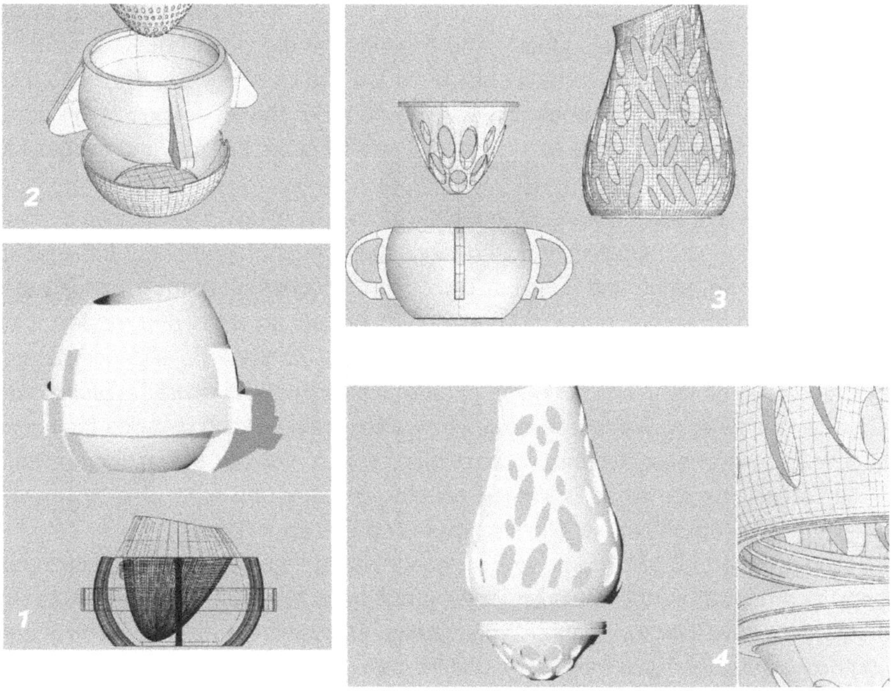

Fig. 9.5 USL smart vase iteration adjustments are shown here to respond to air and plant needs from iterative lab studies. Image by authors assisted by design research student Thelmelis Abreu.

New Developments in the Unit Design

Our three shared design lenses continue to offer perspectives that the team explores together while working through specific project issues. Considerations of transparency, material, and surface still draw the group into reflection and guide our forays across the disciplinary lines of design and science.

The material and surface in the design, such as the development and placement of airflow and air holes in the unit mentioned above, remains a consistent design lens in the process. We continue to develop these features through both physical decision making and lab-based experimentation. As noted earlier, patterns and nature emerged as a significant theme in the development of materials for the unit. Tessellations and transparency are efficient ways to divide and create surface openings using the technology for producing the vase, yet still create a closed sanitary and hospitable environment for the bio-mix development. Experimenting with tessellations and transparency is the engine for forms and surface textures in the most developed unit, shown in Fig. 9.3. Tessellations encourage the bio-mixture to grow and are an efficient way to make patterns when using digital technology to produce outcomes (Iwamoto 2009).

The group continues to explore new forms for the internal and external surfaces of the unit. Our goal for these surfaces is to enhance the growth of both bio-mix and plant, and some examples of newer bio-forms are shown in Fig. 9.6. Exploring this physical form continues to push the team to consider the emerging design as a system of surfaces and forms that must integrate function, biology, and aesthetics efficiently.

Using divergent and convergent thought—while cultivating an awareness of how we are thinking—led to several model designs that brought together our three living systems: bio-mix, plant, and humans. The current conceptual synthesis among the designers and scientists in the group still happens around ideas and models that incorporate bio-mix, plant, and human systems. When modeling ways to incorporate the three living systems into our current phase, we vary how much of the project we consider at a time, as well as the scale of the intervention. Doing so helps the team to develop solutions for discrete pieces of this complex problem. In fact, the unit has at times become as tall as a kitchen counter, or taken the form of a large wall hanging—and versions of it at other scales may yet be considered.

Although the biologists came to the project with a vague sense of architectural systems organization, the designers in the group have learned that bio-systems, by themselves, have their own systems organization. Biologists are trained to look at the contexts of the organisms they study, and the biologists on our team are considering bio-habitats at both micro and macro scales. The biological system within the unit, for instance, led to unintended contamination of the spaces by mold from water retention. The whole team—designers and scientists—needs to understand these

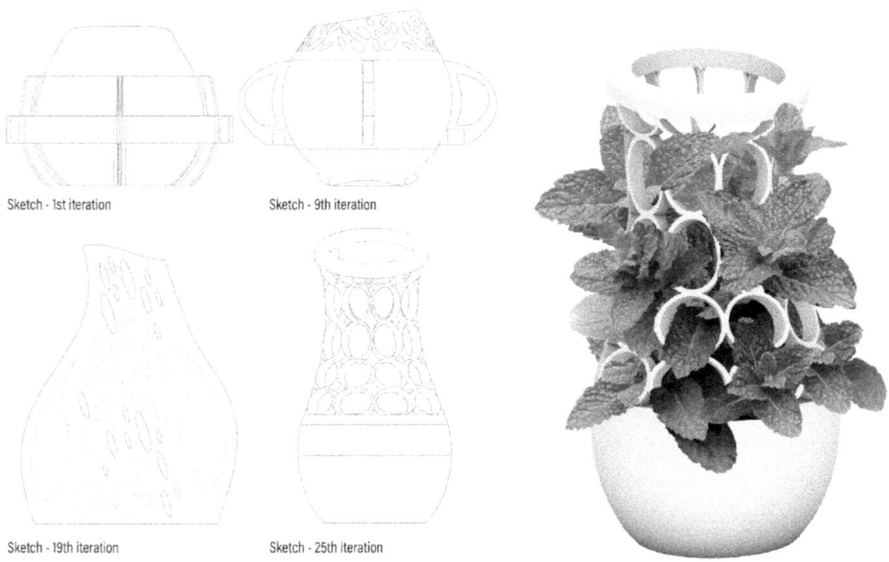

Fig. 9.6 USL smart vase iteration in patterning: patterns drawn from biological forms such as the Voronoi. Image by authors assisted by Design Research student Thelmelis Abreu.

bio-systems in order to figure out how to mitigate mold in the unit to preserve a healthy habitat for the bio-mix, plants, and the humans.

Explorations of overall habitat—how to integrate the growing units into the home, and how to create a healthy habitat for their users—are ongoing. With some input from the scientists, the designers have utilized their knowledge about habitat and living spaces across several scale changes to create various modes of interaction between the living space and the unit. Living space mock-ups have been created and are now being considered in detail to test how people engage with the units.

The rowhome, the main housing type in eastern American cities—which is a structure that can be thought of as a large container for all three systems—has emerged in our current work as the contextual base for sustaining the three living systems. In the process, the team has examined how several prototypes relate to the building façade, and to the growing area in a rowhome, and they have explored large window unit prototypes with water added.

As the group develops an in-depth examination of the way that urban dwellers use their spaces and their kitchens, we have taken some deep dives into rowhome kitchen usage. In urban Philadelphia, many rowhomes in underserved neighborhoods lack access to working plumbing or appliances because of issues of ownership or with rental providers (Wallace and Divringi 2019). One research student, therefore, spent a summer examining mobile ideas in response to the lack of facilities in some rowhouses. One mobile project that came out of this research is a kitchen unit that can be wheeled out onto the porch of a home whose kitchen is not functional.

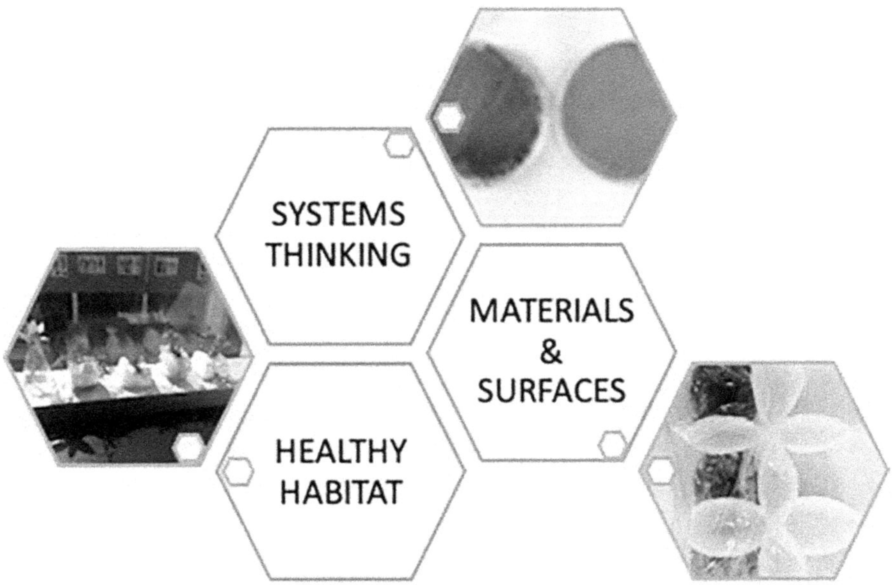

Fig. 9.7 Three primary lenses in the URBN STEAMlab Biodesign Process.

Ultimately, based on our research, the group concluded that smaller portable options for the unit made the most sense.

Luckily, the length of our collaboration, and our shared thinking and making processes, have allowed us to explore many diverse options for the unit. The group modeled a mobile kitchen-sized unit, for instance, that could grow food and also serve as a place to clean and cut vegetables. At the same time, the team attends community events to garner input into the use of such units in local homes. Healthy habitat continues to be a useful approach for the development of our current prototype, and small-scale in-home bio-mix tests are underway, even during the pandemic. Plans for in-home placement continue apace—test units with micro-greens and herbs will be placed in homes in fall 2021, and the data gathered from the round of fall tests will drive a community study in spring 2022.

Conclusion

Design and science can come together when boundaries between the artificial and natural world are blurred. Design and science processes may seem different at first glance, but they share messy problem-solving, systematic visual and verbal pattern vocabularies, reflection and iteration during thinking and doing, and systems thinking.

Although the paradigms and ontology of design and science may differ, they can be successful complements to solving wicked problems. Biodesign work produces bio-driven design solutions that question the complexity of our environment through experimental practice or speculative, future-driven scenarios. Situated at the intersection of biology, design research, and the "wicked" problem of healthy food shortages, the USL project that we present here is a pragmatic solution to a real-world problem that is driven by speculative thought. As a team operating in an urban area, surrounded by the challenges that this implies, a pragmatic world view that includes interdisciplinarity is of tantamount importance. Living spaces are central to the work of the group, as is an interest in working to improve health in underserved communities. A vital theme of our work is to create opportunities that enhance health in underserved communities through collaborative science and design research, while at the same time developing novel processes to carry out these collaborations.

People can create new knowledge about, and solutions to, issues using both scientific and designerly ways of knowing. This process includes utilizing abductive, inductive, and deductive thought, systems-driven synthesis, and design and scientific experimentation. Our use of the double-diamond technique, and systems- and design lens-based approaches, has taught us new ways of organizing skills and interests that lead to solutions that are both entrepreneurial and consumer focused. As the USL group continues our collaborative work to produce a marketable growing unit, we will also build awareness about how we collaborate. USL collaborators will continue to produce biodesign solutions for the urban environment—for urban rowhomes, temporary shelters, and community-based spaces—that are pragmatic, healthy, and forward thinking.

Note

1. This language was related to a similar systematic design system called "shape grammars" that was an early forerunner to the parametric and computational design processes available to architects today through tools such as grasshopper (Menges and Ahlquist 2011).

References

Birch, L. L. and A. K. Ventura (2009), "Preventing Childhood Obesity: What Works?" *International Journal of Obesity (2005)* 33 Suppl 1: S74-81, doi:10.1038/ijo.2009.22 (accessed on June 24, 2021).

Buchanan, R. (1992), "Wicked Problems in Design Thinking," *Design Issues* 8 (2): 5–21. doi:10.2307/1511637 (accessed on June 24, 2021).

Cross, N. (2001), "Designerly Ways of Knowing: Design Discipline Versus Design Science," *Design Issues* 17 (3): 49–55. doi:10.1162/074793601750357196 (accessed on June 24, 2021).

Design Council (2018), "The Design Process: What Is the Double Diamond?" www.designcouncil.org.uk/news-opinion/design-process-what-double-diamond (accessed on June 24, 2021).

dosomething.org (2020), "Volunteer for Social Change." www.dosomething.org/us (accessed on June 24, 2021).

Dudovskiy, J. (2018), "Positivism Research Methodology," *Research-Methodology*, https://research-methodology.net/research-philosophy/positivism/ (accessed on June 24, 2021).

IDEO (2011), *Human-Centered Design Toolkit: An Open-Source Toolkit to Inspire New Solutions in the Developing World*. Second edition. Place of publication not identified: IDEO.

Ikemoto, L. C. (2017), "DIY Bio: Hacking Life in Biotech's Backyard Symposium – Future-Proofing Law: From RDNA to Robots (Part 2)," *U.C. Davis Law Review* 51: 539–68.

Ito, T. (1997), "Three Transparencies – Toyo Ito," *Scribd*. www.scribd.com/document/348327567/Three-Transparencies-Toyo-Ito (accessed on June 24, 2021).

Iwamoto, Lisa. (2009), *Digital Fabrications: Architectural and Material Techniques*. New York, United States: Princeton Architectural Press.

Kern, D. M., A. H. Auchincloss, M. F. Stehr, A. V. Diez Roux, L. V. Moore, G.P. Kanter, and L. F. Robinson (2017), "Neighborhood Prices of Healthier and Unhealthier Foods and Associations with Diet Quality: Evidence from the Multi-Ethnic Study of Atherosclerosis," *International Journal of Environmental Research and Public Health* 14 (11), doi:10.3390/ijerph14111394 (accessed on June 24, 2021).

Kolko, J. (2010), "Abductive Thinking and Sensemaking: The Drivers of Design Synthesis," www.jonkolko.com/writingAbductiveThinking.php (accessed on June 24, 2021).

Kolko, J. (2012), *Wicked Problems: Problems Worth Solving*, Austin: Austin Center for Design, www.wickedproblems.com/read.php (accessed on August 18, 2021).

Menges, A. and Sean Ahlquist (2011), *Computational Design Thinking: Computation Design Thinking*. John Wiley & Sons.

National Institute of Health (1998), "Clinical Guidelines on the Identification, Evaluation, and

Treatment of Overweight and Obesity in Adults—The Evidence Report. National Institutes of Health."

Office of the Surgeon General (US) (2010), *The Surgeon General's Vision for a Healthy and Fit Nation*, Reports of the Surgeon General, Rockville (MD): Office of the Surgeon General (US), www.ncbi.nlm.nih.gov/books/NBK44660/ (accessed on June 24, 2021).

Ogden, C. L., M.D. Carroll, C. D. Fryar, and K. M. Flegal (2015), "Prevalence of Obesity Among Adults and Youth: United States, 2011–2014," *NCHS Data Brief* 219, www.cdc.gov/nchs/products/databriefs/db219.htm (accessed on June 24, 2021).

O'Grady, J. V. and K. V. O'Grady (2017), *A Designer's Research Manual, 2nd Edition, Updated and Expanded: Succeed in Design by Knowing Your Clients and Understanding What They Really Need*, expanded edition, Beverly, MA: Rockport Publishers.

Owen, C. (2008), "Design Thinking: On Its Nature and Use," https://hbr.org/product/design-thinking-on-its-nature-and-use/an/ROT060-PDF-ENG (accessed on June 24, 2021).

Rittel, H. and M. M. Webber (1973), "Dilemmas in a General Theory of Planning," *Policy Sciences* 4: 155–69. doi:10.1007/BF01405730 (accessed on June 24, 2021).

Schön, D. A. (1991), *The Reflective Practitioner: How Professionals Think In Action*. Aldershot: Ashgate.

"Teaching Tolerance – Diversity, Equity, and Justice," (2017), www.tolerance.org/ (accessed on June 24, 2021).

Zeisel, J. and J. P. Eberhard (2006), *Inquiry by Design: Environment/Behavior/Neuroscience in Architecture, Interiors, Landscape, and Planning*. New York: Norton.

10 Designing a Scientific Instrument: Lessons from the Crookes Radiometer[1]

LINA HAKIM

What would happen, I wonder, if we tried to talk about the object of science and technology, the *Gegenstand* [object], as if it had the rich and complicated qualities of the celebrated *Thing*?

Bruno Latour (2004: 233)

The more objects are turned into things—that is, the more matters of fact are turned into matters of concern—the more they are rendered into objects of design through and through.

Bruno Latour (2008: 2)

Introduction

According to sociologist Bruno Latour, the etymology of the word "thing," which used to designate assemblies, or places of gathering and debate in all ancient European languages (*Ding, chose, causa, res, aitia*), allows us to distinguish between an object and a thing (2004: 232–3). The word "object," he explains, refers to what is "out there," "what lies out of any dispute, out of language," unquestionable "matters of fact" that stand apart, objectively and independently, "unconcerned by any sort of parliament, forum, agora, congress, [or] court" (2004: 233; 236).

The word "thing," on the other hand, indicates something that is more rich and complicated—although, importantly, not entirely distinct. As Latour puts it: "A thing is, in one sense, an object out there and, in another sense, an *issue* very much *in* there, at any rate, a *gathering*" (2004: 233). For Latour, a thing is a gathering because of the hybrid assemblage of human and non-human participants that make it, and because of its mutability and openness to interpretation as an "issue that brings people together *because* it divides them" (2004: 237; 2005: 13). "Objects become things," he explains, "when matters of fact give way to their complicated entanglements and become matters of concern" (2005: 31).

In "A Cautious Prometheus," Latour proposes that attending to objects in this way, as complex and engaging "things" that assemble, mediate, and gather, is equivalent to considering them as objects of design (2008). For Latour, objects always become projects in design, which means that rather than appearing as fixed "matters of fact," objects of design are always seen as "complex and contradictory assemblies of humans and non-humans" that are open to change. Objects, therefore, are conceived in terms of their potential and can be rethought, reconfigured, remade, etc. (2008: 6). Taking this idea in combination with Latour's call for thinking of scientific objects as things, this essay explores their implications for the parallel study of science and design (see in particular Latour 2000; 2004; 2005; 2008). Specifically, by looking at the ways in which a scientific instrument carries meaning, I will highlight how scientific practice attends to the "thingness" of things in ways that are characteristic of the practice of design.[2]

The object of this essay is the Crookes radiometer, a nineteenth-century scientific instrument that has resisted being taken for granted as a settled matter of fact. The radiometer's evocative qualities have made it an object of interest and dispute since its debut in the Victorian scientific community. This ongoing concern with the instrument has produced a wealth of resources that emphasize the contingencies and negotiations that shaped it. In particular, scientist William Crookes's detailed account of his radiometer-related research provides rare documentation of the complex play with materials, phenomena, and theories that leads to the establishment of a working device. Such complexities—especially the bifurcations and messiness of the making process—are often omitted from accounts of both science and design that typically (mis-)portray them as linear, solution-oriented, and efficient practices, which is to say that they talk about them in the "language of matters of fact" (Latour 2004: 240). Instead of this "poor proxy of experience and of experimentation"—as Latour describes it—complexities take center stage here, presenting the radiometer as a multifarious "thing" of science and design: an engaging and entangled gathering of socio-material relations that Latour would call "matters of concern" (Latour 2004: 245).

In what follows, I consider the qualities that render the radiometer as this sort of engaging and evocative object, unpack some of the entangled participants and processes that constitute it, trace the multiple and evolving motivations that drove its making, and discuss some of the ways in which it has been understood in the scientific community and beyond. These sections will reveal how the instrument carries meaning in terms of what it is (at the level of materials and technologies), how it was made (experimental opportunity and skilled practice), why it was made (intentions and contexts of investigation), and what it does (from practical understanding of how it functions to metaphorical interpretation of its action). These four dimensions of meaning are, as we shall see, at the root of the artifact's scientific significance and of the kinds of knowledge it carries and produces, but they are also essentially aspects of design. They present the radiometer as a *project* of design—gathering "things" with an attentiveness to details of craft and skill, recognizing the responsive nature of practice, and being aware of the key role of meaning in all its processes—which can be framed in terms of comprehension and interpretation. Approaching the radiometer in this way, as a *thing* of science and design, draws attention to how both science and design are practices that make and think about the world by mobilizing the particularities of experience, arrangements of materials and technologies, and socio-material contexts of action and interpretation.

Evocative Object: What it is

The Crookes radiometer, also known as a "light-mill," typically consists of a partially evacuated glass bulb mounted on a stand that contains a structure of four vanes pivoted on a vertical axis. The vanes hold small movable screens that are usually white on one side and black on the other, all facing the same way. This structure, which resembles an anemometer or wind vane, is known as the instrument's "fly" or "telltale" (Figs. 10.1a and 10.1b). When the instrument is exposed to light, the fly rotates with the white sides leading. This spin intensifies with the length of exposure to light, and it slows down then stops when the light source is taken away.

The Crookes radiometer is a perfect example of what sociologist Sherry Turkle would describe as an evocative object: its aesthetic qualities, the complexity of its materials, and its technological history, render it both emotionally suggestive and thought-provoking (Turkle 2007). The radiometer's glass case, which is a delicate and transparent enclosure, prompts careful attentiveness to its internal workings. Glass vessels are prominent in scientific practice because of their quasi-immateriality: both solid and see-through, they are technologies of containment and observation that can provide a controlled experimental environment. They can also evocatively "disappear" in order to bring attention to what they contain: the actual object of study. Literary critic Isobel Armstrong describes this ambiguity of glass as a solid and invisible medium, and the fluctuation between looking at it and looking through it, as "the dialectic of glass" (Armstrong 2000: 152; 2008: 11). She argues that this tension in glass provokes imaginative and intellectual responses concerned with ideas about mediation and transitivity—ideas that have also informed the making and the interpretation of the radiometer, as discussed later in this essay.

This glass case is evacuated of air to the extent that it was considered to contain an almost perfect vacuum at the time of its invention—which means that it is also a vacuum technology. In this way, the radiometer can be considered to be a descendent of Otto von Guericke's mid-seventeenth-century vacuum pump, and of his spectacular Magdeburg experiments, which launched scientific investigations into the nature and effects of vacuums. The scientist James Clerk Maxwell draws this connection when describing his presentation of the radiometer to Queen Victoria:

> I was sent to London, to be ready to explain to the Queen why Otto von Guericke devoted himself to the discovery of nothing, and to show her the two hemispheres in which he kept it, and the pictures of the 16 horses who could not separate the hemispheres and how after 200 years W. Crookes has come much nearer to nothing and sealed it up in a glass globe for public inspection.
>
> Maxwell 1876, in Brush and Everitt 1969: 112

Fig. 10.1a Radiometer with aluminum vanes and movable screens of clear mica, No. 1043 (No. 11).

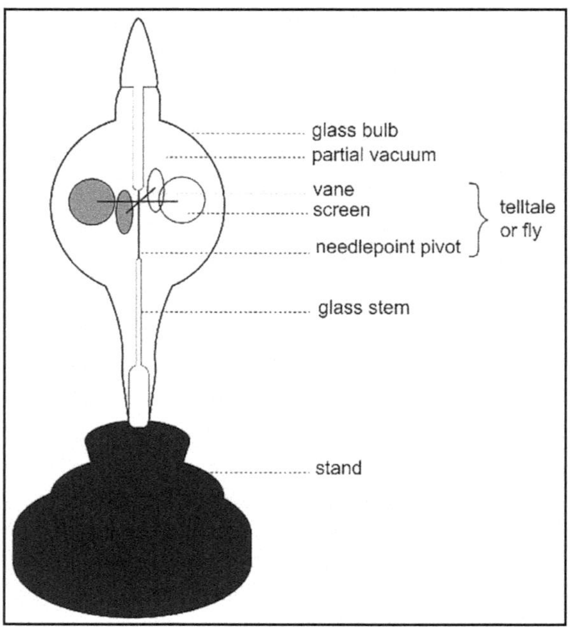

Fig. 10.1b Annotated diagram by the author, 2021.

For Maxwell, the radiometer's most appealing feature was that it presented the unfamiliar medium of a near-perfect vacuum to public and scientific scrutiny.

The radiometer's "fly" functions as a mechanical indicator within this near-perfect vacuum, and its movement can be observed for insights into the properties of rarefied environments, in which air pressure is lower than standard atmosphere. The indicator is an aesthetically appealing element of this evocative object, with its delicately balanced vanes that look like miniature kinetic sculptures enclosed in a glass bulb. This is especially true of the radiometers that were constructed by Crookes and his assistant Charles Henri Gimingham, whose experimentation with scales and structures resulted in an eclectic collection of beautiful artifacts (Fig. 10.2).

The fly's revolving motion, and its mode of activation, are likewise very compelling. The telltale's rapid and inconsequential whirl offers light-hearted pleasure to onlookers, and the mechanism that drives it, which remains mysterious to this day, suggests a seemingly magical conversion of light into kinetic energy. The device at work enthralled early audiences of Crookes's public science demonstrations. Crookes describes, for example, how an outsized radiometer that he exhibited to a large audience became a spectacular public event, its performance experienced as "four disks of light chasing each other round the room" (Crookes 1876a: 353–4).

Fig. 10.2 The Royal Society's collection of Crookes's radiometers and otheoscopes. © Science Museum / Science & Society Picture Library.

In his patent pamphlet, Crookes also acknowledges the allure of the radiometer's constant readiness to perform: "it requires no adjustment, and is always ready to be observed, whilst there is a peculiar charm in using an instrument which is constantly in active work" (1876b: 345). The fact that it doesn't require activation is very closely linked to the most captivating feature of the instrument at work: because of how little energy is needed to sustain a vigorous spin of its fly, it looks as if it could go on indefinitely in ways that would challenge the physical laws of nature.

Crookes draws attention to this suggestive quality in his first report on the instrument, where he writes that the movement observed is "kept up with great energy and regularity as long as the candle burns—producing, in fact, perpetual motion, provided only the radiation falling on the pith be perpetual" (1875: 523). By drawing a connection between his instrument and that most wondrous of impossible devices, the perpetual motion machine, Crookes highlights the extent to which the radiometer's spinning fly captures the imagination.

With its mediating glass case, the near-vacuum it contains, and the delicate telltale that spins inside it when it is exposed to a light source, the radiometer is a very meaningful artifact. The properties and histories of its materials and technologies call for a multifaceted engagement merging thought and feeling—the kind of engagement that defines creative and investigative processes according to the pragmatist view of design (Binder et al. 2011; Östman 2005). In other words, as a result of its evocative qualities, the radiometer demands to be considered as an interesting and complicated "thing" of science and design.

Making Processes (Exploration and Invention): How it was Made

The way in which the radiometer was made is likewise very meaningful because of both the entangled socio-material negotiations involved in making it, and the procedural and techno-material understanding that making it presents. Unpacking how the radiometer was made in this section reveals an emergent, contingent, and open-ended creative process that is shared by science and design. My discussion of this process also highlights the skilled practice and know-how that is necessary for establishing a working device.

In his six-chapter patent pamphlet describing "the construction, action, and uses" of the radiometer, Crookes's preface explains: so "that the action of the instrument may be more clearly understood, a brief account of the researches and experiments which led to the discovery of the Radiometer is first given" (1876b: 4). The pamphlet reports Crookes's investigations and tests in minute detail, no matter how provisional the findings turn out to be, and transparently records the back-and-forth interactions between thoughts and materials that define skilled practice. In this way, his narrative highlights the unpredictability and messiness that characterize the making process: the starts and stops during its non-linear progression, the role played by accidents and chance, the many revisions, adaptations, transformations of targets, and the appreciation of new possibilities that may emerge. By describing this protracted process as it unfolds, Crookes recognizes its significance to the knowledge that the radiometer carries and produces. He is aware, in other words, of how essential it is for the radiometer to be understood as a complex, contingent, and entangled assemblage of materials, actions, and thoughts. And in doing so, he demonstrates how both science and design are embodied, situated, open-ended, responsive, and transformative practices that interweave thinking and doing (cf. Schön's "Reflection-in-Action" 1983, 1987; and Gooding's "Procedural Knowledge" 1990).

Crookes's thinking and making processes began with his interest in the new element, Thallium, which he discovered in 1861 (Crookes 1873). In a paper describing his "very laborious researches" while attempting to determine the atomic weight of Thallium, Crookes noted the odd behavior of the warm samples of the element that he was weighing in an evacuated chamber (1873: 278). His observations suggested an "effect of heat in diminishing the weight of bodies" in "a vacuum as perfect as the mercurial gauge will register" (1873: 287). This was an apparent deviation from what was known about the effect of gravity, and Crookes decided to investigate. Crookes was able to notice this anomalous behavior because he understood the expected

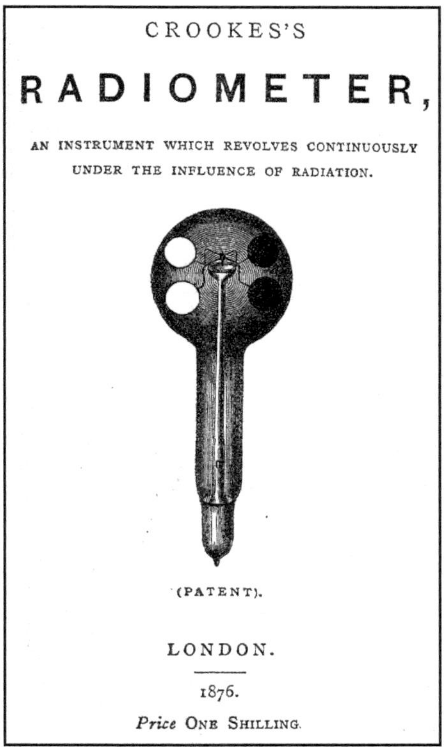
CROOKES'S

RADIOMETER,

AN INSTRUMENT WHICH REVOLVES CONTINUOUSLY UNDER THE INFLUENCE OF RADIATION.

(PATENT).

LONDON.

1876.

Price ONE SHILLING.

Fig. 10.3 Cover of Crookes's *Radiometer, an Instrument which Revolves Continuously Under the Influence of Radiation* [Patent Pamphlet], London, 1876.

ways that heat and gravity functioned in this experimental context. This understanding, and his informed curiosity, led him to pursue further experiments.

Crookes wondered if the effect he had observed was because of an unknown connection between the distinct phenomenal domains of heat and gravity. To investigate, he began by experimenting with the vacuum-balance in which he first noticed the anomaly, but "soon found [it] necessary to investigate the phenomena with smaller and less complicated apparatus" (1874a: 501). Crookes then configured various material set-ups, and carried out "a series of experiments with the view of ascertaining what form of apparatus would be most sensitive to the action sought" (1874a: 505).

His first arrangement consisted of a straw balance with pith balls at each end. The balls were suspended on a horizontal needle inside a wide glass tube that he connected to a Sprengel evacuation pump (Fig. 10.4). Tests with that apparatus using different heat sources, and pumping air out of the container to different levels of evacuation, disproved Crookes's first hypothesis of a connection between heat and gravity. It turned out instead that, in an evacuated space, bodies were *repelled* not *lightened* by heat (1874a: 505–11).

The effect that Crookes observed, which was a movement so sensitive that it could be activated by the heat of a finger "instantly repelling the ball to its fullest extent," was, however, itself worthy of investigation. Adaptation and revision of goals

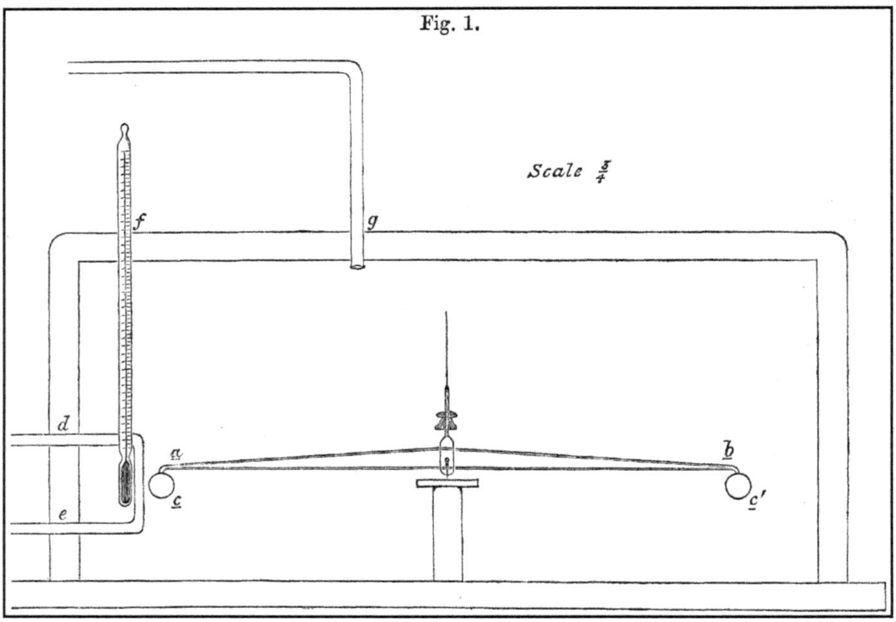

Fig. 10.4 "First Apparatus Tested," in Crookes (1874). "On Attraction and Repulsion Resulting from Radiation," *Philosophical Transactions of the Royal Society of London* 164: 501–27, p.506, Fig. 1.

in response to emergent contexts is an essential aspect of the making process in both science and design, and Crookes decided at this stage to focus on "facts" rather than conjectures (1874a: 511).[3] His next experiments thus aimed to establish the artifactual environment that would best present his observation that, in an evacuated space, bodies were *repelled* by heat. His efforts to find the most fitting apparatus involved an exhaustive evaluation of materials, scales, and arrangements. This immersive exploration is characterized by his biographer William H. Brock as "test it and see, change a variable, and test and see again" (2008: xix). Its entanglement and contingency underscore the radiometer's "thingness" as the output of a messy making process.

While fine-tuning the indicator's movement, Crookes found that an arrangement that allowed for horizontal movement within a glass bulb best illustrated the effect of heat on the movable bodies (1874a: 523; 1875: 520). Tests using various surfaces and coatings showed that a pith surface coated with lampblack was more sensitive to heat than a white one was, so he designed a structure that would rotate by using pith disks that were blackened on one side and kept white on the other (1876a: 339). He then experimented with changing how these rotating bodies were suspended, and replaced the silk fiber—which became twisted and stopped the movement—with a pivot needlepoint on the movable part balanced on a glass stem (1876b: 25). Finally, he increased the number of pith disks from two to four to show the rotation in a more striking fashion. This extensive experimentation ultimately led to a working device that "showed the movement of rotation in a very convenient manner" (1876b: 339).

Crookes's obstinate trial-and-error process shows the connections between speculative experiment and material progress, and exemplifies the negotiations that engage humans and materials in what sociologist Andrew Pickering describes as the "mangle of practice" (Pickering 1993). It also highlights how attentive scientists and designers are towards the materials, contexts, and phenomena with which they are engaged. And it reveals how being attuned to things drives a responsive and non-linear progression towards an initially unknown outcome. This sensitive attunement, and the tentative progression towards an outcome that only gets gradually defined, are characteristic of creative processes across disciplines according to the architect Kyna Leski (2015).

Once he determined the form of this working device, Crookes presented it in the third chapter of his patent pamphlet, and proceeded to list a series of tests and repetitions that confirmed its reliability (Figs. 10.5a and 10.5b). These tests prove that the radiometer reliably and conspicuously presents a remarkable phenomenon: the rotation of its vanes in response to a light source. Even though at this stage it was not at all clear what caused the effect observed, Crookes conclusively demonstrated that the instrument that he had designed *worked*. Establishing that an instrument works means recognizing that it produces what philosopher Davis Baird describes as "working knowledge." This, according to Baird, is a two-fold "thing knowledge" that suggests the "potential production of phenomena through material manipulation,"

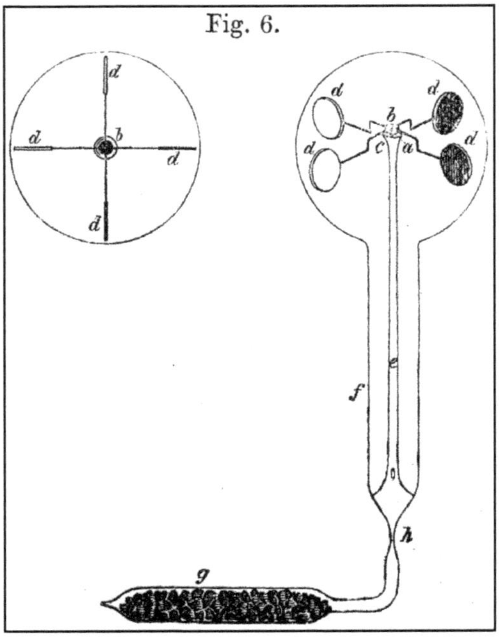

Fig. 10.5a "Radiometer," in Crookes (1876), "On Repulsion Resulting from Radiation–Parts III & IV," *Philosophical Transactions of the Royal Society of London* 166: 325–76, p.339, Fig. 6.

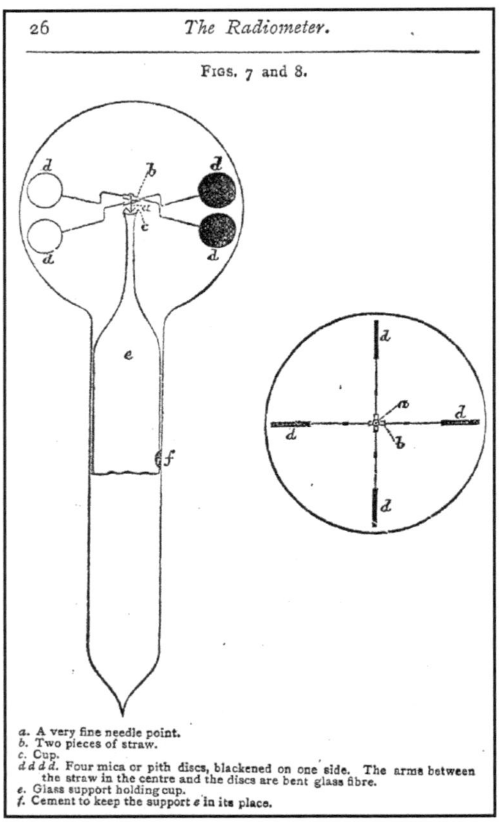

Fig. 10.5b "Radiometer," in *Radiometer, an Instrument which Revolves Continuously Under the Influence of Radiation* [Patent Pamphlet], London, 1876, p.26, Figs. 7 and 8.

and that places these phenomena "in an objective relation to human experience" (2004: 8–9).

Baird's first aspect of "thing knowledge"—the potential production of phenomena through material manipulation—derives from skilled practice: it is the know-how required to construct an artifact that works, and the technical and material understanding that develops in the making process. Like most chemists at the time, Crookes was trained in glassblowing, first as a student at the Royal College of Chemistry from the age of sixteen, then as August Wilhelm Hoffman's assistant there between 1848 and 1853, later passing on his skills to his apprentice.[4] His practice as an analytical chemist and as a chemical physicist meant that he engaged frequently with glass apparatuses. With the help of his assistant Gimingham, whose skills at glassblowing soon outshone his own, Crookes worked on improving some glass devices, such as the vacuum pump, and on developing new ones, such as the radiometer and the cathode ray tube. Crookes's intimate knowledge of his materials, in fact, was critical to designing the radiometer. The detail he provides in publications about the glassware in his experiments demonstrates his understanding of the complexity of glass, and of the specificities required of his glass apparatus to ensure precise observations of the radiometer's effect (1873: 294–5).

Crookes's knowledge about evacuation technologies was also very advanced and played a key role in the development of the radiometer. His paper on Thallium documents the substantial improvements that he and Gimingham made to the Sprengel evacuation pump (1873: 295). In a footnote to his first paper discussing what would come to be known as the radiometer effect, Crookes described different grades of vacuum: "air-pump vacuum," "Sprengel vacuum," and "chemical vacuum" (1874a: 507). His research on "attraction and repulsion resulting from radiation" was conducted in what he called a "chemical vacuum," a vacuum so perfect it could impede the passage of current from an induction coil (Crookes, 1873–1874: 37–41). It was only by achieving this rarefied environment that Crookes could observe and reproduce the radiometer's effect, and, as such, his technical knowledge about vacuum production is encompassed in this device. This knowledge would inform the design of new apparatus by Crookes and Gimingham, as well as by other scientists and makers.

Baird's closely entwined second consideration of "thing knowledge"—the placement of phenomena in an objective relation to human experience—relates to how the material and technical knowledge that goes into making an artifact is, in fact, rendered concrete by that artifact. Because it works, the radiometer demonstrates the possibility of achieving a particular effect through a specific combination of materials and technologies. Moreover, by staging its phenomenon transparently and presenting it to direct experience, the radiometer offers physical understanding that is realized through embodied engagement with the world.

The process of making the radiometer involves a combination of trained understanding (of elements and environments), embodied expertise (in the procedural skills required), and exploratory inventiveness (in experimentation). These complex socio-material relations are gathered in the radiometer, and they shape its meaning as a *made* "thing" of science and design. By considering "how" the radiometer was made, this section has emphasized the role of material agency in the socio-material negotiations that constitute the making process. The next section focuses instead on the "why" that drives this play of resistance and accommodation, and spotlights the role of human agency in the "mangle of practice."

Contexts of Inquiry: Why it was Made

Skilled practice is usually oriented toward more or less defined future goals, and Crookes's process for the radiometer was focused on diverse and evolving goals. His goals were shaped and transformed through an interplay of personal motivations and contexts of investigation. Crookes's emerging intentions during the process of making the radiometer demonstrate how the instrument carries meaning through its promise to be practically or theoretically productive in different investigative contexts.

As discussed earlier in this essay, Crookes's chance observation of the unusual behavior of heated bodies in an evacuated environment triggered his radiometer research. The scientist's curiosity was provoked by what he first believed was evidence of a new force connecting heat and gravity (Crookes 1874a: 501). As a result of this belief, Crookes theorized that the phenomenon he had observed could offer new insight about astronomical motion:

> It is not unlikely that in the experiments here recorded may be found the key of some as yet unsolved problems in celestial mechanics . . . So far as repulsion is concerned, we may argue from small things to great, from pieces of pith up to heavenly bodies.
>
> Crookes 1874a: 527

His first goal in response to this perceived relationship between heat and gravity was to construct a modeling instrument in which small (manipulable) things stand for greater (imponderable) ones. In this device, the almost perfect vacuum he produced would stand for a "stellar void," and the suspended pieces of pith would represent the movement of the planets (1874a: 527). This initial aim for the device was far removed from its eventual purpose in terms of both its operation and the kind of understanding it could offer. His actions show the extent to which envisioned goals are transformed in the making process. As we saw in the previous section, Crookes's further experiments disproved his initial hypothesis, so he decided to resist further speculation and focused his next efforts on a more practical goal.

Crookes's new pragmatic aim was to refine the instrument's telltale in order to construct a device that *worked* well. This kind of applied intentionality—a technical rationale centered on optimizing performance—is characteristic of making processes across disciplines. Optimizing performance is typically characterized by an evolving intentionality that transforms during practice. As part of this transformation, what it means for a device to "work," both conceptually and materially, gets defined and redefined during an open-ended and inquisitive tinkering process. For the radiometer, this designerly process led to an apparatus that performed reliably and compellingly, a situation that allowed Crookes to ascribe the status of instrument to the device and give it a name: "I have proposed for this instrument the name of the *Radiometer*, as it serves to measure the amount of radiation falling upon it by the velocity with which it revolves. It may also be called the *Light-Mill*" (1876b: 340–1).

Crookes's name for the device reveals yet another envisioned goal. Although light-mill may be a more appropriate name for a device that rotates under the influence of light, Crookes settled on *radiometer* instead. He suggests in his patent pamphlet that the radiometer could be used to measure radiation of any kind, with the caveat that it might not provide accurate measurements. He then proposed that it could be used in photographic operations to measure whether the amount of light in a room will expose plates or paper. He finally suggested that the radiometer may work as a "self-registering photometric instrument" and, more vaguely, that it may have a potential use in climatology where light "has been hitherto but very crudely and approximately estimated, or rather guessed at" (1876b: 38).

Crookes's speculations about quantitative uses for the radiometer are likely drawn from an early driving force in its development, rather than from a capacity he had recognized in the device at work. Historian Robert DeKosky argues that Crookes's quest to devise a vacuum that approaches perfection was a major motive for his research program in the early 1870s. In this quest, the radiometer's effect, along with other effects observed in extremely rarefied environments, were seen as potential means to measure the extent of exhaustion of air that was achieved (DeKosky, 1983: 14–15). Crookes's concern with the sensitivity of the radiometer's telltale was partly driven by his desire to accurately measure the degree of vacuum he was able to produce with the apparatuses he was developing.

Another area of concern is also likely to have informed Crookes's intention to devise a highly sensitive device. Crookes was famously interested in spiritualism and séances, and this interest seems to have shaped less conventional aspirations for the radiometer. Crookes was convinced that his skills for detecting and studying subtle physical phenomena gave him the means to study psychic phenomena, which he believed were also, in fact, physical. As part of his research into spiritualism, Crookes proposed the theory of a "Psychic Force," according to which the spiritualist "mediums" he studied were "supposed to possess a force, power, influence, virtue, or gift by means of which intelligent beings are enabled to produce the phenomena observed" (Crookes, 1926: 113). The highly sensitive "mediums" were fickle, however, and historian Richard Noakes suggests that Crookes sought to replace them with stable laboratory instruments that he hoped could also detect and respond to the "Psychic Force" he had identified. Noakes evocatively describes these as "instruments to lay hold of spirits" (Noakes 2002). An article in *The Echo* suggests that, by 1871, Crookes believed that he had developed such an instrument:

> Some recent experiments in my laboratory led me to believe that I have compassed an instrument as purely physical as the thermometer or electroscope, which will enable me to detect the presence of some hitherto unknown form of force or emanation from the fingers of everyone with whom I have tried it.
>
> Crookes 1871: 2

Crookes is referring to experiments that he conducted following his observation of the effects of radiation on bodies suspended in a near-perfect vacuum. These experiments were carried out while he was attempting to determine the atomic weight of Thallium, and are the ones that launched his development of the radiometer. In his first paper on the radiometer's effect, Crookes reports how these effects of radiation take place "[w]hen the finger is rubbed against the exhausted glass tube," suggesting that perhaps something more complex than just heat underlies what he also describes as an "interference" on the suspended bodies (1874a: 512). Crookes had noticed the similarity between this "interference" and some of the manifestations of "Psychic Force" that he had observed when studying "mediums," most notably their ability to move bodies without contact and to alter their weight.[5] This explains philosopher Ian Hacking's claim that the radiometer "was invented as a tool for investigating what has since been called psychokinesis (minds making motion)" (Hacking 1988: 435).

This highly unusual purpose for a scientific instrument demonstrates the range of intentions and contexts of inquiry that came into play during the development of the device. The diversity and evolution of Crookes's envisioned aims for the radiometer expose the optimism that is often at the core of making practices, and they reveal the sense of adventure that things-in-the-making embody. This emergent structure of intentions calls into question depictions of scientific and design practices as problem-solution binaries that progress in a linear fashion. It also helps dispel the myth of functionality as the key driving goal when designing an instrument. Crookes began the chapter in his patent pamphlet on the "useful application of radiometers" with an apology: "This instrument has been too recently constructed to allow for more than a brief notice of the many uses for which it is applicable" (1876b: 36). The vague potential quantitative functions that he briefly listed after this apology were, as we have seen, aspirational afterthoughts at best. This proves that, at least as far as Crookes was concerned, any function the radiometer may have had was of very little relevance to the significance of the instrument.

Crookes's multifaceted hopes for the radiometer encompassed theoretical, practical, technical, scientific, and even psychical frameworks, and provide the artifact with a wealth of potential meanings in association with these different frameworks. "*Form follows meaning*," therefore, is more apropos than the familiar "form follows function" to convey the sense-making process in design, which supports design theorist Klaus Krippendorff's notion that "Design is Making Sense (of Things)" (1989: 14; 9). Krippendorff's proposition amplifies the idea of *sense* through *making* in design, drawing attention to the designer's desire to make something that is recognizable and understandable, as well as something that is new (1989: 9).[6] His proposition also points to the key role of interpretation in design. The importance of this role is addressed in the next section, in parallel to the crucial role that interpretation also plays in science, by looking into the ways that the radiometer carries meaning through *what it does*. The latent meanings envisaged by Crookes while making the radiometer—in combination with meanings that emerge from *how it was made* and to *what it is*—reveal how the radiometer and its activity were understood by its maker and users across a range of socio-cultural and epistemological contexts.

Interpretations: What it does

The processes of interpretation that make sense of what the radiometer does are based on drawing links between the object and the various contexts in which it functions. These processes employ a range of meaning-making tools and skills. They also establish relational networks that foreground the instrument's "thingness" as an object of debate by gathering various players who are in dispute about a range of matters—from explanations of how the device works to conceptions about what it stands for. My examination of these processes and networks highlights the crucial role that interpretation plays in both science and design as practices that make sense of the world.

When they were first exhibited at a Soirée of the Royal Society on April 7, 1875, Crookes's radiometers caused an immediate sensation. This excitement was clearly expressed in a letter by James Clerk Maxwell:

> They whip spirits all to pieces. A candle at 3 inches acts on a pith disk as promptly as a magnet does on a compass needle. No time for air currents and the force is far greater than the weight of all the air left in the vessel. Attraction by a bit of ice very lively. All this at the best attainable vacuum.
>
> Maxwell, quoted in S. G. Brush and C. W. F. Everitt 1969: 109

Enchanted with the instrument's display, and excited by its potential to provide answers about the nature of light and radiation, the Victorian scientific community eagerly proposed explanations for how the radiometer functioned. The conflicting explanations advanced by different members of this community show how personal motivations, professional expertise, and scientific paradigms shaped understandings of the radiometer and its effect.

These debates about the radiometer's function show how advancing a scientific explanation—which is typically thought of as the ultimate "matter of fact"—is actually a process of interpretation where the meaning of a working artifact –now seen as a "matter of concern"– is mutable and context dependent. As a "matter of concern," then, the radiometer lends itself to interpretation as an object of design, befitting Latour's claim that "the more objects are turned into things . . . the more they are rendered into objects of design" (2008: 2). For Latour, meaning is a core feature of design, because "[w]herever you think of something as being designed, you bring all of the tools, skills, and crafts of interpretation to the analysis of that thing" (2008: 4).

Considering the radiometer as a thing in which "matter is absorbed into meaning (or rather as contested meaning)," means drafting these tools, skills, and crafts into making sense of it as an object of design (2008:5).

Maxwell and other scientists dealt with Crookes's radiometer in precisely this way. The repulsion of the radiometer's vanes when the telltale was exposed to a light source was particularly compelling for Maxwell, whose own theory of electromagnetic radiation had predicted a pressure of light. When refereeing Crookes's paper on the radiometer, Maxwell accepted Crookes's explanation that the telltale's rotation was directly caused by radiation, even though the pressure observed in the radiometer was much higher than what Maxwell had predicted. Maxwell declared in his referee report that Crookes had "made a great discovery," but he also pointed out that if such a direct effect of radiation did exist, reflection by the white side of the pith would cause a greater repulsion, forcing the black side to lead in the rotational motion (Maxwell, [1874] 1995: 390–1). The telltale's rotation was, however, in the opposite direction, prompting Maxwell to propose that: "It is not, therefore, because the black surface radiates more, but because it is *hotter*, that it is more repelled" (Maxwell, quoted in R. DeKosky 1976: 39).

Maxwell's moderated support was followed by further oppositional debate on Crookes's theory that the radiometer's rotation was because of a direct effect of radiation. After Crookes's presentation of the radiometer, fellow scientist Osborne Reynolds notably suggested that the rotation was because of evaporation and condensation within the device's glass bulb (Crookes 1875: 544). Following Maxwell's observation about the direction of rotation, Crookes added to his explanation of the telltale's rotation an intermediary process by which light was converted into heat, thereby raising the temperature of the blackened side of the pith screen (Crookes 1876a: 362–3). This edited explanation maintained that radiation was the cause of rotation.

In support of this explanation, Crookes also responded to Reynolds's criticism with experimental evidence that there was very little water vapor in the device (Crookes 1875: 547). In the same volume of *Philosophical Transactions* in which Crookes's paper appeared, however, Reynolds published a paper arguing that the movement of the radiometers' vanes could be explained by the presence of residual gas in the evacuated chamber—the minimal quantity of gas that remained in the radiometer's bulb after it was evacuated to the highest level achievable at that time (Reynolds 1876: 726). Reynolds also referred to an experiment by Arthur Schuster, carried out at his instigation, which offered experimental evidence that "the Force which turns the Mill is not directly referable to Radiation" and that instead "[t]he motion in the light-mill is wholly due to the forces acting between the revolving mill and its enclosure" (Reynolds 1876: 728; Schuster 1876: 718). He proved, in other words, that the telltale's movement was, in fact, a result of the effect of heat on residual molecules of gas. Based on this crucial experiment, Reynolds worked on calculations that led to the explanation, which, although incorrect, is the one most commonly held to this day: that an excess of pressure exerted by gas molecules on the warmer side of each vane is the cause of the observed rotational movement (1876: 730).[7] These negotiations show how a range of objects, experimental processes, theoretical frameworks, and cognitive skills were enrolled to make sense of the radiometer as a "matter of concern" in the context of Victorian science.

After these debates, British scientists working on the radiometer agreed that the rotation depended in some way on residual gas and differences in temperatures, but they could not work out how these caused a stronger pressure on the hotter side. It wasn't until 1879 that the radiometer's phenomenon was finally explained through key contributions by both Reynolds and Maxwell: the difference in temperature between the two sides of the vanes of the radiometer exposed to heat caused the gas molecules closest to the vane's surface to slide, producing tangential stress on the vane and instigating the fly's rotation. By then, with the exception of Crookes's continued research, most scientists were no longer interested in the radiometer itself, and they directed their attention instead to the study of phenomena in highly rarefied environments such as the one presented in its glass case.

This was, however, not the case for Crookes. Although he publicly conceded to the presence of residual gas in the radiometer's glass bulb following Schuster's experiment and Reynolds's explanation, Crookes privately remained suspicious of these theories. He embarked on a divergent course of inquiry, one shaped by his interest in the nebulous area between matter and energy:

> It is a question whether the residual gas in the apparatus, when so highly attenuated as to have lost the greater part of its viscosity, and to be capable of acquiring molecular movement palpable enough to overcome the inertia of a plate of metal, should not be considered to have got beyond the gaseous state, and to have assumed a fourth state of matter, in which its properties are as far removed from those of a gas as this is from a liquid.
>
> Crookes, 1876–1877: 308

Guided by his conception of a fourth state of matter—in which this fourth state has a similar attenuation relationship to gas as gas does to liquid—and driven by his hope that physical and psychical understanding might be simultaneously reachable by means of the radiometer, Crookes developed a new understanding of the medium enclosed in the radiometer bulb. In his 1879 lecture "On Radiant Matter," Crookes introduced his new theory and quoted Michael Faraday to explain why he chose to name this fourth state "radiant matter":

> If now we conceive a change as far beyond vaporization as that is above fluidity, and then take into account also the proportional increased extent of alteration as the changes rise, we shall perhaps, if we can form any conception at all, not fall far short of radiant matter.
>
> Faraday, quoted in Crookes 1879: 3

Crookes then described experimental arrangements that he devised to investigate the properties of radiant matter and listed his main findings about it: it can exert powerful directed mechanical action, it causes repulsive action between molecules, and it can be deflected by magnetic action (1879: 17–27).[8]

These properties bore a striking correspondence to events that Crookes had identified in his parallel researches into spiritualism and psychic phenomena. This led him to speculate that he had constructed a device that could effectively replace the spiritualist medium's body in its mediation between the physical and psychical worlds.[9] As a spectator at one of Crookes's early demonstrations of the radiometer, the scientist Francis Galton describes how "when the finger is *approached* the needle moves, sometimes [?] by attraction, sometimes by repulsion," and points out that "different people have different power over the needle" (Francis Galton, letter to Charles Darwin [28 March 1872], quoted in Noakes 2002: 145–7). The physical medium inside the radiometer's bulb appeared, in these circumstances, to have effectively replaced the spiritualist medium's body by rendering spiritual energy concrete and by acting as an intermediary between the seen and unseen.

In the same way as the more conventional scientific context did, the cultural complex at the interface of spiritualism and science offered a particular set of material and contextual tools that Crookes utilized to make sense of how the radiometer works as an object of design. Even though he ultimately concluded that the phenomena observed within the radiometer resulted from heat rather than bodily emanations, the subtle medium in the radiometer's bulb retained its clairvoyant affordance as a so-called fourth state of matter that promised to reveal new worlds to science. In conclusion to his lecture "On Radiant Matter," Crookes explains how "[i]n studying this Fourth State of Matter":

> We have actually touched the border land where Matter and Force seem to merge into one another, the shadowy realm between Known and Unknown, which for me has always had peculiar temptations. I venture to think that the greatest scientific problems of the future will find their solution in this Border Land, and even beyond; here, it seems to me, lie Ultimate Realities, subtle, far-reaching, wonderful.
>
> Crookes 1879: 30

The radiometer is thus seen as a bridge between material reality, as perceived by our senses, and the zone of energy and forces that scientific experiments can reveal; it is a technology that inhabits the space between known and unknown in which Crookes's explorative practice thrived. At this symbolic level, the radiometer's meaning extends from its cultural significance to more subjective understandings that are built by drawing personal associations with the object over time.

This is the way in which evocative objects come to hold cultural and personal significance according to Turkle (2007). Instruments are often personally meaningful objects for their designers, and for Crookes the radiometer represented the intellectual and enlightened character of his practice. The device is the central and most prominent motif in his coat of arms, which was conceived when Crookes was knighted in 1897 for his scientific achievements (Fig. 10.6a). His chosen motto, inscribed on a scroll underneath it, reads "Ubi crux ibi lux" (Where the cross is, there is light). Although this line most obviously refers to the Maltese cross component of his cathode ray tube—which it features twice as a green icon at the upper corners of

Fig. 10.6a Sir William Crookes Armorial Ex Libris Bookplate, taken from Crookes's copy of G. Hilton Scribner (1883), *Where Did Life Begin? A Brief Inquiry as to the Probable Place of Beginning and the Natural Courses of Migration There from of the Flora and Fauna of the Earth; A Monograph*, New York: C. Scribner's Sons. Available online through the Wellcome Collection at: https://wellcomecollection.org/works/cj8uydh9 (last accessed 17/05/2021).

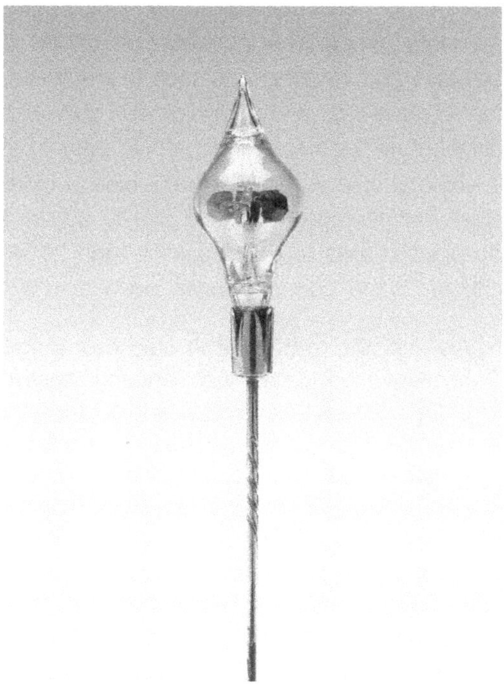

Fig. 10.6b Small radiometer mounted on pin and lead base. © Science Museum / Science & Society Picture Library.

the design—his biographer, Brock, also reads it as wordplay by the inventor. He interprets it to mean "Where Crookes is, there is light"—the light of knowledge provided by the radiometer, its whirling vanes representing "the black of scientific ignorance fleeing from the white of a new understanding of fundamental physics"[10] (Brock 2008: 222).

A tiny otheoscope, a variation on the radiometer fashioned at "about ½ inch in diameter mounted as a scarf pin" that he joyfully describes in a letter to his son, likewise demonstrates the device's symbolic significance for Crookes (Fig. 10.6b) (Crookes, letter to [his son] Henry [April 12, 1877], quoted in Fournier D'Albe 1923: 261). This working miniature performs its phenomenon as an intimate event held literally close to the heart, presenting the radiometer as an emblem of what could be described as Crookes's scientific virtue, affirming that he possessed the skills, attributes, and character required for the practice of science.

In a similar vein, the radiometer was a meaningful object lesson as an illustration of principles in a dispute between Crookes and the physiologist, and active anti-spiritualist, William Benjamin Carpenter. Carpenter first used it to point out what he saw as a contrast between the "admirable series of scientific investigations which led up to that invention," and the "thoroughly unscientific course" of Crookes's research into spiritualism (Carpenter 1877: 243). In "Another Lesson from the Radiometer," his reply to Carpenter in the literary periodical *The Nineteenth Century*, Crookes elaborates on the significant lessons offered by his radiometer research. He explains that the radiometer exemplifies the importance of being attentive to and following up on "residual phenomena," the kinds of phenomena that current theories "cannot account for" and that constitute "hints" that may lead to "the discovery of new elements, of new laws, possibly even of new forces" (Crookes 1877b: 886). Such a sensitive and inquiring approach to the world, Crookes argued, requires "finished manipulative skill" and a "disciplined mind," both attributes that are demonstrated in his work on the instrument (1877b: 886). In this lesson, the radiometer embodies intellectual curiosity and the principles of skilled practice, and illustrates Crookes's epistemic courage, which led him to pursue knowledge about a heterodox topic despite the risks to his career (cf. Kidd 2014). Crookes's article concludes with a criticism of the "common sense" advocated by Carpenter as a defense of ignorance:

> Are scientific men never to step over a rigid line, to refrain from investigation because it would clash with common-sense ideas? [. . .]. Can the wildest dreams of the spiritualist ask credence to anything more repugnant to "common sense" than the hypothesis imagined by science, and now held to account for the movements of the Radiometer? In the glass bulb, which has been exhausted to such a degree that "common sense" would pronounce it to be quite empty, we must conceive there are innumerable smooth elastic spheres, the molecules of residual gas, dashing about in apparent confusion, with sixty times the velocity of an express train, and hitting each other millions of times in a second. Will the "common sense of educated mankind" consider this rational doctrine?
>
> Crookes 1877b: 887

The radiometer here becomes a lesson about what can be gained from embracing what is considered strange and unusual as potential opportunities for new understanding, a call to pursue explorative and interpretative practices at the periphery of accepted standards.

The meanings that are revealed by what the radiometer unravels, the discourses, actions, and participants—human and non-human, real and imaginary, abstract and concrete—that are brought together to make sense of the device. This process emphasizes the various "matters of concern" that the radiometer draws together, the many people it gathers "*because* it divides them," and the different tools with which it is understood (Latour 2005: 13). My discussion of how the radiometer was debated and construed highlights how scientific practice, like design, engages different networks of meanings by different communities in various contexts that extend beyond those of its intended users. It has shown, in other words, how the radiometer is a "thing" of design as well as a "thing" of science, made to be interpreted, practically and symbolically, by its makers and its users.

Conclusion

Following Crookes's lead, and with special attention to Bruno Latour's notes about design in "A Cautious Prometheus?" perhaps there is a further lesson to be drawn from the radiometer, one specifically concerned with the parallel study of science and design. Approaching the Crookes radiometer as a design "project," this essay foregrounds some key characteristics that are shared by the practices of science and design. They are both skilled practices that carefully attend to, and thoughtfully engage with, the physical and imaginative properties of materials, technologies, and phenomena. They both possess an attitude to practice that Latour considers to be a form of "humility" because it involves recognizing that science and design are responsive, diverging from established narratives of truth and originality that depict them as foundational (2008). And they are both meaning-making practices, enrolling various processes of understanding and interpretation in order to make sense of the world.

By attending to the diverse ways that the Crookes radiometer carries meaning, I have also shown that both science and design are meaningful practices of making and thinking *because they deal with things.* They are meaningful in light of the fact that they produce evocative and engaging artifacts as opposed to detached objects "out there." They are meaningful because they gather human and non-human participants in the "mangle of practice" rather than involving a systematic goal-oriented process. They are meaningful because they mediate between different intentions and contexts of investigation instead of following a linear, rational, or function-led progression. And they are meaningful because they are open to interpretation and debate as "matters of concern" as opposed to stable or defined "matters of fact." In this essay, then, the Crookes radiometer becomes an object lesson in an interdisciplinary philosophy of science and design.

Notes

1. I am very grateful for Prof. Leslie Atzmon's stimulating feedback and discussions, which have helped me clarify and organize my ideas when preparing this essay.
2. The notion of "thingness" is taken from Heidegger, and is the functional equivalent to Latour's "gathering" and "matters of concern," and to Pickering's "mangle of practice" (Heidegger 1967; Latour 2004; Pickering 1993).
3. Crookes decided at this stage to focus on observable "facts" rather than conjectures, and he based the direction of his next investigations on empirical phenomena rather than theoretical speculation (1874a: 511).
4. Crookes was closely involved with the nineteenth century's emergent technical glass chemistry, publishing the results of collaborative research between scientists and glass manufacturers in his *Chemical News* and later being an active member of a British government-sponsored program for the investigation of glass (Brock 2007).
5. These "Spiritual Phenomena" are described by Crookes as "Class I: *The Movement of Heavy Bodies with Contact, but without Exertion*" and "Class III: *The Alteration of Weight of Bodies*" (Crookes 1926: 94; 97).
6. Krippendorff's proposition is meant to be a corrective to the meaning of design that amplifies the aspect of "applying a technical-functional rationality to the material world, at the expense of the *sense* that was to be achieved thereby" (1989: 9). The parenthesis in Krippendorff's proposal is further meant to highlight, in line with the position taken in this essay, "that we cannot talk about things that make no sense at all, that the recognition of something as a thing is already a sense-derived distinction, and that the division of the world into a subjective and an objective realm is therefore quite untenable" (1989: 9).
7. The problem with Reynolds's explanation is that the kinetic theory of gases would argue for a state of equilibrium instead of the movement observed.
8. These were instrumental variations on the radiometer: the "cup radiometer" (because of the chance observation of a radiometer with a crumpled vane spinning the wrong way), the otheoscope (a variation "in which a movable fly is caused to rotate by the molecular pressure generated on fixed parts of the apparatus"), and the "electrical radiometer" (in which the fly acts as a cathode), which led to the development of the cathode ray tube and provided the equipment for the discovery of X-rays (1895) and later the electron (1897) (Crookes 1876–1877: 308 [footnote]; 311).
9. The most notably similar occurrences are "Class I: *The Movement of Heavy Bodies with Contact, but without Exertion*" and "Class III: *The Alteration of Weight of Bodies*" (Crookes 1926: 94; 97). See Noakes on how Crookes's research into spiritualism and psychic phenomena may have contributed to his speculations on the fourth state of matter (Noakes 2004a; 2004b; 2008; 2013).
10. Brock is here referencing Philip Pullman's description of a device in *Northern Lights* that is named "photo-mill" in the novel and is unmistakeably a radiometer (Pullman 1995: 149).

References

Armstrong, I. (2000), "Technology and Text: Glass Consciousness and Nineteenth-Century Culture," in K. Flint and H. Morphy (eds.), *Culture, Landscape and Environment: The Linacre Lectures*, 149–75, Oxford: Oxford University Press.

Armstrong, I. (2008), *Victorian Glassworlds: Glass Culture and the Imagination*. Oxford: Oxford University Press.

Baird, D. (2004), *Thing Knowledge: A Philosophy of Scientific Instruments*. Berkley: University of California Press.

Benjamin, W. (1999), "Cultural History of Toys" [1928], in M. W. Jennings, H. Eiland, G. Smith (eds.) *Walter Benjamin: Selected Writings Vol.2 1927–1934*, 115–16, Cambridge: Harvard University Press.

Binder, T., G. De Michelis, P. Ehn, G. Jacucci, P. Linde, and I. Wagner [as A. Telier] (2011), *Design Things*. Cambridge: MIT Press.

Brock, W. H. (n.d.) (2004), "Crookes, Sir William (1832–1919)," *Oxford Dictionary of National Biography*, www.oxforddnb.com/view/article/32639 (accessed on August 6, 2013).

Brock, W. H. (2007), "The Royal Society's Glass Workers' Cataract Committee; Sir William Crookes and the development of sunglasses," *Notes and Records of the Royal Society* 61 (3): 301–12.

Brock, W. H. (2008), *William Crookes (1832–1919) and the Commercialization of Science*. London: Ashgate.

Brush, S. G. and C. W. F. Everitt (1969), "Maxwell, Osborne Reynolds, and the Radiometer," *Historical Studies in the Physical Sciences* 1: 105–25.

Carpenter, W. B. (1877), "The Radiometer and its Lessons," *Nineteenth Century* (1): 242–56.

Crookes, W. (n.d) (1879), *On Radiant Matter: A Lecture Delivered to the British Association for The Advancement of Science, at Sheffield, Friday, August 22, 1879,* London: E. J. Davy.

Crookes, W. (1871), "Mr. Crookes's 'Psychic Force,'" *The Echo*, November 10, 1871: 2.

Crookes, W. (1873), "Researches on the Atomic Weight of Thallium," *Philosophical Transactions of the Royal Society London* 163: 277–330.

Crookes, W. (1873–1874), "On the Action of Heat on Gravitating Masses," *Proceedings of the Royal Society of London* 22: 37–41.

Crookes, W. (1874a), "On Attraction and Repulsion Resulting from Radiation," *Philosophical Transactions of the Royal Society of London* 164: 501–27.

Crookes, W. (1874b), "Remarks on a paper by Professor Osborne Reynolds 'On the Forces Caused by Evaporation from and Condensation at a Surface,'" *Chemical News* 30: 24–5.

Crookes, W. (1875), "On Repulsion Resulting from Radiation – Part II," *Philosophical Transactions of the Royal Society of London* 165: 519–47.

Crookes, W. (1875–1876), "On the Movement of the Glass Case of a Radiometer," *Proceedings of the Royal Society of London* 24: 409–10.

Crookes, W. (1876a), "On Repulsion Resulting from Radiation – Parts III & IV," *Philosophical Transactions of the Royal Society of London* 166: 325–76.

Crookes, W. (1876b), *Crookes's Radiometer, An Instrument Which Revolves Continuously Under the Influence of Radiation* [Patent Pamphlet].

Crookes, W. (1876–1877), "Experimental Contributions to the Theory of the Radiometer – Preliminary Notice," *Proceedings of the Royal Society of London* 25: 303–14.

Crookes, W. (1877a), "On Attraction and Repulsion Resulting from Radiation – Preliminary Note on the Otheoscope," *Proceedings of the Royal Society of London* 26: 176–9.
Crookes, W. (1877b), "Another Lesson from the Radiometer," *Nineteenth Century* 1: 879–87.
Crookes, W. (1926; 1874), "Notes of an Enquiry into the Phenomena called Spiritual during the years 1870–73," in W. Crookes, *Researches in the Phenomena of Spiritualism*, 88–114, London: The Psychic Bookshop.
DeKosky, R. K. (1976), "William Crookes and the Fourth State of Matter," *Isis* 67 (1): 36–60.
DeKosky, R. (1983), "William Crookes and the Quest for Absolute Vacuum in the 1870s," *Annals of Science* 40 (1): 1–18.
Fournier D., E. E. Albe (1923), *The Life of Sir William Crookes*. London: T. Fisher Unwin Ltd.
Gooding, D. (1990), *Experiment and the Making of Meaning: Human Agency in Scientific Observation and Experiment*. New York: Springer.
Hacking, I. (1988), "Telepathy: Origins of Randomization in Experimental Design," *Isis* 79 (3): 427–51 [Special Issue on Artifact and Experiment].
Heidegger, M. (1967), *What Is a Thing?* trans. W. B. Barton, Jr., and Vera Deutsch, Chicago: Henry Regnery Company.
Heidegger, M. (1971), "The Thing," in M. Heidegger, *Poetry, Language, Thought*, 163–84, trans. Albert Hofstadter, New York: Harper Collins.
Kidd, Ian. (2014), "Was Sir William Crookes epistemically virtuous?" *Studies In History and Philosophy of Biological and Biomedical Sciences* 48 (A): 67–74.
Krippendorff, K. (1989), "On the Essential Contexts of Artifacts or on the Proposition That 'Design Is Making Sense (Of Things),'" *Design Issues* 5 (2): 9–39.
Latour, B. (2000), "When things strike back: a possible contribution of 'science studies' to the social sciences," *British Journal of Sociology* 51 (1): 107–23.
Latour, B. (2004), "Why Has Critique Run out of Steam? From Matters of Fact to Matters of Concern," *Critical Inquiry* 30 (2): 225–48.
Latour, B. (2005), "From Realpolitik to Dingpolitik or How to Make Things Public," in B. Latour and P. Weibel (eds.), *Making Things Public: Atmospheres of Democracy*, 14–41, Cambridge, Massachusetts Karlsruhe, Germany: MIT Press, ZKM/Center for Art and Media in Karlsruhe.
Latour, B. (2008), "What is the Style of Matters of Concern?" in N. Gaskill & A.J. Nocek, *The Lure of Whitehead*, 92–127, University of Minnesota Press.
Latour, B. (2008), "A Cautious Prometheus? A Few Steps Toward a Philosophy of Design (with Special Attention to Peter Sloterdijk)," www.bruno-latour.fr/sites/default/files/112-DESIGN-CORNWALL-GB.pdf (accessed on 8 February 2019).
Leski, K. (2015), *The Storm of Creativity*. Cambridge: MIT Press.
Maxwell, J. C. (1873), *A Treatise on Electricity and Magnetism*. London: Macmillan and Co.
Maxwell, J. C. (1879), "On Stresses in Rarified Gases Arising from Inequalities of Temperature," *Philosophical Transactions of the Royal Society of London* 170: 231–56.
Maxwell, J. C. (1995), "Report on Mr. Crookes' Paper on the Action of Heat on Gravitating Masses," [1874], in E. Garber, S. G. Brush and C. W. F. Everitt (eds.), *Maxwell on Heat and Statistical Mechanics: On "Avoiding All Personal Enquiries" of Molecules*, 390–1, London: Associated University Presses.
Noakes, R. (2002), "Instruments to Lay Hold of Spirits: Technologizing the Bodies of Victorian Spiritualism," in I. R. Morus (ed.) *Bodies/Machines*, 125–63, Oxford: Berg Publishers.

Noakes, R. (2004a), "The Bridge which is between Physical and Psychical Research: William Fletcher Barrett, Sensitive Flames, and Spiritualism," *History of Science* 42: 419–64.
Noakes, R. (2004b), "Natural Causes? Spiritualism, Science, and the Supernatural in Mid-Victorian Britain," in N. Brown, C. Burdett and P. Thurschwell (eds.), *The Victorian Supernatural*, 23–43, Cambridge: Cambridge University Press.
Noakes, R. (2008), "The 'World of the Infinitely Little': Connecting Physical and Psychical Realities circa 1900," *Studies in the History and Philosophy of Science* 39 (3): 323–33.
Noakes, R. (2013), "Exceptional Phenomena: A Congruence of Psychical and Physical Sciences, circa 1870–1930," paper presented at the "Psychical Research in the Histories of Science and Medicine" conference, January 25, 2013, University College London.
Östman, L. E. (2005), "A pragmatist theory of design: The impact of the pragmatist philosophy of John Dewey on architecture and design," Doctoral dissertation, Karlstad University, Sweden.
Pullman, P. (1995), *Northern Lights*. London: Scholastic.
Reynolds, O. (1876), "On the Forces Caused by the Communication of Heat between a Surface and a Gas; And on a New Photometer," *Philosophical Transactions of the Royal Society of London* 166: 725–35.
Reynolds, O. (1879), "On Certain Dimensional Properties of Matter in the Gaseous State. Part I & II," *Philosophical Transactions of the Royal Society of London* 170: 727–845.
Schön, D. A. (1983), *The Reflective Practitioner: How Professionals Think in Action.* London: Temple Smith.
Schön, D. (1987), *Educating the Reflective Practitioner.* San Francisco: Jossey-Bass.
Schuster, A. (1876), "On the Nature of the Force Producing the Motion of a Body Exposed to Rays of Heat and Light," *Philosophical Transactions of the Royal Society of London* 166: 715–24.
Turkle, S. (2007), *Evocative Objects: Things We Think With*. Cambridge: MIT Press.

11 From the Laboratory to the Studio: Microorganisms in Art and Design

CHRISTINE MARIZZI AND NURIT BAR-SHAI

Introduction: Art and Design Informed by Science

In the seventeenth century, Robert Hooke and Antoni van Leeuwenhoek independently pioneered microscopy and discovered the world of microbes—living creatures too small to see with the naked eye. Hooke published impressive illustrations in his 1664 book *Micrographia*, including the first image of a microbe. Van Leeuwenhoek designed the first single-lensed microscopes, allowing him both to see and to experiment with microbes (which he originally referred to as animalcules from the Latin term *animalculum*, meaning "tiny animal"). Although both scientists intended to depict what they saw under the microscope (Fara 2009), Hooke's double-spread micrographic illustrations forever changed people's perspective on the world by revealing previously unseen minuscule living creatures in dramatic renderings (Henderson 2016).

Alexander Fleming's Lifesaving "Germ Art"

This perspective on microbes inspired the work of future scientists, who approached microbes simultaneously as biological and visual. In the early 1930s, Sir Alexander Fleming, the Nobel Prize-winning scientist who discovered the antibiotic penicillin, used colorful microbes to create "germ-paintings" depicting scenes from everyday life, such as faces, soldiers, and ballerinas. An avid painter and a lifelong member of the Chelsea Arts Club in London, Fleming is credited with being one of the first people to use pigmented bacteria on agar in Petri dishes as an alternative to conventional visual media such as pastels or watercolors (Dunn 2011). Doing so required some training and microbiological experimentation. Fleming expanded the natural bacterial color palette from yellow and creamy white by using microbes that produce various pigments. He then synchronized their growth to ensure that all of the different bacteria grew into the images he desired, which posed a great technical challenge for Fleming. Some have speculated that Fleming's visual training allowed him to observe his science experiments in unorthodox ways.

When painting with microbes, Fleming also apparently did not follow the strict standard sterile technique, which minimizes unwanted environmental microbe growth. It happened that one of his painting plates (Petri dishes) that was sitting next to an open window became contaminated with mold. Instead of tossing the contaminated plate, Fleming observed that the bacteria in proximity to this mold were dying. He followed up on these findings and identified the mold as a member of the *Penicillium* genus (Dunn 2011). Fleming found this mold could effectively destroy all Gram-positive pathogens, which are bacteria that are responsible for diseases such as scarlet fever, pneumonia, and meningitis, among others (Tan and Tatsumura 2015). Fleming's discovery set the wheels in motion for the development of one of the most useful drugs in medical history—the penicillin group of antibiotics (Garrod 1960). Utilizing biological systems in conjunction with art and design techniques—such as drawing with microbial "living ink"—requires visualization, as in Fleming's practice. But it also requires a basic understanding of the biology of the organisms, in this case microscopically small microorganisms.[1]

Hacker Community Science Practices: (The) Growth Medium is the Message

Bringing non-scientific making techniques together with biological processes can spur novel ideas and outcomes, as Fleming's work did. This approach can likewise be effective for designers and artists, or even ordinary citizens, who work in this way. Agar is a gelatin-like substance that serves as structural support and food for microorganisms that grow on its surface. It is a robust mixture of Agarose (a complex polysaccharide, generally extracted from certain red seaweeds) and Agaropectin. Agar is also easy to obtain, inexpensive, temperature stable, and it cannot be digested by bacterial enzymes.

That modern scientists can grow bacteria in pure culture (as opposed to a wild mixture that is unsafe), and study them, is a fascinating story. Using culinary arts processes, the German-American Fanny Hesse discovered agar as a medium for culturing microorganisms in the 1880s, and consequentially transformed microbiology (Agapakis 2014; Smith 2005). Fanny was married to microbiologist Walter Hesse, who joined Robert Koch's lab in Germany. She also worked as an assistant and illustrator in Koch's microbiology lab (Agapakis 2014).

At the time, microbiologists used broth, potato starch, potato slices, or animal-derived gelatin as gelling agent for culturing bacteria. Gelatin has the severe disadvantage of liquefying on warm days, though, turning culturing bacteria into a gooey mess. Moreover, some bacteria possess enzymes that are capable of digesting gelatin. Fanny Hesse solved the problem of making particularly suitable plates. A friend of hers introduced her to an algae-based gelling agent called agar, which Fanny used at home in her kitchen to make exquisite edible jellies. Her husband noticed that her jellies were impeccable even on hot summer days, and Fanny suggested that her husband replace gelatin with agar, which turned out to be a perfect ingredient for Petri dishes in the laboratory.

Alerted by Walter Hesse, Robert Koch used this new solid media to isolate and discover the etiologic agents of diseases including tuberculosis, plague, malaria, and cholera—which brought about a huge advance in clinical microbiology. Robert Koch received the Nobel Prize in Physiology and Medicine in 1905 for his investigations on and discoveries about tuberculosis. He also identified many other bacterial infectious diseases such as anthrax and cholera, and pioneered many of the basic principles and techniques of modern microbiology. Without Koch's assistant, Julius Petri, who invented the Petri dish, and Fanny Hesse's agar kitchen skills, these discoveries might not have happened. Koch did not credit Hesse in any of his publications, and the suggestion never resulted in financial benefit for the Hesse family (Agapakis 2014; Hesse and Grschel 1992). As a pioneer dilettante scientist, however, Fanny Hesse's innovative problem solving became common practice in microbiological research. Fanny Hesse's "hacker" approach and motivation is today part of what is called community science practices in which public participation and collaboration in scientific research increases scientific knowledge. We present some recent community science projects later in this essay.

Composite Growth Medium

The visual qualities of agar turn out to be quite versatile for science as well as for art and design. New versions of agar-based growth media have allowed scientists to expand their understanding and representation of microbes. Microbiologists use various methods, including "differential agar," to distinguish between microbial species visually on the basis of biochemical differences. Differential agar contains specific ingredients that reveal microbial species by allowing them to be differentiated visually from other microbes. Red blood agar—which is red because it contains red blood cells—is an example of a differential agar medium that distinguishes bacterial species by their ability to break down red blood cells. This capability is particularly useful for classifying *Streptococci* species. Another differential medium, MacConkey agar, indicates microbes that can ferment the sugar lactose by a pink color change in both the microbes and the agar. Non-fermenters will appear normally colored or colorless on the plates.

Building on the idea that differential agar can change the color of the cultured bacteria, some contemporary artists and designers have expanded the color palette of the agar "canvas" itself. The standard agar plate is a lysogeny broth (LB) medium solidified with agar that is transparent and yellow. Tarah Rhoda of the School of Visual Arts in New York added charcoal to the standard LB medium to create a black background for her microbial culture, and William Shindel used LB whitened with titanium dioxide at Genspace in New York. Simon Park (2019) utilized red cabbage as a Ph indicator in his homemade manuals.[2] Finally, co-author of this essay, Christine Marizzi, used mica powders and food coloring to create shimmering plates in all colors of the rainbow to motivate high-school workshop participants to engage with scientific practices.

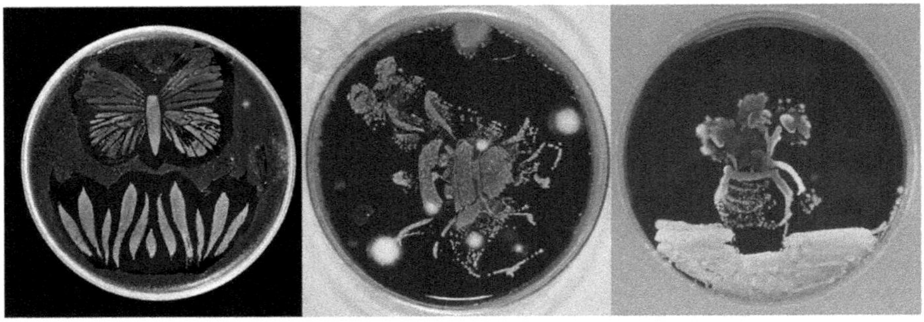

Fig. 11.1 ASM Agar Art Contest projects created at DNA Learning Center's Saturday DNA workshops led by Christine Marizzi. Left: Science teacher Allison Granberry, *A Bumble Bee, Bacteria, and Mold: Could this be Art in the Making?* Center: Ten-year-old Kate Lin, *The Magnificent Butterfly*. Right: High-school student Jenny Xu's submission *Sunflowers*. Image courtesy of Christine Marizzi.

Living Art and Design

Contemporary living art and design incorporate both agar's capability to change color and traditional scientific processes, along with hacking and innovation processes. These hybrid creative approaches are found in the work of bioartists, biodesigners, and in community lab practices that have adopted standard scientific techniques used in microbiology to create "agar art," "microbial art," "microbial painting," or "bacteria painting."[3] In this process, microbes are applied onto solidified agar as "paint," and are then allowed to grow into a living artwork. Living art and design projects interrogate critical issues such as time, death, and decay. Some bioartists and biodesigners explore the morality and safety of genetically modified organisms, while others repurpose bacteria to be used for medical procedures or to mitigate environmental issues.

Amy Chase – Kitchen Culture

Escherichia coli (*E. coli*)[4]—which is a commonly used organism in scientific research—is an especially popular species for bioartists and biodesigners. In 2009, artist Amy Chase Gulden collaborated with molecular biologist Kristin Baldwin to create the project *The many (still) lives of E. coli*, which integrated traditional bacteria culture techniques with art and design practices. Guided by Baldwin, Gulden prepared the bacteria culture in a homemade DIY (Do It Yourself) lab, using an oven top to prepare the agar medium and an electric blanket with a thermometer as an incubator in which the bacteria grew overnight (Gulden 2019). Using a technique that she called "growing impressions," Chase used brushes to hand paint genetically modified blue *E. coli* bacteria onto standard agar plates. The bacteria grew into blue patterns that were

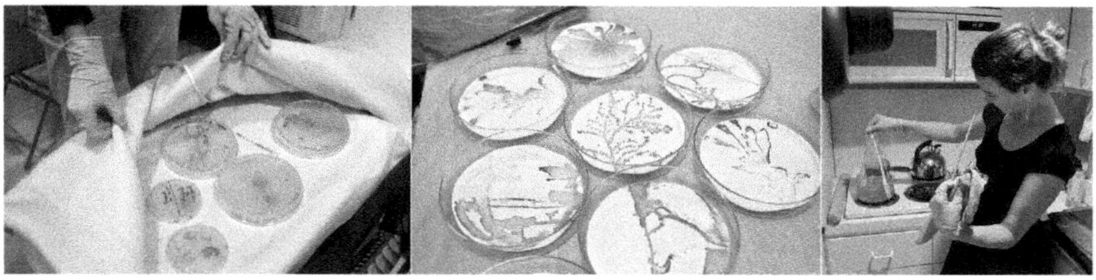

Fig. 11.2 Artist Amy Chase Gulden and molecular biologist Kristin Baldwin (2009). Homemade lab setup for *The many (still) lives of E. coli*, bacterial growth printed on paper. Images used by permission of Amy Chase Gulden.

guided by the brushstrokes, and the patterned bacteria were then printed onto fitted circular paper by transfer. The printed bacteria, which were fixed onto paper circles that fit within Petri dishes, were exhibited in a gallery.

Like Fanny Hesse, Chase appropriated modern kitchen science and repurposed her kitchen into a temporary lab space. She likewise refitted her tools and common items, such as brushes and an electric blanket, to be used to grow her project. Chase was able to create a new framework to generate and grow her images by bringing together scientific techniques with visual practices.

Kac's Bacterial Paintings and Living Time

Utilizing living organisms as material in art and design necessarily introduces time as a factor. Biological systems naturally encompass time-based processes, such as self-assembly, self-organization, collective collaborations, and mass communication. Driven by processes such as evolution, emergence, growth, and reproduction, biological systems produce inherent themes of control, life, death, and decay. Thus, when incorporating biological systems into their practices, designers and artists often reflect upon human-environment / human-nature interrelationships, prompting them to think critically about life itself.

For his project *Specimen of Secrecy About Marvelous Discoveries* (2006), for example, pioneer bioartist Eduardo Kac created a biotopes series, a living artwork that continually evolved during the exhibition in response to both the work's "internal metabolism" and the exhibition space's environmental conditions. The microorganisms in this project were used as living paint, which was applied on nutrient-rich media. As the work traveled the world, the different venues and local visitors affected its "internal ecology" (Kac 2011), which had an effect on the piece.

Kac's work changed partly in response to the artists' initial settings, partly in response to the (gallery) environmental conditions, and partly as an expression of the organisms as they grew. Because they are complex living creatures, work using bacteria transforms constantly. "Living" projects, therefore, contend with notions of growth, evolution, and decay over time and how these changes happen in response to environmental conditions—which are likewise a crucial part of biological processes.

Edgar Lissel Growing Photographs Using Light Sensitive Bacteria

Change is a time-based process. Change and time, however, can be transient as well. German Photographer Edgar Lissel uses light-sensitive bacteria to grow his images and capture change over time, fixing fleeting moments of life created using living microorganisms. His content includes ephemeral moments, objects, or places, as well as still life motives and ruins. In particular, Lissel creates stunning self-assembly bacterial photographs using Petri dishes with cyanobacteria, which obtain energy from light.[5] In his *Project Bakterium* series, Lissel used a camera obscura to produce "living" photographs (Lissel 2008) in which photographic images projected light and shadow areas onto the agar. These projections allowed cyanobacteria microbes to

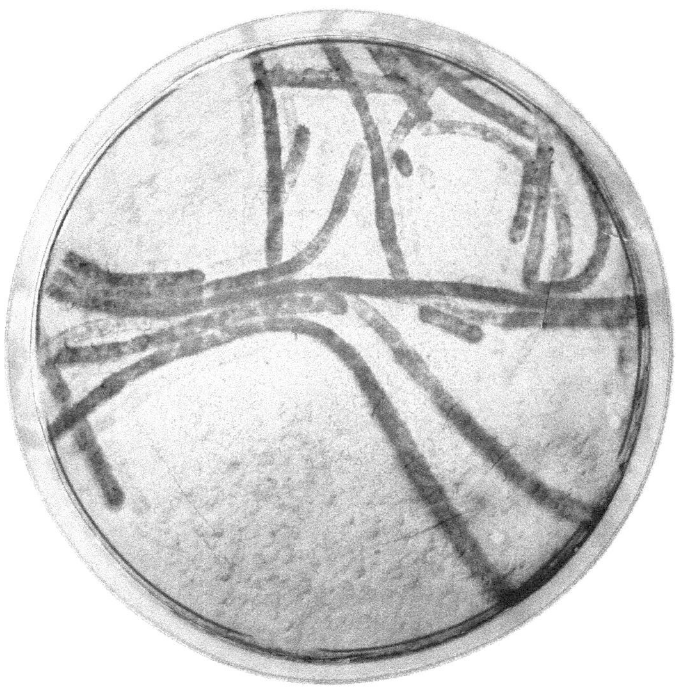

Fig. 11.3 Edgar Lissel, *Project Bakterium: Self-Portrayal* (1999, 2001). Image used by permission of Edgar Lissel. Bacteria reconstruct their own micro-image. Using what Lissel calls "photographing myself – my invisible self in time," the bacterial image grows within the project as opposed to taking and fixing a photographic of the same subject.

proliferate only in the areas that were exposed to light. Like light sensitive photo-paper, life recording unfolds in these projects—and as the bacteria "outgrow," dynamic photographic images form and change.

In *Self-Portrayal* (1999, 2001), Lissel took photomicrographs of individual bacterium and projected them onto Petri dishes that contained cyanobacteria. As a consequence, the bacterial cultures reproduced their own micro-image "cellfies." In *Domus Aurea* (2005) Lissel's microbes reconstruct the outlines of the original *Domus Aurea* fresco image in Rome. Lissel chose cyanobacteria because this fresco had been damaged by the light-sensitive cyanobacterium *Leptolyngbya*,[6] which had, unfortunately, metabolized and faded the fresco. Lissel came up with a constructive process in response. He illuminated the ruin, and projected a negative image of the deteriorating fresco onto plaster panels in order to guide the bacteria culture, which eventually grew to replicate the outlines of the original fresco image. In *Domus Aurea*, Lissel captures both the notion of change over time and the transience of time. The piece also expands upon themes of creation, destruction, and reconstruction, while equating human-made and naturally occurring entities.

The Vanishing Painting of Marta de Menezes—Decomposing Color

Other bioartists and biodesigners are also interested in both transience and the convergence of making processes with natural processes. For her project *DECON: Deconstruction, Decontamination, Decomposition, 2007* (de Menezes 2019), Portuguese artist Marta de Menezes collaborated with Dr. Ligia O. Martins of the Instituto de Biologia Química e Biológica in Lisbon, Portugal. Martins works on the degradation of toxic textile dyes using the bacterium *Pseudomonas putida*. Informed by bioremediation strategies—in which microorganisms remove pollutants or toxins—de Menezes utilizes this soil bacteria in her work to break down and decompose toxic dyes that are used in the textile industry, which is one of the most polluting industries in the world.

De Menezes used textile dyes in her custom-made Petri dishes that reproduced artist Piet Mondrian's constructivist paintings in shape, color, and form. The basic geometric structures in the piece were filled with agar, and pigmented with textile dyes in primary colors. She then applied living *Pseudomonas putida* provided by Martin's laboratory; the microbes deconstructed the artwork as they metabolized the dyes, and eventually the dyes in the piece faded.

Pushing against the idea of traditionally fixed imagery, de Menezes' project existed only for a finite time, foregrounding notions of process and progress, life and death, life and decay, and decay as matter. Moreover, the project raised social and environmental awareness about toxic materials such as textile dyes. At the same time, this project applied biotechnology—which uses cellular and biomolecular processes to produce technologies and products—and it utilized bacteria as a novel

Fig. 11.4 Marta de Menezes and Dr. Ligia O. Martins, *DECON: Deconstruction, Decontamination, Decomposition* (2007). Image by Marta de Menezes. Work by Marta de Menezes and Dr. Ligia O. Martins @ Instituto de Biologia Química e Biológica in Lisbon, Portugal, 2007. Image used by permission of Marta de Menezes and Dr. Ligia O. Martins.

resource for remediation. De Menezes' critical work illuminated her collaborator's (Dr. Ligia Martin) scientific practices outside of the laboratory, and produced innovative ways of working with a combination of biological systems and art and design techniques. De Menezes shared her practices as well. She collaborated with scientists to found the Cutivamos Culture citizen community biolab in Odemira, Portugal, which promotes biolab-based educational and outreach programs to the general public.

Hacking Traditional Microbial Culture Approaches

Like de Menezes, bioartist and researcher (and co-author of this essay) Nurit Bar-Shai—also a co-founder of the Genspace community biolab in Brooklyn, NY—is interested in bacterial behavior. The immediate environment, such as the nutrient medium upon which bacteria feed and swarm, has the ability to alter their color[7] or behavior. Bacteria change how they act in response to temperature, oxygen levels, moisture, antibiotics, or nutrient composition. These behaviors are part of a microbial survival mechanism, in which versatile native adaptation offers them better opportunities for growth and proliferation. Utilizing the environmental modifications that affect microbial cultures expands the tools available to scientists, and to biodesigners and bioartists.

Bar-Shai's Modifications of Traditional Lab Practices to Alter Highly Motile Bacterial Behavior

For her work, co-author of this essay Nurit Bar-Shai has modified environmental factors for *Paenibacillus* bacteria. *Paenibacillus* are soil microorganisms with advanced social motility—which is the coordination of movement of single cells in response to external growth factors or biochemical signal stimuli that results in the formation of complex architectural patterns. The *Paenibacillus vortex* bacteria that were discovered in the early 1990s by biophysicist Eshel Ben-Jacob's group at Tel Aviv University are considered to be among the "smartest bacteria," with "collective intelligence" behavior and an advanced social communication system. Acting much as complex multi-cell organisms would, this single-cell organism makes collaborative decisions as a colony.

Inspired by Ben-Jacob's research on the social life of bacteria, Bar-Shai questioned the methodologies that scientists used to culture the microbes, and she manipulated the natural high-motility growth patterns of various strains of *Paenibacillus* bacteria through artistic interventions.

Bar-Shai's interest in structural and environmental impact on bacterial collective behavior—and thus bacterial pattern formation—led to a series of experiments in which she linked bacterial morphogenesis and designed ecosystems. *Paenibacillus* are naturally colorless microorganisms. In order to see them, Bar-Shai dyed the transparent bacteria with a pigment that fixes the bacteria and makes their patterns visible. Instead of using flat 2D standard round plates that are used in labs for research, for these experiments Bar-Shai redesigned the bacteria environment. She cultured *Paenibacillus* in custom-made Petri dishes, including geometric-shaped plates.

Bar-Shai also tinkered with the agar substance, defining areas on the agar in order to choreograph—to induce or hinder—form-dependent bacterial growth. She also experimented with agar concentrations to achieve different degrees of substrate stiffness and softness. Instead of 2D, flat agar surfaces, Bar-Shai adapted molding,

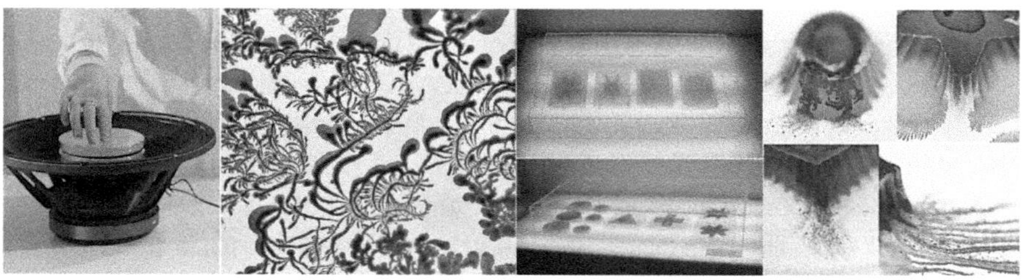

Fig. 11.5 Nurit Bar-Shai, *Objectivity* [tentative] (2009–ongoing). A series of experiments and interventions inspired by Professor Eshel Ben Jacob's research. This project investigates the complex networks and communication systems of intelligent bacteria. Images courtesy of author.

3D-printing, and sculpting techniques to create 3D agar structures that initiated microbial behaviors that resemble those that take place in nature or inside our bodies. Finally, Bar-Shai applied sound to the solidifying agar, which affected the complexity of the bacterial growth patterns. The resultant piece was a product of interplay between Bar-Shai's criteria and bacterial behavior. The piece's expression was influenced by growth factors, as well as biophysical and biomechanical properties, such as surface topographies or stiffness, structural forms and shapes, gravity, and area boundaries or sound vibrations.

Synthesizing Bacteria: Genetically Modified (GM) Colors

The circumstances that Bar-Shai used to modify bacterial behavior, such as external environmental stimuli, are further enhanced by utilizing genetic engineering. Genetic engineering, in which an organism's genes are manipulated using biotechnology, has changed microbiology practices significantly. Genetic manipulation has expanded the toolboxes of scientists, and consequently those of bioartists and biodesigners.[8] *E. coli* bacteria, which are naturally white, for example, have been genetically engineered for research purposes to synthesize and secrete a variety of colored pigments, including fluorescent proteins.[9] One famous example is the green fluorescent protein (GFP) gene, which is found naturally in the Pacific jellyfish *Aequorea victoria*. When GFP is produced and "excited," a fluorescent green light is emitted; this ability to glow is called "bioluminescence." Using genetic engineering, scientists introduced the GFP gene into specific host organisms for genetic tracking. In addition to GFP, fluorescent color proteins have been developed or discovered that are yellow, orange, and red, or blue and purple hues. Although the yellow fluorescent protein (YFP) is a genetic mutant of GFP, the red fluorescent protein (RFP) originated from a completely different organism, the sea anemone *Discosoma sp*. Multicolored fluorescent proteins have become a common biological tool used by scientists to tag and track where and when certain genes are expressed in cells—as well as to track tissues or whole organisms in real-time—without the need to sacrifice the host organism.[10]

GFPixel Living Portrait

Bioartists and biodesigners have reflected critically upon various aspects of the new GFP gene technology. The piece, GFPixel / Living Portrait (2001) by Austrian artist Gerfried Stocker and molecular biologist Reinhard Nestelbacher (Stocker and Nestelbacher 2008), is a project that is made with 4,000 standard-size Petri dishes.

This work was cultured using *E. coli* bacteria, some of which were genetically modified with GFP. The Petri dishes were arranged as a portrait, like pixels on a digital screen. Each plate represents a pixel, an arrangement that merges a digital visual vocabulary with living organisms. The Petri dishes that expressed the GFP gene emitted green visible light when the dishes were exposed to UV light—meaning that the GFP gene was "switched ON." Those without the GFP gene didn't emit green light and were "switched OFF." The portrait looked digital, but in fact it lived and died during the exhibition. This work illustrates certain similarities and distinctions between altering living organisms using genetic modification and digital representations, and between binary code and DNA engineering. Both GFPixel and actual digital imagery deal with how the mind perceives color and form that are composed of smaller units of color. Living organisms used to create imagery, however, require more sophisticated and specific optimal environments to produce an image that will ultimately decay.

E. Chromi—The Colorful Palate of Biotechnology

The ways that GM bacteria produce color in certain environments also informed the project *E. chromi*. Designers Alexandra Daisy Ginsberg and James King—who collaborated with scientists from Cambridge University for the 2009 International Genetically Engineered Machine (iGEM) competition[11]—created *E. chromi* from GM[12] bacteria (Ginsberg 2009).

The team genetically engineered *E. coli* bacteria to produce a variety of fluorescent proteins in red, yellow, green, blue, and violet. The team designed standardized sequences of DNA that produced these fluorescent proteins, which they called BioBricks,[13] and then combined several of them and inserted them into *E. coli* bacteria. This genetic engineering enabled the bacteria to manufacture the red, yellow-, green-,

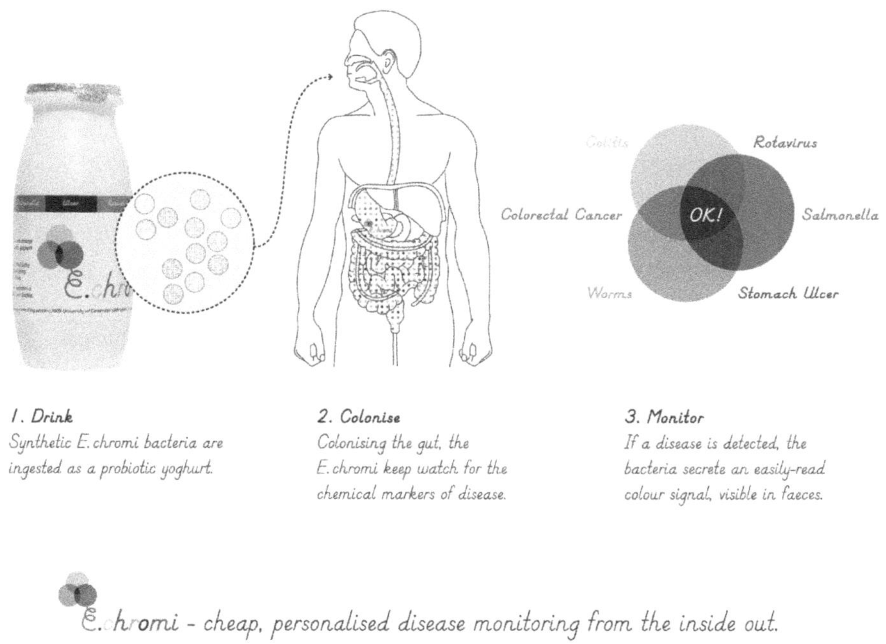

Fig. 11.6 Alexandra Daisy Ginsberg, James King, and Cambridge iGEM 2009 team (2009). System diagram for *E. chromi* Scatalog. Image used by permission of Alexandra Daisy Ginsberg.

blue-, and violet-colored proteins—but only if a certain "trigger" molecule, such as a toxin, was present in their environment. Bacteria that turned red, for example, would indicate a certain toxin. The designers also imagined speculative future scenarios in which *E. chromi* could reshape our everyday lives. In one scenario, this genetically engineered bacteria could be used for personalized medicine as a food additive when added to a probiotic drink. The GM microbes would then indicate gastrointestinal disease by producing different colors—basically "color-coding" a person's feces to indicate disease.

Biotechnology: GM Color Manufacturing

> We can use cells as tiny factories, and DNA as a programing language that can control living systems.
>
> **Natalie Kuldell, Massachusetts Institute of Technology (MIT)**

In recent years biotechnology has been incorporated into various aspects of our everyday lives—it has become part of food, medicines, biofuels, and biomaterials. Our perceptions of microbes, the biotechnologist's "unseen helpers," is changing as well. Perceived in the past as disease-causing "germs," microbes are beginning to be understood instead as "factory workers." We are also starting to appreciate DNA as a language that can control living systems, and biotechnology laboratories as new sorts of factories.

The implications of biotechnology for the human body, and on our environment, are well studied and highly regulated by government agencies. The fact that biotechnology can interfere with nature, however, is a great concern for most people. A prominent, and very controversial example, is bioartist Eduardo Kac's GFP bunny, named Alba (Kac 2001). Kac collaborated with French geneticist Louis-Marie Houdebine, whose team had already created several transgenic rabbits in the laboratory by splicing the GFP gene into the animal's genome. Kac visited Houdebine's lab and planned to adopt one of these rabbits as a pet for his family in the United States, and to exhibit it. Kac's intentions called attention to the promises of this new biotechnology—which borrows biological parts, repurposes them, and shares them between species—and to bring awareness of what would be known as post-naturalism.[14] But, when Kac announced the project publicly, it was understandably seen as controversial. Public protest was spearheaded by animal rights activists who were concerned about the application of biotechnology only for artistic purposes, while the art community argued against the exhibition of a lab animal as an art object (Clüver 2010; Collins 2003; Kac 2011). In the end, the GFP bunny was banned from the public domain.

Biotechnology in Public Domain

Although the GFP bunny was banned, GM organisms have already been available for purchase as pets: for example, since 2003 at PetCo stores in the United States.[15]

The commercial distribution of GMO into the public domain has been widely debated and has raised serious concerns.

Bioartists Shiho Fukuhara and Georg Tremmel have pushed back against indiscriminate acceptance of genetically modified organisms. They created a piece called *Common Flowers, Flower Commons* (2009), which was in response to the genetically modified Florigene MoonDust carnation. Developed by Suntory Flowers—the largest Japanese beer manufacturer, and a producer of GM blue carnations and blue roses since 1996 (Chandler 2007b; Chandler 2007a)—the purple MoonDust carnation was the first GM plant sold in stores. Protesting against commercial GM blue carnations, bioartists Fukuhara and Tremmel used DIY plant-tissue culturing techniques to develop a method for breeding these manufactured sterile blue carnations, and then open-sourced this method.

Using standard plant tissue culture techniques, at first, Fukuhara and Tremmel open-sourced the native sterile plant, which grew roots so that it could later be planted in public parks. They then reversed Suntory's GM process for the Florigene MoonDust carnation through "back-breeding," "restoring" its original white color. By "wiping out" Suntory's genetically modified purple pigment, the bioartist duo questioned the practice of patenting living organisms, as well as the human interventions and community science practices around GM organisms. Their reverse and open-source bioengineering process challenged authorship for, and the implications of, unregulated GM distribution into the environment (Cloutier 2017).

Fig. 11.7 Shiho Fukuhara and Georg Tremmel (BCL), *Common Flowers, Flower Commons* (2009). The artists cloned Suntory's first commercial GM product and then open-sourced protocols for making blue carnations with "kitchen technologies." Images used by permission of Georg Tremmel.

Full STEAM Ahead! Collaborations, Outreach, and Activism

Although biotechnology has made its way into our everyday lives, there is little scientific literacy among the general public about genetic engineering practices. Community science labs are working to address these issues. As we discussed above, people are also justifiably concerned about certain biotechnological techniques and outcomes. The mission of many community science labs is to educate people about issues around biotechnology, and to share knowledge through public participation in outreach programs that foreground hands-on experience. These community labs often utilize a STEAM educational approach, which uses science, technology, engineering, the arts, and mathematics to guide inquiry. Those who run community science labs feel that this methodology helps people both to understand concerns about, and participate in, the development of new biotechnology.

The *NYC Biome MAP*—Urban Biome Drawn Through Art Science and Public Collaborations

In 2015, this essay's co-authors, Dr. Christine Marizzi and Nurit Bar-Shai, collaborated on the *NYC Biome MAP*—an accessible and inclusive platform for the public to engage with science. This platform also incorporates Marizzi and Bar-Shai's concerns about problematic issues around emerging biotechnology.

The *NYC Biome MAP* is more than a unique visual project; it exemplifies a collaboration that involves STEAM-based outreach activity. The *NYC Biome MAP* illustrates how pairing art and design with science can serve as an entry point for those who are hesitant about or intimidated by technical fields—or for those who are passionate about tinkering with biotechnology. By drawing on the nexuses among art, design, and science, this project reached many citizen scientists across the NYC metropolitan area.

The *NYC Biome MAP* project was crafted in response to New York City's New Museum's IDEAS CITY Festival 2015. Their call for projects asked contributors to reflect upon the festival's theme, "The Invisible City," in homage to author Italo Calvino's 1972 literary masterpiece of the same name. We approached the theme by making the invisible visible through working with microbes that are too small to be seen with the naked eye. We invited the general public to take part in the project and designed a two-part outreach workshop. In the first part, participants learned about bacterial culturing, sterile technique, and how to work with live microbes as bio-regenerative paint. In the second part, participants discussed DNA technology and observed microbes grown on agar plates in a laboratory setting.

The first part of this project took place at the New Museum's IDEAS CITY Festival street fair in New York's Lower East Side. Each participant selected an area from a grid map of NYC that was designed for the project, with a Petri dish prepared with sterile laser-cut stencils that matched the topography of the assigned grid map area. The stencils covered parts of the Petri dish, directing the bacterial growth in the

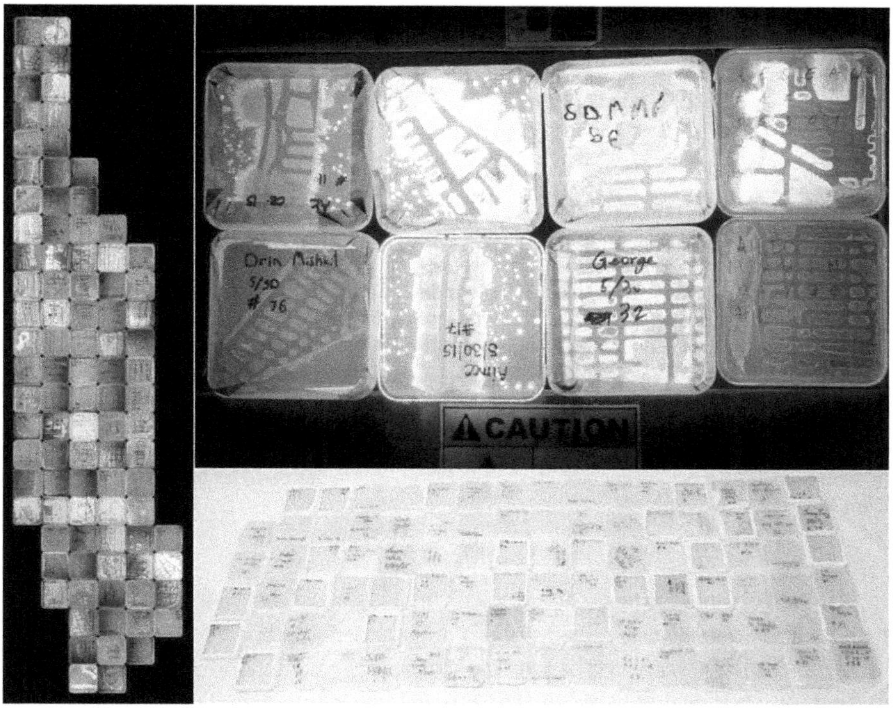

Fig. 11.8 Christine Marizzi and Nurit Bar-Shai, the *NYC Biome MAP* (2015). Incorporating many personal biomes into a collective biome, The *Urban-Biome MAP of New York City* is a collaborative bio-painting artwork. Images courtesy of authors..

image of particular NYC streets and waterways. Participants learned how to safely use colorful *E. coli* bacteria as paint, and applied them onto the exposed agar areas in their dishes. Finally, participants learned how to seal their dishes to prevent contamination. The Petri dishes were returned to the laboratory so that the bacteria could incubate.

As a guest activity, Marizzi and Bar-Shai invited the "Swab Squad" team of the PathoMap research group to introduce their "SwabKit." PathoMap,[16] which is led by physiologist and biophysicist Chris Mason of Weill Cornell Medical School, aims to map New York City's microbiome. Their SwabKit contains tools that participants can use to swab and sample the urban microbiome, and then crowd-source microbial samples from around the city.

The second part of the activity took place at the Genspace community lab during which participants observed how the bacteria on their topographical Petri dishes had grown. For the majority of the participants, this was the most memorable part because they had never seen live bacteria on a Petri dish. This hands-on experience was enriched by background material on, and a lively discussion about, perceptions of genetic engineering and microbes.

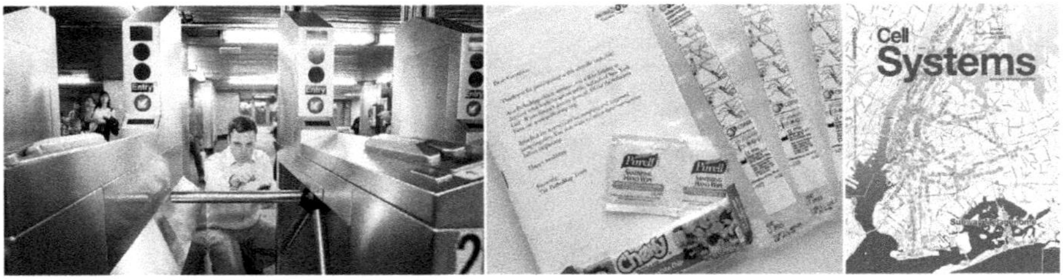

Fig. 11.9 Christopher E. Mason, PhD and team with Genspace, *PathoMap* (2010). Dr. Mason and his team at Weill Cornell Medicine study the microbiome and metagenome in New York City's subways, parks, and waterways. The Brooklyn community-science lab Genspace worked with Mason's group to identify extremophiles at the polluted Gowanus Canal in Brooklyn. Extremophiles are microoganisms, typically archaea, which live in environments with extreme temperatures, acidity, alkalinity, or chemical concentrations. PathoMap (pathomap.org) results appeared in several scientific publications, with citizen-scientists co-authors. Image used by permission of Ebrahim Afshinnekoo. (a) Source: Wall Street Journal (b) Image by: Ebrahim Afshinnekoo (c) Image by: Cem Meydan.

The participants' experience suggests that agar art, as well as other kinds of bioart and biodesign, can support public engagement, and help to address possible social, moral, and scientific issues about biotechnology. The *NYC Biome MAP* participants were especially interested in how biotechnology had revolutionized drug production—for example, how genetically engineered bacteria produce human insulin to treat people with Type I diabetes. Most had a vague idea how this process could work, but the informative dialogue with the workshop instructors helped to create a deeper understanding of biotechnology. Participants also learned how the scientific community is mapping microbial ecosystems in urban environments, and how citizen scientists can get involved in this work.[17]

After participants' plates were documented and safely disposed, the *NYC Biome MAP* team assembled the images into the final collective artwork: a map of Manhattan. Microbes reside everywhere—on land, in water, and in the air—but, of course, they are too small to be seen with the human eye. NYC is a melting pot of cultures, both human and microbial, and each citizen actually has his or her own personal microbiome. Collectively, New Yorkers shape NYC's urban biome, and this unseen microbial world has a significant impact on them. Making the microbial community of NYC visible, the project renders a poetic representation of the city's collective microbiome—and thus its shared ecology.[18]

Drawing on notions of our personal microbiomes and our city's collective urban-biome, the *NYC Biome MAP* project has achieved critical acclaim. It came in second in the American Society for Microbiology (ASM) Inaugural Agar Art Contest in 2015, and was covered in several news articles worldwide. The BBC World Service's flagship discussion program, *The Forum*, invited Christine Marizzi to discuss the project process—including its outreach component—on their program *Microbes and Humans: The Science of Living Together* (BBC World Service 2016). The now iconic

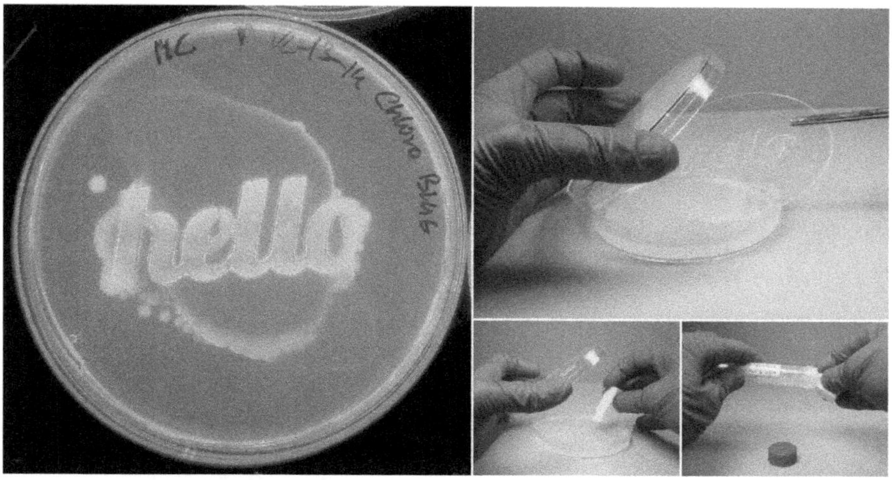

Fig. 11.10 Nurit Bar-Shai, Instructions on how to grow *Hello World* with GFP bacteria. Open-source BioArt Kit that is offered as an Instructable workshop online. Photography by Nurit Bar-Shai. Image courtesy of author.

NYC Biome MAP is showcased annually at the ASM Microbe Conference Agar Art Gallery and was presented at the United Nations General Assembly 74 in New York at a program about antimicrobial resistance. These events led to a partnership for Bar-Shai and Marizzi with the ASM to develop an open case study about the project for the public to follow, as well as workshops and an open-source "BioArt kit" (Bar-Shai 2016).

Microbial Landscapes

In 2016 Marizzi and Bar-Shai collaborated on another investigation on the microbes that are found in our environment. Microbial life impacts the entire biosphere, which is the global ecological system that includes all living organisms. It has been estimated that there are more microbes on earth than stars in the sky, and microbial life plays a primary role in metabolic processes and regulating biogeochemical systems in virtually all environments on earth. Eukaryotes are organisms in which genetic material is organized into chromosomes found within the cell nucleus. Animals, plants, algae, and fungi are all eukaryotes; most microorganisms are prokaryotes. The immensity of microorganism production is such that—even in the total absence of eukaryotic life, including us(!)—microoganisms, and their biological processes, would continue unchanged.

However, we humans collectively shape the global microbiome by our lifestyle choices, and this unseen microbial world also has a powerful impact on us. When Marizzi and Bar-Shai attended the prestigious Aspen IDEAS Festival, they designed and ran a project about the microbiome for the Spotlight Health Festival attendees, which involved painting with colorful GM *E. coli* microbes. Participants worked

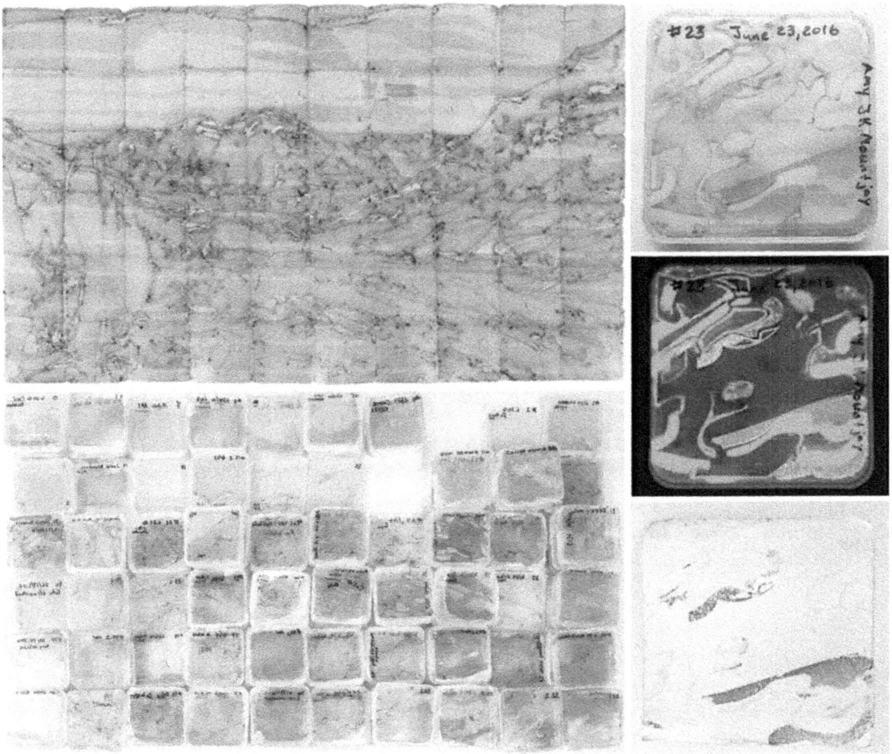

Fig. 11.11 Christine Marizzi and Nurit Bar-Shai, *Microbial ecologies* (2016) This collaborative bio-painting and printing project with 150 plates draws links between the microbiome and our environment, Aspen IDEAS Spotlight Health Festival. Images courtesy of authors.

together to create a collaborative microbial piece that depicted Aspen's iconic Maroon Bells mountain peaks (Marizzi 2016). The project consisted of individual pieces that were completed by participants, making up a collective image of the iconic Maroon Bells landscape. During the process, participants learned from Marizzi and Bar-Shai about microbial ecology, including the relationship between the microbiome and its environment, and about human interventions in nature.

[s]Teaming Up

Biotech institutions and companies are reaching out to artists, designers, and citizen scientists. In 2014, iGEM added new submission categories, including a Community Labs track (2014.igem.org/Tracks/Community_Labs) and an Art and Design track (2018.igem.org/Competition/Tracks/Art_Design). ASM, which was founded in 1899, held its first Agar Art Contest in 2015, which was received with great enthusiasm. The collaborative nature of the *NYC Biome MAP* inspired ASM's science outreach program, and the organization formed partnerships with ten community labs In the United States and Europe ("ASM Agar Art Contest" 2018; Alfano 2018). ASM also established separate contest categories for experts and non-experts, including a category for children. In 2018, ASM's partner-hosted workshops were attended by more than 400 participants, many of them with little to no prior experience with microbiology (Lontok, Marizzi, and Sturm 2018; Lontok 2018). Public outreach, such as the initiatives above, suggests that hands-on projects at the intersection of art and design and science can foster positive notions of microbiology: designing with microorganisms, and the visible manifestation of invisible microbes, may mitigate the "ick factor" around microorganisms. These outreach projects may also reveal the beneficial microorganisms that we rely upon for life and wellbeing. Working with microorganisms could help the public to understand the human body and our environments better.

Artists and designers and community scientists at large, are in a unique position to explore the social, moral, and ethical questions that are related to new biotechnologies through creativity and public engagement. Thoughtfulness, openness, and listening-ability have earned some bioartists, biodesigners, and community scientists the trust of people in their communities. And what they are hearing is a call to democratization: breaking down walls in order to increase access to science and get the public engaged. Alternative education methodologies and collaborative opportunities, combined with public participation and making practices, could make art, design, and science collaborations common practice. We need more spaces with open access to life-sciences—such as community biolabs—near libraries and fabrication laboratories (fablabs). These kinds of spaces, all of which endorse both art and design and science and technology practices, have the potential to become social hubs with substantial impact on the public. Their initiatives inform citizens about the roles that art, design, and science can play together, and inform scientists about unconventional ways to incorporate art and design with science.

Acknowledgments

We would like to thank all our supporters and collaborators, especially Genspace NYC staff and volunteers. Special thanks to Ali Schachtschneider and Martha Molina Gomez for their design and production work, and Emily Dilger and Katherine Lontok of ASM, the New Museum's IDEAS CITY Festival team and the Aspen IDEAS Festival team for their collaboration and support. The *NYC Biome MAP* was partially funded by Genspace NYC and the DNA Learning Center at Cold Spring Harbor Laboratory.

Notes

1. One important aspect is safety. Microbes are usually maintained in pure culture, ensuring that laboratory strains are not introduced into the surrounding environment. However, microbes in pure culture are not easy to obtain and require very specific conditions to survive and grow. Preparing media and sterilizing equipment to culture bacteria, and responsible cleanup afterwards, requires careful and long-term planning. In addition, any

microbes used as a scientific or artistic tool must be handled according to their appropriate Biosafety Level (BSL) and following standard microbiological practice, such as presented in the American Society of Microbiology (ASM) Biosafety Guidelines (Emmert 2013) and the National Institutes of Health (NIH) Guidelines (NIH 2013). Interested citizen scientists and artists, therefore, regularly seek BSL-certified community laboratories to work with safe microbes under expert supervision.

2. Simon Park discusses in "Microbial Art" (2019) that red cabbage is a natural pH indicator and turns green/yellow if the bacteria generate alkaline conditions, or red if they produce acid. Purple indicates neutral conditions.
3. Bacteria can be naturally pigmented, such as *Micrococcus luteus* (yellow), *Rhodococcus rhodochrous* (pink), *Sarcina aurantiaca* (orange-yellow), and *Serratia marcescens* D1 (red). The biological function of these pigments ranges from aiding in basic metabolism to communication, UV-protection, and defense mechanisms (Soliev, Hosokawa, and Enomoto 2011; Malik, Tokkas, and Goyal 2012; Wynn-Williams et al. 2002; Tuli et al. 2015; Liu and Nizet 2009; Ligia, Karen, and Susan 2017).
4. Naturally occurring in the lower intestines of humans, *E. coli* is widely used as a research tool and model organism to study disease. Professional and community labs usually carry several *E. coli* derivatives from its wild-type *E. coli* strain K-12, which was originally isolated from a patient at Stanford University in 1922, but cannot effectively colonize the human gut (Bachmann 1996). In evaluating biological agents that pose theoretical health risk to humans, the NIH assigns K-12 to Risk Group 1 (RG 1): Agents that are not associated with disease in healthy adult humans (NIH 2013).
5. Cyanobacteria, which are 3.5-million-year-old photosynthetic microbes, are among the oldest known living creatures on earth—and the earliest organisms that obtained energy from light.
6. A morphotype strain of the light-sensitive *Cyanobacteria*, found at the excavated site of the Domus Aurea in Rome, was responsible for the destruction of its frescoes.
7. For example, a Winogradsky column provides a view into the versatile culture of various anaerobic microorganisms that are found in the ground, and which express diverse colors as a response to their environmental conditions as they metabolize.
8. Scientists often use fluorescent proteins, such as GFP, to understand the function of cells and proteins in living organisms. Proteins are the chemical tools of life – they control most of what happens within a living cell. Every human being functions thanks to the well-oiled machinery of thousands of proteins, such as enzymes, antibodies, and peptide hormones. If a protein malfunctions and cannot be corrected, illness and disease often follow. Therefore, it is fundamental for the biosciences to map out the role of various proteins in model organisms and humans. For the discovery of this groundbreaking GFP technique, Roger Tsien, Osamu Shimomura, and Martin Chalfie were awarded the 2008 Nobel Prize in Chemistry (www.nobelprize.org/ 2008).
9. Transformation is the process by which foreign DNA is introduced into a cell. Transformation provides an understanding of the molecular basis of the DNA being studied, and it also provides a highly effective tool for manipulating bacteria to manufacture a desired protein.

10. To do this, the gene coding for a protein of interest is first fused with the gene for a fluorescent protein (like GFP) and transformed into a model organism. When the corresponding protein is produced, it will be tagged with GFP. This causes the protein of interest to glow inside the model organisms' cells when the cell is irradiated with ultraviolet light, and allows scientists to track its location in real time under a microscope. It also allows the observation of the spatial overlap between two (or more) different fluorescent labels—each with a separate emission wavelength, and therefore a specific color—to see if the different "targets" are located in the same area of the cell or not. For example, an interaction of a green-labeled protein 1 with a red-labeled protein 2 would produce a yellow color signal. These "co-localization" studies make characterization of unknown proteins in live cells possible and also help to uncover where interactions between proteins occur.
11. See iGEM, http://igem.org.
12. A GM organism (also abbreviated as GMO) is any organism whose genetic material has been altered using genetic engineering techniques.
13. Rethinking biological parts as Lego blocks, BioBricks are standard DNA parts that are used as building blocks to design and assemble a larger synthetic biological circuit with purposed functions. These BioBricks could then be engineered into a living cell.
14. Postnaturalism is the study of organisms that have been intentionally and heritably altered by humans by means such as selective breeding or genetic engineering. It is studied at the Center for PostNatural History in Pittsburg.
15. GM drugs have been in use for human consumption since the early 1980s (insulin in bacteria). Plants have been genetically modified since the early 1980s, beginning with antibiotic-resistant tobacco. The first commercially grown GM crop approved for sale as a food product in the US was the FlavrSavr tomato. Specifically designed to have a prolonged shelf life, it was sold in stores around the US for human consumption from 1994 to 1997. Other crops (such as rice, wheat, corn, soy and cotton, etc.) were genetically optimized for pesticide, herbicide, and pathogen resistance as well as fast growth, and better nutrient profiles (Baulcombe et al. 2014).

    The GloFish (www.glofish.com) is a zebrafish (*Danio rerio*) that was genetically modified with GFP in 1999, and later with other fluorescent proteins producing colors such as red, magenta, yellow, and blue. Originally developed as a simple way to detect the presence of environmental toxins in drinking water, it was the first whole GM organism that entered the market and was distributed to the public domain. Recent additions to the GloFish collection are tetra (*Gymnocorymbus ternetzi*) and a shark (*Epalzeorhynchos frenatum)* genetically modified to fluoresce (Knight 2003; Blake 2005).
16. See http://www.pathomap.org/.
17. See the Human Microbiome Project (https://hmpdacc.org/), American Gut Project (http://humanfoodproject.com/americangut/), Urban Barcode Project (www.dnalc.org/websites/urbanbarcodeproject.html), PathoMAP (www.pathomap.org/), and community science courses (www.genspace.org/classes/ and https://sites.google.com/biobus.org/exploreathome/home).
18. The term microbiome was coined by Joshua Lederberg of Rockefeller University (Lederberg and McCray 2001). The microbiome is the ecological community of all

microorganisms that share our body spaces. The human body contains up to three times more microbial cells than human cells, although the entire microbiome weighs only about three pounds (Sender, Fuchs, and Milo 2016). The total number of genes associated with the human microbiome exceeds the total number of human genes. We are outnumbered indeed! Perhaps then, when we look into our own bacteria, we might see the reflection of ourselves. Who we are is directly linked to what we eat, where we were raised, how much sleep we get, how much sunlight exposure we receive, who we associate with, or how often we are in close contact with animals and plants. All those directly affect the microbial community in and on our body, which, in turn, could affect our wellness and even our decision-making processes. Based on our individual biome, companies (like DayTwo www.daytwo.com), can map our microbiome and personalize our diet. These companies claim that they will revolutionize the future of our nutrition and wellness.

References

Agapakis, C. (2014), "The Forgotten Woman Who Made Microbiology Possible," *Popular Science*, www.popsci.com/blog-network/ladybits/forgotten-woman-who-made-microbiology-possible (accessed on June 25, 2021).

Alfano, A. (2018), "Microbial Masterpieces Win ASM Agar Art Contest," www.cshl.edu/microbial-masterpieces-win-asm-agar-art-contest/ (accessed on June 25, 2021).

"American Society for Microbiology (ASM) Agar Art Contest," (2018), www.asm.org/index.php/public-outreach/agar-art (accessed on June 25, 2021).

Bachmann, B. J. (1996), "Derivations and Genotypes of Some Mutant Derivatives of Escherichia Coli K-12," in F. C. Neidhardt (ed.), *Escherichia Coli and Salmonella: Cellular and Molecular Biology*, 2460–2488, Washington, D. C.: ASM Press, www.asmscience.org/files/Chapter_133_Derivations_and_Genotypes_of_Some_Mutant_Derivatives_of_E._coli_K-12.pdf (accessed on June 25, 2021).

Bar-Shai, N. (2016), "How to Grow 'Hello World' With GFP Bacteria," Instructables, www.instructables.com/How-to-Grow-Hello-World-With-GFP-Bacteria/ (accessed on June 25, 2021).

Baulcombe, D., J. Dunwell, J. Jones, J. Pickett, and P. Puigdomenech (2014), "GM Science Update: A Report to the Council for Science and Technology," *A Report to the Council for Science and Technology* 3: 1–49, https://assets.publishing.service.gov.uk/government/uploads/system/uploads/attachment_data/file/292174/cst-14-634a-gm-science-update.pdf (accessed on June 25, 2021).

BBC World Service, (2016), "Microbes and Humans: The Science of Living Together," www.bbc.co.uk/programmes/p03yvzfv (accessed on June 25, 2021).

Blake, A. (2005), "GloFish – The First Commercially Available Biotech Animal," *Aquaculture Magazine* 31 (6): 17–26.

Cardoso, L., Y. F. K. Karen, and G. K. Susan (2017), "Microbial Production of Carotenoids A Review," *African Journal of Biotechnology* 16 (4): 139–46, www.researchgate.net/publication/315057397_Microbial_production_of_carotenoids_A_review (accessed on June 25, 2021).

Chandler, S. (2007a), "Genetic Modifications in Floral Crops: Research to Marketplace," in R.E. Litz, R. Scorza (eds.), *Acta Horticulturae* 738, 37–50, Leuven, Belgium:

International Society for Horticultural Science, doi:10.17660/ActaHortic.2007.738.2 (accessed on June 25, 2021).
Chandler, S. (2007b), "Practical Lessons in the Commercialization of Genetically Modified Plants – Long Vase-Life Carnation," in P.E. Read (ed.), *Acta Horticulturae* 764, 71–81, Leuven, Belgium: International Society for Horticultural Science, www.actahort.org/books/764/764_8.htm (accessed on June 25, 2021).
Cloutier, M. (2017), "Bioart as a Space for Identity Conceptualization: Figuring the Human Body under the Scope of Biotechnologies," *Leonardo* 50 (5): 330, doi:10.1162/LEON_a_01489 (accessed on June 25, 2021).
Clüver, C. (2010), "'Transgenic Art': The Biopoetry of Eduardo Kac," in Lars Elleström, *Media Borders, Multimodality and Intermediality*, 175–86, New York: Springer. doi:10.1057/9780230275201 (accessed on June 25, 2021).
Collins, D. (2003), "Tracking Chimeras: The Eighth Day of Eduardo Kac," in S. Britton and D. Collins (eds.), *The Eighth Day: The Transgenic Art of Eduardo Kac*, Phoenix: The Institute for Studies in the Arts, Arizona State University, www.asu.edu/cfa/art/people/faculty/collins/Kac/ (accessed on June 25, 2021).
De Menezes, M. (2013), "DECON: Deconstruction, Decontamination, Decomposition," https://martademenezes.com/art/life-cycles/decon/ (accessed on June 25, 2021).
Dunn, R. (2011), "Painting With Penicillin: Alexander Fleming's Germ Art," *Smithsonian*, www.smithsonianmag.com/science-nature/painting-with-penicillin-alexander-flemings-germ-art-1761496/ (accessed on June 25, 2021).
Emmert, E. A. B. (2013), "Biosafety Guidelines for Handling Microorganisms in the Teaching Laboratory: Development and Rationale," *Journal of Microbiology & Biology Education* 14 (1): 78–83, https://journals.asm.org/doi/10.1128/jmbe.v14i1.531 (accessed on June 25, 2021).
Fara, P. (2009), "A Microscopic Reality Tale," *Nature* 459: 642–644, www.nature.com/articles/459642a (accessed on June 25, 2021).
Garrod, L. P. (1960), "Relative Antibacterial Activity of Three Penicillins," *British Medical Journal* 1 (5172): 527–29, doi:10.1136/bmj.1.5172.527 (accessed on June 25, 2021).
Ginsberg, D. (2009), "*E.chromi*: Living Colour from Bacteria," www.daisyginsberg.com/projects/echromi.html.
Gulden, A.C. (2019), "Working with Live Painting Materials," http://amychasegulden.blogspot.com/2007/08/working-with-live-painting-materials.html#links (accessed on June 25, 2021).
Henderson, F. (2016), "'The most ingenious book that ever I read in my life' Pepys and Micrographia," www.rmg.co.uk/stories/blog/most-ingenious-book-ever-i-read-my-life-pepys-micrographia (accessed on June 25, 2021).
Hesse, W., and D. H. M. Grschel (1992), "Walther and Angelina Hesse-Early Contributors to Bacteriology," https://jornades.uab.cat/workshopmrama/sites/jornades.uab.cat.workshopmrama/files/Hesse.pdf, *ASM News* 58 (8): 425–8 (accessed on June 25, 2021).
"iGEM." (n.d.), http://igem.org/Main_Page.
Kac, E. (2011), "Bio Art: From Genesis to Natural History of Enigma," *Imagery in 21st Century* 1997: 57–80.
Knight, J. (2003), "GloFish Casts Light on Murky Policing of Transgenic Animals," *Nature* 426 (372), www.nature.com/articles/426372b (accessed on June 25, 2021).

Lederberg, J. and A.T. McCray (2001), "''Ome Sweet 'Omics – A Genealogical Treasury of Words," *The Scientist* 15 (7): 8, https://lhncbc.nlm.nih.gov/LHC-publications/pubs/OmeSweetOmicsAGenealogicalTreasuryofWords.html (accessed on June 25, 2021).

Lissel, E. (2008), "The Return of Images: Photographic Inquiries into the Interaction of Light," *Leonardo* 41 (5): 438–45, doi:10.1162/leon.2008.41.5.438 (accessed on June 25, 2021).

Liu, G. Y., and V. Nizet (2009), "Color Me Bad: Microbial Pigments as Virulence Factors," *Trends in Microbiology* 17 (9): 406-13, https://pubmed.ncbi.nlm.nih.gov/19726196/ (accessed on June 25, 2021).

Lontok, K. (2018), "Science Outreach: Engaging, Inspiring and Welcoming Nonscientists," *Microcosm* 2 (Winter): 64–6.

Lontok, K., C. Marizzi, and T. Sturm (2018), "The Many Dimensions of Agar Art," *SciArt* 34, www.sciartmagazine.com/december2018contents.html (accessed on June 25, 2021).

Malik, K., J. Tokkas, and S. Goyal (2012), "Microbial Pigments: A Review," *International Journal of Microbial Resource Technology* 41 (4): 361–5, www.researchgate.net/publication/324835431_Microbial_Pigments_A_review (accessed on June 25, 2021).

Marizzi, C. (2016), "Painting with Microbes Brings an Unseen World Into Public View," Labdish blog, Cold Spring Harbor Laboratory, www.cshl.edu/labdish/painting-with-microbes-brings-an-unseen-world-into-public-view/ (accessed on June 25, 2021).

NIH, (2019), "(Major Actions Sect III-A) NIH Guidelines for Research Involving Recombinant or Synthetic Nucleic Acid Molecules," NIH Guidelines, March: 1–4. https://osp.od.nih.gov/biotechnology/nih-guidelines/ (accessed on June 25, 2021).

Nobel.org. (2008), "The Nobel Prize in Chemistry 2008: 'for the Discovery and Development of the Green Fluorescent Protein, GFP,'" Nobelprize.org 38 (10): 2821–2, doi:10.1039/b917331p (accessed on June 25, 2021).

Park, S. (2019), "Microbial Art," www.microbialart.com/microbes/ (accessed on June 25, 2021).

Sender, R., S. Fuchs, and R. Milo (2016), "Are We Really Vastly Outnumbered? Revisiting the Ratio of Bacterial to Host Cells in Humans," *Cell* 164(3): 337–40, *doi:10.1016/j.cell.2016.01.013* (accessed on June 25, 2021).

Smith, A. (2005), "History of the Agar Plate," Laboratory News, www.labnews.co.uk/article/2029646/history_of_the_agar_plate (accessed on June 25, 2021).

Soliev, A. B., K. Hosokawa, and K. Enomoto (2011), "Bioactive Pigments from Marine Bacteria: Applications and Physiological Roles," *Evidence-Based Complementary and Alternative Medicine*, doi:10.1155/2011/670349 (accessed on June 25, 2021).

Stocker, G. and R. Nestelbacher (2008), "GFPixel Living Portrait," www.turbulence.org/blog/archives/002071.html (accessed on June 25, 2021).

Tan, S. Y., and Y. Tatsumura (2015), "Alexander Fleming (1881–1955): Discoverer of Penicillin," *Singapore Medical Journal* 56 (7): 366-7, doi:10.11622/smedj.2015105 (accessed on June 25, 2021).

Tuli, H. S., P. Chaudhary, V. Beniwal, and A. K. Sharma (2015), "Microbial Pigments as Natural Color Sources: Current Trends and Future Perspectives," *Journal of Food Science and Technology* 52(8): 4669-78, doi:10.1007/s13197-014-1601-6 (accessed on June 25, 2021).

Wynn-Williams, D. D., H. G.M. Edwards, E. M. Newton, and J. M. Holder (2002), "Pigmentation as a Survival Strategy for Ancient and Modern Photosynthetic Microbes under High Ultraviolet Stress on Planetary Surfaces," *International Journal of Astrobiology* 1 (1): 39–49, doi:10.1017/S1473550402001039 (accessed on June 25, 2021).

12 Making Justice with Biodesign: A Pedagogical Approach

DEEPA BUTOLIYA

Introduction

Biodesign is an emerging, futures-driven field within design. Building and expanding upon the existing practice of biomimicry, and situated in between bioart and biotechnology, biodesign generates both working artifacts and visions for the future that incorporate facets of design and biology. Speculative biodesign, which imagines possible futures as a "catalyst for change," considers imaginaries for a future in which we co-exist and co-create with agents of and processes in nature (Dunne and Raby 2013: 33). In these ways—through both real-world biodesign artifacts and speculative biodesign projects—biodesign explores design-enabled futures in which living organisms, biological processes, and living systems play a significant role in the lives of people.

I teach a biodesign class at Stamps School of Art and Design at the University of Michigan. When designing this class, I began by asking: in what ways can we incorporate social justice into biodesign? Using this framework, in this essay I present my social-justice oriented approach to the pedagogy of biodesign in a design studio setting. I define a social-justice oriented approach for my students as a design mindset that takes into consideration how biodesign solutions, artifacts, systems, and services can be valuable for communities. My social-justice oriented approach also teaches students how to interrogate and critique potentially repressive power structures and practices, for example, police-enforced facial recognition that may be used to target vulnerable people.

As part of the class coursework, students submit projects to the Biodesign Challenge (BDC).[1] The Biodesign Challenge is "an education program and competition" whose intention is to shape the "first generation of biodesigners." BDC is an acclaimed platform in which an international group of students, designers, artists, and researchers present and discuss emerging biodesign ideas and projects. University and high-school teams submit projects ahead of the annual BDC summit, which is held each year in June. The winning team is chosen by a dedicated panel of judges over several rounds of judging and presentations. BDC is also a thriving community where many like-minded people meet to further the field of biodesign.

Design and Science

The University of Michigan is a major liberal arts university, which is also a leading research university in the sciences and technology. Working in this environment, I have come to realize that it is crucial to explore the role that science both has played and continues to play in design. For much of the twentieth century, the ways that science influenced design was driven by principles and practices that came out of the Design Methods Movement (DMM) of the 1960s and 1970s. The founders of this movement attempted to "use techniques . . . to make design more 'scientific' in . . . industrial design, architecture, and town planning" (Langrish 2016).[2] According to design theorist Nigel Cross (1993), the central precept of DMM was to apply scientific methods to the design process. The paramount goal of the movement was to produce objective and rational design that would play a critical role in industrial optimization. Cross explains that "the origins of the emergence of the new design methods in the 1950s and 1960s lay in the application of the novel scientific methods to the novel and pressing problems of the second world war" (1993: 16). Cross's claim reaffirms that the "scientification" of design in DMM concentrated on corporatization, militarization, and industrialization.

Around the beginning of the twenty-first century, some new work began to explore how biological processes could prompt innovative design solutions. In *Design Paradigms* (2000), design researcher Warren Wake suggests that plant, animal, and human bodily systems and structures could serve as paradigms for design problem solving and form development. The concept of biomimicry (Benyus 1997) similarly suggests drawing inspiration from the design of the natural world. Biomimicry has led to innovations such as Velcro, which was designed to mimic the microstructures found in the Gecko lizard's feet. Like design based in the DMM, biomimicry tends to focus on the manufacture of designed products rather than on analyzing and addressing critical needs of communities.

Biodesign, though, frequently confronts the instrumental, objective-driven ways that DMM and biomimicry use science to produce designed artifacts as final outcomes. Over the past two decades, biodesigners have instead explored biology and design together as a combined medium for investigating and addressing sociopolitical concerns, and often for enabling equity and justice. Post-normal science (PNS) is a contemporary approach to science that has direct relevance to sociopolitical-oriented biodesign—and to my biodesign course. In PNS, the stakes and uncertainties of scientific decisions are considered in their social contexts. PNS also disavows the traditional model of applied science in which the goal is to be value-free and to work toward certainty in scientific outcomes (Ravetz and Funtowicz 1999). Finally, PNS also engages with extended peer-review communities, such as citizen scientists. In citizen science everyday people participate in scientific research. This approach is particularly useful in uncertain social and political situations because those who are impacted by the uncertain situations can be involved in crafting solutions to them. I introduce PNS and the social impacts of citizen science in my class.

Speculative and Critical Biodesign and Design Fiction

Many biodesigners challenge traditional design methods, products, and services, and provoke public debate through speculative and critical design (SCD) practices. Speculative design "envisions solutions for a speculative future . . . at a societal scale" and this kind of work "goes beyond the creation of an individual object for use" (Craswell 2020). Critical biodesign uses "speculative design proposals to challenge narrow assumptions, preconceptions, and givens about the role products play in everyday life" (Dunne and Raby 2013). SCD, then, is primarily focused on critical aspects of designed artifacts and on speculating shared futures—for example investigating the futures of emerging biotechnologies (Dunne and Raby 2013). I developed my biodesign course using SCD as an investigative design framework.

Speculative and critical biodesigners frequently collaborate with scientists—some of whom are synthetic biologists—to explore how scientific discoveries and emerging biotechnologies may be manifested in responsive artifacts and systems. Synthetic biology involves redesigning organisms by engineering them to have new abilities.[3] Working with synthetic biologists establishes and clarifies the roles that biological organisms play in emerging biotechnologies, and the effects that these technologies may have on people and other organisms. SCD methodology, then, compels designers—and the scientists who work with them—to debate and discuss the pros and cons of specific biotechnologies. This flexible approach also works well when designers analyze and modify existing biotechnologies. In short, working with scientists using a "futures" lens encourages biodesigners to create flexible solutions that are shaped by biotechnology—which utilizes biological systems or living organisms to develop products and systems—and synthetic biology that also addresses lived realities.

Despite the exciting possibilities that biodesign and biotechnology projects can offer, however, it is crucial to question controversial aspects of emerging biotechnologies. CRISPR, for example, is a technology that is used to edit genes and has revolutionized approaches in health and medicine. Its gene-editing function promises to cure diseases in the future.[4] But this technology can have negative implications—for instance, individuals may disturb the natural order by predetermining gender, physical traits, or athletic abilities in their offspring. This problematic use of CRISPR could create further disparities for marginalized communities and groups. And new studies suggest that CRISPR may cause more damage to genes than previously thought (Wu 2020). Many researchers who carry out bioscience research or develop biotechnologies, in fact, do not fully take into account the negative social, political, economic, and ethical impacts of their projects. It is therefore imperative to interrogate biotechnological developments and also to create alternative imaginaries in biodesign that are rooted in social justice, especially for those at the margins of society.

SCD, as I have already noted, uses artifacts and systems to question both narrow assumptions about, and the futures suggested by, these artifacts and systems. In the related practice of design fiction, a term often attributed to Julian Bleecker of the Near Future Laboratory, artifacts are designed based on emerging technologies. Bleecker defines design fiction as "making things that tell stories" (Bleecker 2009). Science fiction writer Bruce Sterling writes that design fiction involves "the deliberate use of diegetic prototypes to suspend disbelief about change" (Sterling 2009). Design fiction is like science fiction in that it helps us to imagine future scenarios.

About the Course

I bring all of the ideas, concerns, and deliberations considered above to my biodesign class. My students use speculative and critical biodesign, design fiction processes, and the principles of PNS to explore the implications of future biotechnological solutions for marginalized communities. I first taught biodesign in the winter semester of 2019, and the course followed the standard timeline of a sixteen-week university course for three credit hours. I divided the class into four teams. There were three teams with five students and one team with four.

Each team produced a design project to be entered into BDC 2019, and one team among the four was actually selected as a finalist for the final round of BDC projects. For the finalist team, the work on the project extended beyond the regular term into the summer of 2019. During the summer, the team members refined their ideas and then exhibited and presented their work at MoMA (Museum of Modern Art) in New York and at the biodesign summit, which was held at The New School in New York City in June 2019. The class was also supported by the expert advice of Dr. Marcus Ammerlaan, who is a Collegiate Lecturer in the Department of Molecular, Cellular, and Developmental Biology at the University of Michigan.

Course Description

I run my biodesign course as a design research studio that focuses on the intersection between design and science, and biodesign and biotechnology. It is taught as a hybrid of a design seminar and design studio: as a mix of reading, dialogue, and making. The first half of the class includes lectures, weekly readings and discussions, project overviews, and workshops on relevant topics—such as pivotal examples of biodesign work, and readings about PNS and the principles of synthetic biology—that demonstrate how students might explore the implications of biotechnology.

The readings, discussions, and workshops also help to prepare students for BDC. BDC introduces students to the ways that biotechnology, biology, and design intersect. Over the course of the winter 2019 semester, BDC organizers worked hand-in-hand with me on my course curriculum, and paired students with expert consultants, such as synthetic biologists, ecologists, life-cycle analysts, and other relevant subject-matter experts. At mid-semester, my students broke up into small teams to begin producing designs that address social, cultural, economic, and political issues. The support from the BDC people, and the expert guidance they offered in evolutionary biological sciences—along with the expertise of University of Michigan scientist Dr. Marcus Ammerlaan—enabled the class to ground their work solidly in scientific research.

What the Students Brought to the Class

The seventeen students were seniors and juniors who were divided into four teams based on their backgrounds. Many of them had art and design experience, but there were a few students who came from other academic backgrounds, such as biology. We made an effort to distribute students among the groups based on skills and experience, gender, and compatibility. Besides their academic backgrounds, the students brought enthusiasm and a willingness to explore new territory: for the artists and designers this new territory was biology, and for life sciences students it was design.

What I Brought to the Class

In my own research, which I call critical *jugaad*, I investigate futures for marginalized people (Butoliya 2016). I coined the term critical *jugaad* to describe the DIY making and doing processes of people in marginalized communities who engage in these activities in order to build resilience in the face of adversity. These practices present an alternative way of being in the world that is represented by the Hindi term *jugaad*, which means a quick fix, making, making-do, and survival. *Jugaad* typically infers a pragmatic workaround and a subaltern social practice (Rai 2015).

In my work, I adapt SCD approaches to fit the modes of being and doing in *jugaad*.[5] SCD are good for analyzing emerging technologies in the West, and possibly to suggest alternatives—but these methods seldom provide alternatives for peoples and technologies at the margins of Eurocentric capitalist Western societies (Tonkinwise 2014). Critical *jugaad*, on the other hand, brings to the forefront the making, making-do, and survival-oriented designerly practices of the Global South. My adaptation of speculative and critical approaches spotlights how these South Asian practices differ from the Western understanding of design.

Critical *jugaad* is a frugal, decolonial, and inclusive equal of SCD. In Western SCD human and animal waste is represented as a possible future energy source. Speculative designers Anthony Dunne and Fiona Raby's project, *Our Energy Futures* (2005), for example, includes a "poo lunch box" with two compartments—one for lunch and one for fecal waste. This poo lunch box speculates about futures in which children bring their fecal waste back home after lunch because it is a valuable energy source. This idea is not new outside of the West. Animal waste has been used as sustainable biomass fuel in agrarian countries such as India for centuries. Western speculative projects, such as the poo lunch box, overlook the fact that in developing countries many everyday people have come up with various methods for using waste as fuel. In my critical *jugaad* research, I point out these sorts of Western oversights.

I also present critical alternatives to their solutions, for instance, how modern-day Hindus buy cow dung biofuel online on Amazon for rituals. As critical *jugaad*, this practice appropriates and subverts current capitalist systems of e-commerce. My critical *jugaad* research has also helped to solidify my interest in teaching students to create biodesign artifacts, systems, and services that are decolonial and produced using social-justice considerations.[6]

The Process

My biodesign class emphasizes how process is a crucial part of design. The associations between design and biology are fairly limited in conventional knowledge systems. Traditional designers don't use biology in their work, and scientists typically don't use design, except occasionally as a visual style add-on to data or new instruments. To begin designing for future critical issues, students must first unlearn their preconceived notions of what design is and how design and the biological sciences relate to one another.

They also need to develop an awareness of biases in the Western paradigm for imagining futures, in which most designed artifacts and systems are centered on privileged, middle-class, cis-gendered, heteronormative white males. The students analyzed case studies, seminars, and readings on bioart, biotechnology, synthetic biology, citizen science, and PNS. They examined, in critical fashion, the work of bioartists and biodesigners, such as Agi Haines, Alexandra Daisy Ginsberg, and Neri Oxman. I highlighted biases and disparities in design practice in this speculative design and critical design work and in course readings. I shared and demonstrated ways that they could operate with an ethos of social justice while creating solutions for existing problems and issues in predominantly Western societies. Then, as students chose their project topics, I asked them to consider who they felt was marginalized in the problems that they were exploring. Finally, we began our initial brainstorming and idea-generation exercises collectively as a class, which reinforced our plan to compete in a healthy and supportive environment.

Idea Generation and Making Games

Generative-design methods and tools encourage flexibility, critical thinking, and creative ideation by enhancing cognitive abilities for designers, and helping to stimulate a large number of design possibilities. Interdisciplinary and generative problem solving in design, co-design, human-computer interaction, and data collection commonly use innovative methods, such as cultural probes, to enhance ideation (Gaver, Dunne, and Pacenti 1999). In cultural probes, participants observe and reflect upon experiences around a design solution and the community in which it will be used. Cultural probes are tools that are designed to engage with the community. I had the student teams design generative tools that were also cultural probes. They studied certain existing tools and methods from the speculative design literature for analyzing and generating ideas: *The Thing from the Future* by Situation Lab,[7] *Syn Bio Tarot Reading*,[8] and *Instant Archetypes: A New Tarot For The New Normal* by Superflux studio.[9] *The Thing from the Future* is a collaborative card game designed by Stuart Candy that uses prompts from the card decks to generate visions of alternate futures. *Syn Bio Tarot Reading*, which is designed by UK-based Superflux studio, is a design-fiction-generating card game that is based on synthetic biology and emerging biotechnologies. *Instant Archetypes* is developed by Superflux as well. It is a strategy game in which "Anyone looking for a deeper understanding of the world around them can use the cards to explore their place within it. The cards are a useful way for shifting perspectives, problem solving, block dislodging, strategy finding: a toolkit for exploring plural futures" (Superflux).

Studying these existing methods invigorated and motivated the students' thinking, and helped the teams to create games that generated new ideas (and also could be used by others to help generate ideas). As we explored ideas for this course, and crafted the conceptual spaces in which the projects would develop, the teams also shared the brainstorming tools and methods that they used with the other teams. The tools were tested by all of the teams and by outside experts as well. Many of these tools were small projects that dealt with topics such as futures of food, communication, energy, and transportation—some of which I present below.

Figs. 12.1 and 12.2 illustrate tools that were designed by teams in the biodesign class that can help other design teams to come up with innovative solutions. Fig. 12.1 shows a card game for futures of food, and Fig. 12.2 depicts a game in which participants select word prompts randomly from different category bags. Participants create a combination of those words, one from each category bag.

In this tool-creating part of the project, students designed useful artifacts that contributed to both their group research and the meta-design process for their whole project. In retrospect, this part of the project, which actually requires a semester's worth of work, took time away from developing the final team projects. However, it was worth the time. The tool creating was a critical part of the project; it sparked enriching discussions about, and added conceptual rigor to, the final outcomes.

Final Concepts

After investigating critical theories and concepts, and reflecting upon and critiquing ideas that the students generated, the class finalized four concepts. The students produced artifacts and scenarios to visualize their concepts, and websites to explain their concepts and outcomes to others. Biologist Marcus Ammerlaan helped to ensure that the students investigated biological-science-based ideas and applied them to their design projects in well-reasoned ways. One of the requirements for the BDC competition is to use existing scientific evidence as proof of the project concept. Ammerlaan worked with each of the design teams, validating their scientific concepts, and bringing in his own graduate students for co-ideation sessions to help make sure that my students employed scientifically robust ideas. Dr. Ammerlaan's experience

Fig. 12.1 A biodesign card game, *Food for 1,000 years*. Designed by students from the biodesign class at the Stamps School of Art and Design, University of Michigan, 2019.

with a competition that is similar to BDC (the iGEM competition) allowed him to guide my students in ways to approach competitive situations.[10]

Project 1: *Ecto*

Team members: Kelsey Segasser, Yiyi Gao, Justine Abbo, Liana Smale, Henry Olstein

Ecto is a speculative design concept for a wearable device that addresses the social issues of opioid addiction and over-prescription. In the project mission statement, the student team proposed to "revolutionize the pain medication industry" in order "to

Artifact:

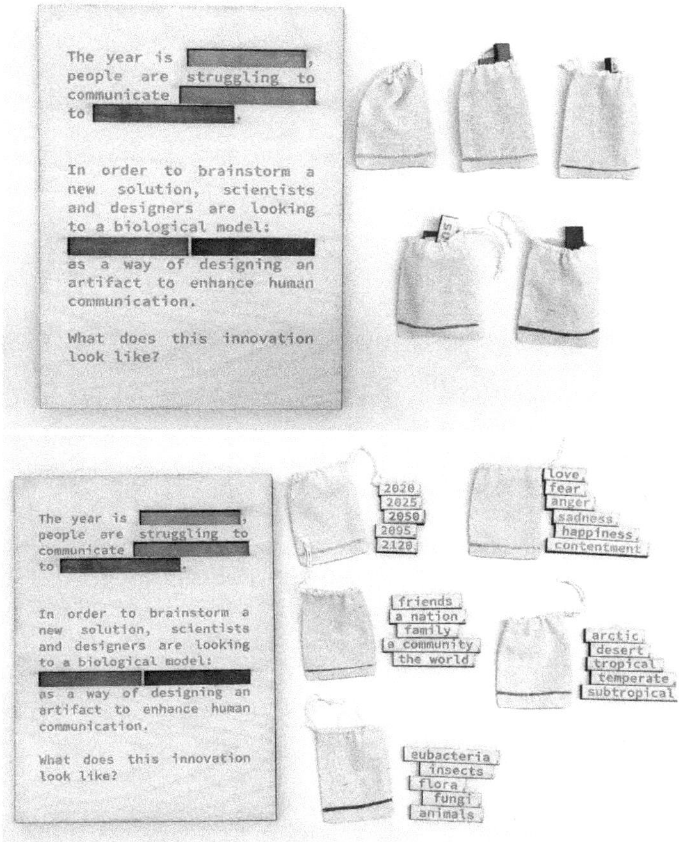

Fig. 12.2 A board game designed for exploring biodesign futures, designed by students from the biodesign class at the Stamps School of Art and Design, University of Michigan, 2019.

eliminate addicting drugs & habits." Their concept was for a device that administers a biobased alternative to addictive pain medication, especially opioids.

This project takes a critical stance on recent medical malpractices, such as over-prescription of opioids and the life-threatening side effects of such practices. The design team adopted the familiar form language of popular wearable devices, while simultaneously introducing the provocative concept of using a genetically modified insect as an agent of pain management. Their wearable device houses a genetically engineered tick that can administer a non-addictive, long-acting dose of a toxin-TTX (Tetrodotoxin). TTX is a small-molecule blocker of voltage-gated sodium channels on nerves that conduct pain signals. This speculative design idea builds on existing scientific research carried out at WEX Pharmaceuticals Inc. of Vancouver, BC—a

Fig. 12.3 ***Ecto.*** **A speculative biodesign wearable concept by team Ecto students in the biodesign class at the Stamps School of Art and Design, University of Michigan. Submitted to the BDC 2019.**

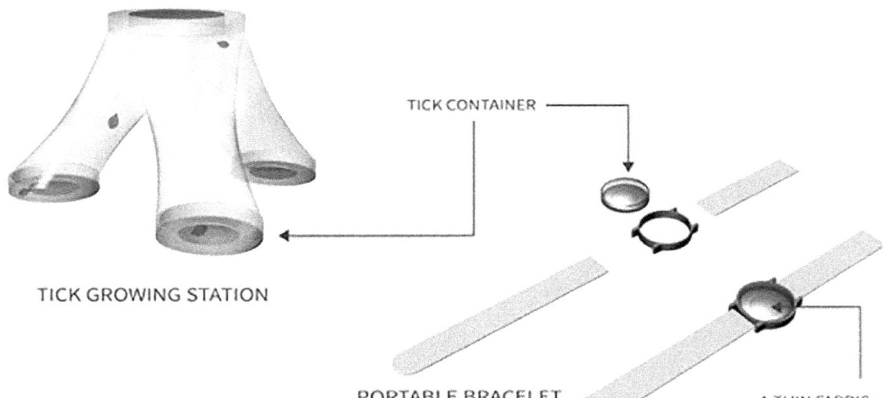

Fig. 12.4 Illustrations for the tick housing station and wearable device concept for ***Ecto.*** **Designed by students from the biodesign class at the Stamps School of Art and Design, University of Michigan, 2019.**

company that is testing the toxin TTX as a potential pain-relief treatment.[11] Unlike morphine, TTX doesn't cause opioid-like side effects.

The ticks in this project are engineered to bite when they sense that nociceptors, a sensory receptor for pain stimuli, have been stimulated, and the material at the bottom of the casing is a fabric that enables the tick to administer its dose. Those who use this alternative pain management wearable device can breed and grow the ticks in their own homes, as illustrated in Fig. 12. 4. Using an artifact and accompanying website, the design creates a scenario that provokes viewers to imagine using living, visible biological agents in medical treatment and life support. Their website elaborated the team design concept.

Project 2: *Chitonet*

Team members: Gillian Yerington, Chelsea Markus, Siena Mckim, Sarah McNamara

This project was more community based than the other three project ideas presented in the class because it used biodesign to deal with the invasive species problems in the local ecosystem. This project employed community participation, which included input on methods and ideas from the community, as a way to create equitable futures, and the project was based on a slow-science paradigm. Slow science is a movement that encourages slow and thoughtful scientific work and pushes back against quick, simplistic solutions to complex problems.[12]

The *Chitonet* project participants—who questioned the implications of our relationships to invasive species—proposed a fishing net made from cattails, which is an invasive grass found in Michigan lakes. The fishing net is coated with chitosan, a naturally derived coating material made from crayfish shells, which is yet another invasive species found in Michigan lakes (see Fig. 12.5). This project also devises a fictional nonprofit that teaches local fishermen to make these sustainable nets and avoid plastic fishing nets, and it reconnects people to the local environment and traditional fishing practices. The team used the chitosan-coated net, and information

Fig. 12.5 ***Chitonet*. This net, which is made from cattails coated with chitosan, is part of a biodesign proposal that was conceptualized by a student team from the biodesign class at the Stamps School of Art and Design, University of Michigan. Submitted to the BDC 2019.**

from interviews with experts in the Michigan Department of Natural Resources about existing practices and environmental policies, as proof of the scientific basis of their project concept. This project demonstrated that biodesign is about creating design artifacts that are provocative, strange, or artistic. But this project also proclaims that a well-rounded concept should consider the context of the idea and its impact on the local community while being sensitive to ecology and culture. Interviews with local fishermen, and learning about their current practices, broadened the assumptions that designers might have held about the situation and its contexts.

Project 3: *Lumiflage*

Team members: Olivia Keener, Joseph Mandel, Madalyn Lelli

Lumiflage is a speculative cosmetics line, shown in Fig. 12.6, that uses bioluminescence to protect against unwanted facial recognition surveillance technologies. This speculative artifact is a critical and discursive design object that prompts scrutiny of government use of surveillance technologies to track citizens. The team presented the concept in a video document that explains how using *Lumiflage* cosmetics can prevent detection by surveillance technologies (Lumiflage cosmetics video: www.youtube.com/watch?v=yhv-F5I9Sfl). This video also offers critical commentary on the contemporary lack of privacy, and describes how to take back digital privacy by using *Lumiflage* cosmetics.

Fig. 12.6 Concept prototypes for *Lumiflage*, a speculative biodesign proposal from a team of students from the biodesign class at the Stamps School of Art and Design, University of Michigan. Submitted to the BDC 2019.

The students worked with scientists at the University of Michigan who were researching how to synthesize luciferin. Luciferins are light-emitting compounds that come from bioluminescent organisms. They give off light in the presence of luciferase enzymes, which catalyze the light-emitting reaction. Using these same scientific processes to convert luciferin by luciferase to light emitting chemicals, *Lumiflage* cosmetics give off a bioluminescent glow when the luciferase is activated by a flash of light or the movement of facial muscles. This glow disrupts the ways surveillance technologies use facial recognition. The rigor of this group's scientific investigation was bolstered by existing research at the University of Michigan.

Project 4: *Gnosis*

Team members: Jessie Adler, Robert Gotham, Michael Moon, Alexa Smith, Evan Vollick-Offer

Project *Gnosis* developed condoms that are coated inside with a natural lubricant that detects sexually transmitted infections (STIs) in male users. The lubricant is made from flavin biochrome, which is isolated by boiling red cabbage. When it is worn, the lubricant-coated condom changes color if it comes into contact with acidity from the STIs Herpes Simplex Virus 1 and 2, syphilis, chancroid, and genital warts. All of these STIs are more acidic than unaffected genital skin, as illustrated in Fig. 12.7.

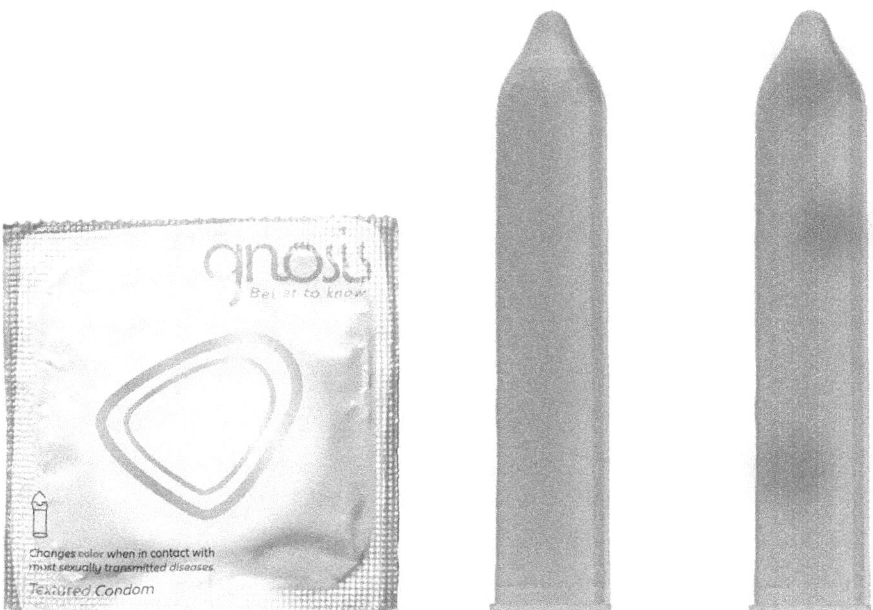

Fig. 12.7 Prototype of packaging for STI-detecting condom, *Gnosis*, and illustration for the concept by a team of students from the biodesign class at the Stamps School of Art and Design, University of Michigan. Submitted to the BDC 2019.

Gnosis was selected as a finalist in the BDC competition because it attended to the social stigma around STIs. This project was provocative and impactful. It proposed a solution that is based on simple biotechnology processes, and it offered contextual awareness through community engagement workshops that educated university students about sexual health. The team also sought external input from university student groups about ways to structure sexual health education for *Gnosis* workshops. As finalists, the *Gnosis* team presented the idea at the BDC Summit, and exhibited an installation of the concept at the final juried show. Fig. 12.8 shows the *Gnosis* team display at the BDC 2019 exhibition.

Gnosis was also selected by guest critics and me as the class finalist based on the originality of the group's scientific concept and their attention to the product's sustainability: the condoms were recyclable and the lubricant was made from bioavailable compounds. This group's provocative project proposed a condom used for detection rather than prevention, a move that helps to reduce stigmas around STIs and encourages users—especially young adults—to have agency over their sexual health. Although this project idea did not win the BDC competition, it generated several conversations at the BDC Summit among participants and attendees.

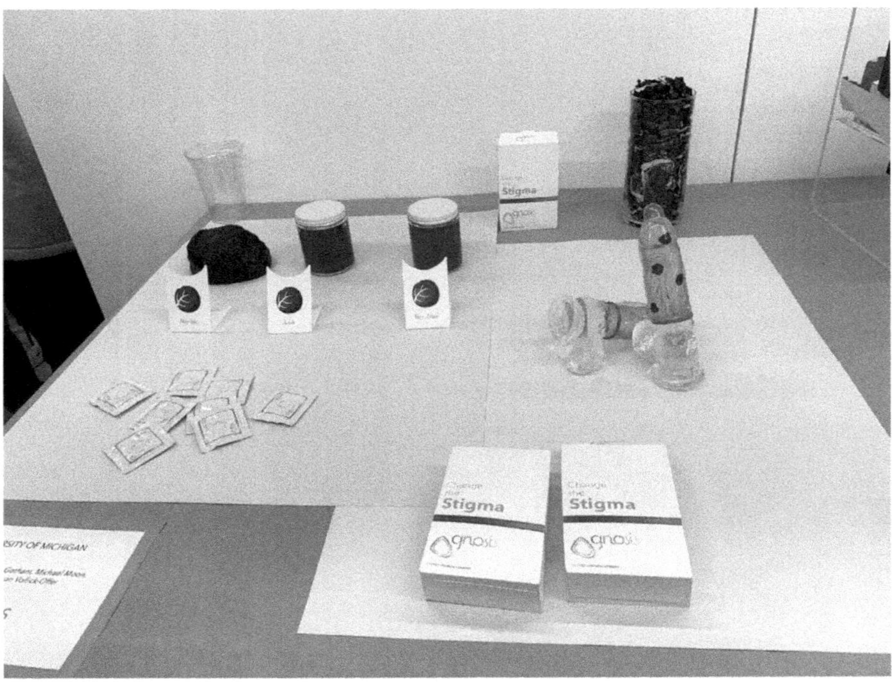

Fig. 12.8 ***Gnosis*** **concept from the biodesign class in an exhibition at The New School, New York, BDC 2019.**

Conclusion

Delving into the pedagogy of biodesign suggested many possibilities to both the students and I for exploring the potential of this creative practice. Teaching biodesign also presented many challenges, in particular, questions about how design students should engage with scientific protocols in the absence of scientific training, and with limited lab facilities. My class met these challenges with resilience, which led to an open design mindset that included reaching out to relevant experts and tapping suitable resources at our research-oriented university.

Merely creating interesting critical and speculative design is not enough. Biodesigners and biodesign-oriented practices must have a real impact on users and communities. The projects *Chitonet* and *Gnosis*, in particular, embodied these principles—one dealing with the issue of invasive species, and the other with diagnosing and destigmatizing STIs in certain communities. These examples, and the various learning, critiquing, thinking, and making processes that we carried out in my biodesign course, reflect just some possible ways to encourage a strong ethos of social justice in biodesign work. A self-reflective, critical, adventurous, and thoughtful approach to teaching biodesign students will turn out (bio)designers who contribute to a just and equitable society. These biodesigners will understand how critical it is to create artifacts that are, at once, productive implements for social justice and creative prospects for biodesign. Finally, I would like to share with aspiring biodesigners the possibility that their projects can be the medium for both resolving and addressing social issues and questioning the future of emergent biotechnologies. The radical engagement of design and science in such projects yields avenues for further explorations in the field, and enables students to query the futures that we may well inhabit.

Notes

1. The competition has been held since 2016, and the website has winning and non-winning entries from all those years of inspiring new ideas and minds. The group is led by Daniel Gruskin and associates who also provide course materials, hold webinars from experts, and guide instructors and students through the entire process. For more information visit the Biodesign Challenge website. See www.biodesignchallenge.org.
2. Bruce Archer, John Chris Jones, Christopher Alexander, and Horst Rittel are considered to me the founders of the Design Methods Movement, although there were certainly other people contributing as well.
3. Definition from https://www.genome.gov/about-genomics/policy-issues/Synthetic-Biology.
4. CRISPR is an acronym for Clustered Regularly Interspaced Short Palindromic Repeats. For more information, see crisprtx.com/gene-editing/crispr-cas9.
5. https://medium.com/post-normal-design-post-speculative-critical/speculative-and-critical-design-futures-and-imaginings-from-the-margins-fall-2017-699531ead23.

6. Decolonial artifacts are based on the notion of decoloniality, which examines the relationships between Frankfurt School critical theory and the paradigm of modernity/coloniality (Bhambra 2014). Decoloniality asks: "What transformations are needed in the 'critical theory' project if gender, race, and nature were to be fully incorporated into its conceptual and political framework?" (Mignolo 2007: 449). In decoloniality theory, Mignolo situates modernity in a radical paradigm of coloniality, which he claims is bigger than, and inseparable from, concepts of imperialism and colonialism (2007).
7. http://situationlab.org/futurething-print-and-play-edition/.
8. https://superflux.in/index.php/work/synbio-tarot-reading/.
9. https://superflux.myshopify.com/.
10. iGEM is "an independent, non-profit organization dedicated to the advancement of synthetic biology, education, and competition, and the development of an open community and collaboration," https://igem.org/Main_Page (accessed on May 25, 2021). See essays in this volume by Geaney, and also Bar-Shai and Marizzi.
11. For more information see https://wexpharma.com/clinical-trials/ongoing-trials/.
12. For more information, see http://slow-science.org/.

References

Benyus, J.M. (1997), *Biomimicry: Innovation Inspired by Nature*. New York: Morrow.

BDC, biodesignchallenge.org (accessed on May 20, 2021).

Bhambra, G.K. (2014), "Postcolonial and decolonial dialogues," *Postcolonial Studies* 17 (2): 115–21.

Bleecker, J. (2009), "Design Fiction: A Short Essay on Design, Science, Fact, and Fiction," Near future laboratory, https://blog.nearfuturelaboratory.com/2009/03/17/design-fiction-a-short-essay-on-design-science-fact-and-fiction/ (accessed on June 25, 2021).

Butoliya, D. (2016), "Critical Jugaad," In *Ethnographic Praxis in Industry Conference Proceedings* 2016 (1): 544, https://doi.org/10.1111/1559-8918.2016.01118 (accessed on June 25, 2021).

Craswell, P. (2020), "What is Speculative Design?" *The Design Writer,* https://thedesignwriter.com.au/what-is-speculative-design/ (accessed on June 2, 2021).

Cross, N. (1993), "A history of design methodology," in de Vries, M. J., N. Cross, D.P. Grant, (eds.), *Design Methodology and Relationships with Science*, 15–27, New York: Springer.

Dunne, A. and F. Raby (2013), *Speculative Everything: Design, Fiction, and Social Dreaming*. Cambridge: MIT Press.

Funtowicz, S. O. and J.R. Ravetz (1993), "Science for the post-normal age," *Futures* 25 (7): 739–55.

Gaver, B., T. Dunne, and E. Pacenti (1999), "Design: Cultural Probes," *Interactions* 6 (1): 21–9.

Langrish, J.Z. (2016), "The Design Methods Movement: From Optimism to Darwinism," Proceedings of the Design Research Society Conference 2016, https://static1.squarespace.com/static/55ca3eafe4b05bb65abd54ff/t/574f0971859fd01f18ec63c1/1464797554420/222+Langrish.pdf (accessed on June 25, 2021).

Mignolo, W.D. (2007), "Delinking: The rhetoric of modernity, the logic of coloniality and the grammar of de-coloniality," *Cultural Studies* 21 (2–3): 449–514.

Rai, A.S. (2015), "The affect of Jugaad: Frugal innovation and postcolonial practice in India's mobile phone ecology," *Environment and Planning D: Society and Space* 33 (6): 985–1002.

Sterling, B. (2009), "Design fiction," *Interactions* 16 (3): 20–4.

Superflux, https://superflux.myshopify.com/collections/frontpage/products/instant-archetypes-a-new-tarot-for-the-new-normal (accessed on June 1, 2021).

Tonkinwise, C. (2014), "How We Intend to Future: Review of Anthony Dunne and Fiona Raby, *Speculative Everything: Design, Fiction, and Social Dreaming*," *Design Philosophy Papers* 12 (2): 169–87.

Wake, W.K. (2000), *Design Paradigms: A Sourcebook for Creative Visualization*. New York: John Wiley & Sons.

Wu, K. (2020), "Crispr Gene Editing Can Cause Unwanted Changes in Human Embryos, Study Finds," www.nytimes.com/2020/10/31/health/crispr-genetics-embryos.html (accessed on June 3, 2021).

Nicola, "Instant Archetypes: A New Tarot For The New Normal" (2021), Superflux blog, https://superflux.myshopify.com/collections/frontpage/products/instant-archetypes-a-new-tarot-for-the-new-normal (accessed on May 26, 2021).

Part 4 Data Manifestation in Design and Science

13 Data Manifestation: Climate Change Data in the Home and on the Body

KARIN VON OMPTEDA

Introduction

Our everyday interactions with data visualization are usually through the line, bar, and pie charts that have become ubiquitous in our visual culture. These methods of displaying quantitative data are still the most frequent sorts of information graphics, even after being used for more than 200 years (Mollerup 2015). Our understanding of climate change data, therefore, is still primarily communicated through these traditional data visualization methods, for example, the line charts of global temperature rise and sea level rise occurring over the past century (e.g., NASA 2021a). Although we may acquire some sense of climate change through such charts, this form of transmitting information isn't fully effective for comprehending a process that occurs on a global scale, changes at a relatively slow pace, and isn't immediately tangible.

Charts also fall far short for connecting global, long-term, and intangible information to our everyday lives, and for helping us to engage with this information emotionally. Engaging people emotionally on climate change can be difficult: "People actually work to *avoid acknowledging disturbing information*," sociologist Kari Marie Norgaard writes in the *Oxford Handbook of Climate Change and Society*, "to avoid emotions of fear, guilt, and helplessness" (2013: 400). In this essay, I present a method of data visualization that endeavors to circumvent triggering avoidance responses, while it also makes climate change relatable to people by engaging them through sensorial data objects.

Data visualization is traditionally viewed as a scientific tool for analysis, and is commonly understood to be neutral, dispassionate, and objective (Vande Moere and Patel 2010). Over the past decade, however, artists and designers have been engaged in data visualization as a creative practice. *Artistic* data visualization, a term that was coined in 2007, is a subjective practice that employs a range of varied visualization techniques to express points of view and to persuade (Viégas and Wattenberg 2007). During this expansion of the ways that data can be visualized as a creative practice, there has also been a proliferation of unexpected forms of information representation, especially representations that have been taken off of the page or screen (von Ompteda 2019a).

Data manifestation takes an artistic or designerly approach to the communication of quantitative information, and does so through the creation of objects, installations, and sensory experiences (von Ompteda 2019a). In contrast to standard data visualization, such as charts and graphs, data manifestation utilizes a process that is visual, but also physical, sensorial, and experiential. I developed a data manifestation practice beginning in 2010 while teaching at the Royal College of Art (United Kingdom), and then, since 2014 at OCAD University (Canada). Other academics working in this field use other terms to describe the physical manifestation of data: Yvonne Jansen and colleagues (2015: 3228) have proposed the term *data physicalization*, which they define as “a physical artefact whose geometry or material properties encode data.” Andrew Vande Moere (2008: 469) has used the term *physical visualization* to describe “contemporary methods that map and materialize abstract data as physical artifacts.” I continue to use the term *data manifestation*, however, because of its sensorial open-endedness (von Ompteda 2019b). The term data manifestation places less emphasis on the *physical* mapping of data, and better encompasses projects that communicate data through a range of sensory experiences (e.g., communicating data through sound).

What unique opportunities does data manifestation offer to the challenge of communicating climate change data? Almost 100 graphic design students at OCAD University have taken on this question, answering it in their own ways through the creation of design pieces using data manifestation techniques. In my third-year graphic design course, which I taught between 2017 and 2019, the students were assigned the following brief: Create a meaningful connection between people and climate change data. The resultant projects have been exhibited internationally (*Design and Science* 2019) and are featured in two recent publications (von Ompteda 2019b; von Ompteda 2022). Summarizing a report in the British medical journal *The Lancet*, *Wired* writer Adam Rogers points out that to “actually spur people to do something about it . . . Show them how climate change affects them personally” (2018). This essay presents two of my students’ projects, each of which explores contexts in which people need to personally engage with complex information. Cecilia Salcedo-Guevara brings climate change data into the home through the design of a blanket, and Justin Yoon places it on the body through the design of a t-shirt.

Climate Change and Data Manifestation

The average global temperature has risen approximately 1.18 degrees Celsius since the late-nineteenth century (NASA 2021b). Although the Earth’s climate has changed throughout history, the current warming trend is significant because it has been attributed to human activities and is occurring at a rate that is approximately ten times faster than rates of warming which followed the ice ages (NASA 2021b). Increasing global temperatures have already had observable effects on the environment, including “loss of sea ice, accelerated sea level rise, and longer, more intense heat waves” (NASA 2021c). Although the “effects on individual regions will vary over time and with the ability of different societal and environmental systems to mitigate or adapt to change," (NASA 2021c), the Intergovernmental Panel on Climate Change (IPCC) has concluded that "the net damage costs of climate change are likely to be significant and to increase over time" (IPCC 2007, 17).

As clear as the above descriptions seem to be, climate change is difficult for people to comprehend because, as noted above, it is vast, slow moving, and intangible. Science writer Phil Plait explains that "humans have a miserable sense of scale. We see what's immediately around us, and have difficulty extrapolating to the greater world" (2016). And Jeremy Talbot writes on the Global Climate Adaptation Partnership blog:

> When talking about the climate system, we have to realize that we are not dealing with something tangible. Climate is not to be seen outside the window; climate is not the weather. It is a collection of data and patterns in a statistical construct.
>
> Furthermore, climate is not here and now. Its only possible way to be perceived is through recognition of patterns, by computer modeling and, most importantly, through representations.
>
> 2015

How could the practice of data manifestation be utilized to address these communication challenges? Vande Moere (2008) and Jansen et al. (2015) offer insight into potential opportunities afforded by the physical manifestation of data. Those most relevant to the projects presented within this essay—a blanket and a t-shirt—are described below.

Design informatics scholar Andrew Vande Moere (2008) suggests considering alternatives to screen-based data visualization, ones that involve ways that humans experience and interpret the world around them. He highlights the inherent capabilities of material objects to communicate meaning by the natural affordances they possess. For example, the blanket and t-shirt described in this essay are personal objects that we wrap around or wear on our bodies. By communicating data through these objects, an intimate relationship with climate change can be engineered. In this way, everyday objects offer interesting possibilities for data communication, because of their immediate closeness and emotional connotations to end users (Vande Moere 2008).

Jansen et al. (2015) point out that while vision is the dominant sense, it is not the only way that we explore the world. When data is physicalized, all senses can be involved in information gathering, and each sense has unique characteristics that can be leveraged for a range of sensory input (Jansen et al. 2015). Designer and journalist Lena Groeger explains in *Scientific American* that "Our five senses—sight, hearing, touch, taste and smell—seem to operate independently, as five distinct modes of perceiving the world." "In reality," she continues, "they collaborate closely to enable the mind to better understand its surroundings" (2012). In this way, "Physicalizations can take advantage of these additional sensory channels to convey a larger range of meanings than a simple visual display" (Jansen et al. 2015, 3229). As I show in this essay, the blanket and t-shirt engage their users visually, as well as through touch and smell.

So-called sensory design, such as the data objects I discuss here, are designed intentionally to bring about an embodied emotional reaction in those who use them. Designer and curator Ellen Lupton describes how sensory designers must: "consider interaction of bodies and things. What sound does a chair make when it scrapes along the floor? How hard does a button need to be pressed to register a response? How much does a surface flex when we push against it?" (2018: 11). Sensory design moves beyond just sight, as in charts and graphs, to shape "our behavior, our emotions, our truth" (2018: 18–19).

Lupton's characterization reveals how and why the two sensorial data objects discussed here can communicate more effectively than flat charts and graphs do. "Sensory experience hits us from all directions" and it has the power to affect us deeply (2018: 10). Charts and graphs are flat and abstracted; we don't experience their content and their meaning within our own bodies. In the two case studies I present, graphic design students at OCAD University engage the senses with complex information on climate change. Through an analysis of these "data objects," I consider the sensorial methods that these designers use to create meaningful connections between climate change data and their audiences.

The students at OCAD University were asked to consider the following questions to aid them in their task and stimulate their creativity.

How might you:

- Create an emotional connection?
- Create a personal or intimate connection?
- Create connections through metaphor and analogy?
- Translate the global to the human scale?
- Translate the future to the now?
- Translate the distant to the proximate?
- Communicate a "slow" process with a sense of urgency?
- Relate the abstract to our everyday lives and experiences?
- Make scientific data feel human?
- Make the complex understandable, simple, and/or relatable?
- Make the invisible tangible?
- Translate the abstract to something visual, tactile, and/or experiential?

The data objects that I present here embody many of the above prompts. They make data tangible, understandable, and translate it to a human scale. Importantly, these objects locate climate change data within our everyday lives, bringing this information into our personal, intimate, and private realms. Lastly, the objects are designed to elicit a visceral response and engage people emotionally in ways that are hard to ignore.

Global Warming in the Living Room

As discussed earlier, people have trouble relating abstract data that is communicated in a bar chart to their everyday lives. But, what if that information was lying on the living room couch? The first project that I present in this essay, entitled *The Comfort of Ignorance*, brings climate change data into the home. Cecilia Salcedo-Guevara spent more than twenty hours weaving global temperature data (NASA 2021a)[1] into a beautiful wool blanket. She photographed the blanket and presented it as the perfect accessory for any living room (Fig. 13.1). Or is it? It announces: "Our Global Temperature Is Rising" on its leather tag. The bar chart that is woven into the blanket

Fig. 13.1 **Cecilia Salcedo-Guevara, *The Comfort of Ignorance*, 100% wool yarn, leather, 2017. Design, photography, and image permission by Cecilia Salcedo-Guevara.**

illustrates global surface temperature from 1965 to 2015, relative to average temperatures from 1951 to 1980 (NASA 2021a)—which is referred to as temperature "anomaly" data.

Salcedo-Guevara's intention was to communicate the dramatic changes in temperature over time. She presented temperature anomaly data in five-year increments (NASA 2021a) as shown in Table 13.1 below. Temperature anomaly data shows how much warmer or colder any one year is compared to the long-term average temperature (NASA 2021a). Salcedo-Guevara translated the data into stitches: the width of each bar was assigned an arbitrary value of five stitches; the height of each bar—which communicates the temperature data—was determined by translating each 0.01 degree Celsius into one vertical stitch. The 2015 temperature anomaly value of 0.87 degrees Celsius, for example, translated into a height of 87 vertical stitches, the final and highest bar on the graph (see Fig. 13.1). The differences across years are indeed dramatic, ranging from low and even negative values (i.e., colder than the long-term average) on the left side of the graph, to progressively higher values on the right, all of which are positive values (i.e., warmer than the long-term average).

Table 13.1 Global temperature anomaly data presented in five-year increments

Year	Average Temperature Anomaly (°C)
1965	-0.10
1970	0.02
1975	-0.02
1980	0.27
1985	0.12
1990	0.44
1995	0.46
2000	0.42
2005	0.69
2010	0.71
2015	0.87

Blankets keep us warm and, in this way, the project represents climate change in another quite clever way. The layers of greenhouse gases in our atmosphere are referred to as a "thermal blanket" around the Earth (NASA 2018). These gases absorb heat and warm the surface of the planet to a life-supporting temperature (NASA 2018). Human activities since the industrial revolution, including the burning of fossil fuels, such as coal and oil, have increased the concentration of atmospheric carbon dioxide by more than a third (NASA 2018). Most climate scientists agree that this expansion of the "greenhouse effect" is the main cause of global warming (NASA 2018). The NASA Global Climate Change website, in fact, describes how greenhouse gases—carbon dioxide, methane, and nitrous oxides—as heat-trapping gases can be "thought of as a blanket wrapped around the Earth, which keeps it toastier than it would be without them" (NASA 2021d). Salcedo-Guevara's blanket thus represents data on global temperature increase through a graph, but also by warming the person who interacts with it.

Whether or not users are affected by this data lies at the core of this project. Salcedo-Guevara explains that she was interested in "the oblivion and ignorance of our society regarding climate change, and how comfortable we often are with it." The blanket is deceptively comfortable while it lies innocuous and inviting on the living room couch. When users wrap it around their body, though, the 100% wool yarn makes itself known by causing a prickling sensation. If a user buries their face in natural wool, the familiar smell and fibers tickle the nose—the prickling sensation on the face can actually feel unbearable. This wool blanket surrounds its users; it literally engulfs them in the discomfort of this data. Furthermore, the warmth of this blanket at a time of global temperature increase accentuates this discomfort; imagine a warm prickly blanket on a warm winter day.[2]

The Comfort of Ignorance is both a metaphor and a litmus test for who we are in this period of climate change, and it provokes some pointed questions. First, are you aware of what is going on around you? If so, are you willing to remain engaged with the issue despite the discomfort of this reality? Will you wrap yourself in this blanket the next time you watch television, or will you stop noticing it on your couch?

The choice of materials and fabrication methods provide further layers of meaning for users. The blanket is made entirely of natural animal-based materials: 100% wool yarn and a leather tag. These materials reflect human subjugation of the natural world—in this case the animals that we breed, shave, and slaughter. The craft methods further emphasize this relationship of control: the weaving of the yarn (i.e., the hair of a sheep) and the burning into the leather to create the text on the tag (akin to branding livestock) in order to create a human-made object.

This object should comfort users, instead it makes them uncomfortable on a whole range of levels. A blanket, which is typically created for human comfort, rebels against its role. In this way, the blanket embodies the conflict between humanity and the natural world; our struggle for control. We cannot interfere, and expect no interference in our own activities. We cannot engage in endless extraction and evisceration of the Earth's resources and expect no consequence. The natural world is not so simple to control.

Climate Change on my T-shirt

One major effect of global temperature increase is sea level rise (NASA 2021b). Although some information about sea level rise has become common knowledge, the details of how much rise there is and how fast it is occurring have not. As discussed above, it is natural for people to think: does this really affect me?—or else not to think about sea level rise at all. The second project that I present in this essay is Justin Yoon's *Sea Level Rise T-Shirt*, which brings climate change data about sea level rise onto the body.

Sociologist Sherry Turkle discusses what she calls "evocative objects," which are familiar objects that we love and, whether we realize it or not, also often use to feel and think.

> We . . . consider objects as useful or aesthetic, as necessities or vain indulgences. We are on less familiar ground when we consider objects as companions to our emotional lives or as provocations to thought. The notion of evocative objects brings together these two less familiar ideas, underscoring the inseparability of thought and feeling in our relationship to things. We think with the objects we love; we love the objects we think with.
>
> Turkle 2007: 153–6

Both *The Comfort of Ignorance* and *Sea Level Rise T-Shirt* are evocative objects. Like Salcedo-Guevara's cosy-looking blanket, Yoon's simple and generic t-shirt is inviting to users because of its familiarity (Fig. 13.2). It is ordinary—white, 100% cotton, crew neck. It is so commonplace, so non-threatening in its banality that a person would be forgiven for overlooking it entirely—that is, except for the bottom of the shirt, which has layers of colored fabric sewn onto it. This fabric is sea level rising up your t-shirt.

Global sea level has risen by approximately eight inches in the past century (NASA 2021b). Although this rate may appear slow, it is accelerating every year (NASA 2021b). Rising sea levels are caused primarily by two factors related to temperature increase: "the added water from melting ice sheets and glaciers and the expansion of seawater as it warms" (NASA 2021e). Through increased flooding, sea level rise has, and will continue to have, an effect on global populations.

Yoon's t-shirt illustrates sea level rise data from 1995 to 2017 measured in millimeters (NASA 2021e). Although information is available dating back more than 100 years, it is not surprising that he chose to depict this date range since this sea level rise has occurred within his lifetime. Yoon's t-shirt communicates cumulative sea level rise during five-year periods. He used NASA's sea-height data from January of the years 1995, 2000, 2005, 2010, and 2015 (NASA 2021e), and through

Fig. 13.2 Justin Yoon, *Sea Level Rise T-Shirt*, cotton fabric, 2017. Designer: Justin Yoon. Photography: Nicole Torres and Karin von Ompteda. Image used by permission of Justin Yoon.

simple subtraction he calculated the sea level rise over these five-year periods (see Table 13.2). The first column of data in Table 13.2 is taken directly from the NASA website (NASA 2021e), and the second and third columns contain his calculations. Yoon wanted to present information that was as up to date as possible, therefore he also included sea-height data from July 21, 2017 (two months before this project was assigned). Yoon also designed a paper tag, which hangs from a string, fastened to the t-shirt with a safety pin. The first three columns of data from Table 13.2 are printed on one side of the tag, assisting users by helping them to understand how the data is mapped to the t-shirt.

Yoon chose to communicate his data through a plain white t-shirt, which he employed as a metaphor for society's indifference to the effects of climate change.

He explained that whenever we look at the ocean, it looks the same; we do not notice the sea level rising as it is not visible to us. If people look closely at both the t-shirt (and the ocean), however, they will discover this information. Yoon communicated sea level rise data through layers of fabric sewn into the base of the t-shirt. He sewed the garment at the dry cleaner where he was working part time. There are five layers of fabric, the height of each one representing a millimeter of sea level increase over the measured time period (see Table 13.2, final column). His final sea level rise calculation reflects approximately a two-and-a-half-year span, which is the reason why the final representation of data (i.e., piece of fabric) is of a lesser magnitude than the ones which precede it.

Yoon thoughtfully utilized aspects of the substance from which his data was taken—water—for the fabrication of his piece. He hand-dyed the pieces of fabric with a blue water-based dye in order to achieve a water-like texture. The uneven coloring of the fabric, with areas absorbing more and less dye (see Fig. 13.2), reference the surface of the blue sea with its patterns of light and shadow. Yoon also created a gradient such that the layers of cloth are progressively lighter in color from the base of the t-shirt upward. Through the sewn fabric layers, uneven dying, and color gradient, Yoon attempted to suggest ripples and waves of the sea, like a tide coming in on the body.

One of the most successful aspects of this project is how clearly it communicates the data. The height of each layer of fabric measured in millimeters directly corresponds to the height of global sea level rise during that time period. Like Salcedo-Guevara did, Yoon translated data to a human scale. Although Salcedo-Guevara's piece communicates temperature through a bar chart that necessarily presents data at an arbitrary scale, Yoon's data is presented at a clear 1:1 scale (i.e., 1mm of fabric represents 1mm of sea level rise). In Yoon's project, sea level rise data, which is global and abstract, is connected directly to everyday experience and the body through the t-shirt design.

Looking at the t-shirt and wearing it are two very different experiences (see Fig. 13.2).[3] This garment was designed to induce anxiety in its wearer. The layers of

Table 13.2 Sea level rise data and calculations (1995 to 2017)

Date	Cumulative Sea Level Rise Since 1993 (mm)	Sea Level Increase Since Previous Measured Date (mm)	T-Shirt Layer (ordered from bottom to top)
July 21, 2017	85.7	9.6	5
January 1, 2015	76.1	20.0	4
January 23, 2010	56.1	13.2	3
January 24, 2005	49.2	17.0	2
January 7, 2000	25.9	18.0	1
January 28, 1995	7.8	-	

fabric at the base of the t-shirt invite tactile exploration of the data, with the opportunity for users to literally feel sea level rise during these time periods with their fingers. The threat of sea level rise is made personal for the user, as the layers appear to be rising up the body. The final layer of fabric (i.e., layer five) may seem the least threatening layer because it is shorter than the others, however, an awareness that this layer represents a partial time period brings home the notion that sea level rise is in progress, and still rising.

T-shirts and Carbon Footprint

Some other t-shirts are locations for self-expression—for example, the wearer's political beliefs, personal humor, or favorite bands. Yoon's t-shirt, on the other hand, provides an opportunity for self-reflection. Although this t-shirt could help to spread awareness about climate change, its true purpose is to create a context for individuals to contemplate the major changes in sea level rise that are occurring globally, and to get wearers to consider how they or others might be affected by those changes. Like Salcedo-Guevara's blanket, this t-shirt is meant to be worn. It asks its wearer to bring threatening phenomena into a traditionally comfortable place. Like *The Comfort of Ignorance*, Yoon's t-shirt asks the wearer: who are we in all of this?

This question takes on even more relevance because this sea level rise data has been presented on and attached to a garment that represents a significant carbon footprint, and one important area where individuals can decrease their own. Recent statistics estimate that the apparel sector accounts for approximately five percent of global greenhouse gas emissions (Bauck 2017). Although this number may seem small, five percent is approximately the impact of the aviation sector for greenhouse gases produced by plane travel (Bauck 2017).

When individuals consider how to decrease their carbon footprint, they often consider air travel because of its high emissions (*Nature Climate Change* 2018). Everyday choices, however, including reducing consumerism, can also have an impact on carbon footprint (*Nature Climate Change* 2018). Fast fashion—a term used to describe clothing production on shorter timeframes—has become increasingly common (*Nature Climate Change* 2018). In fast fashion, new garments appear in stores every few weeks, as opposed to the traditional seasonal spring/summer and autumn/winter collections that are released twice a year. People are purchasing sixty percent more clothing than in the year 2000, and they are wearing them less before disposing of them (*Nature Climate Change* 2018), which produces a larger individual carbon footprint.

Like Salcedo-Guevara's blanket, Yoon's cheap ten-dollar t-shirt reflects the human subjugation of the natural world; in this case, the cultivation of plants through industrial agriculture to create our human-made products. A connection is made in the project between our own choices and the larger global consequences, which are brought right back to the individual for consideration. Yoon's object has been designed for people to face their own accountability. In this way the project combats the helplessness and apathy that individuals feel regarding climate change. It asks us to consider our agency.

Discussion and Summary

By mapping data onto a blanket and t-shirt, climate change can be transported into our personal spaces and into our everyday lives. These projects exemplify the capability of physical and sensorial manifestations of data to expand potential locations and contexts for the engagement with information (Jansen et al. 2015). This is particularly important in the case of climate change, as people typically perceive it as a distant threat and something that happens in remote places (Brügger et al. 2015). Psychological boundaries also exist with regard to scale, as the magnitude difference between an individual life and a global process is profound (von Ompteda 2019b); and people may not believe that their own behavior can make any sort of impact (Gifford 2011). Through data manifestation, climate change can be communicated in the present, brought to our locations, and transformed to a human scale (von Ompteda 2019b).

When data is physically manifested, the communication potential of materials can be harnessed. *The Comfort of Ignorance* employs leather and wool yarn, and *Sea Level Rise T-Shirt* is made of industrially produced cotton. Through the use of these materials, both projects embody complex metaphors of humanity's subjugation of the natural world, analogous to human-driven climate change. Data mapping metaphors are often not immediately understandable but are meant to be discovered (Vande Moere 2008). Although some methods of data communication may not be related to the data, others reference it either directly or indirectly. Yoon's t-shirt, for example, uses water-based dyes. In this way, the textiles that will communicate sea level rise are plunged into the very substance they are communicating about—water. The evidence of this process also remains in the water-like texture that can be seen on the layers of fabric sewn into the garment.

The hand-crafted nature of both data objects also serves a communication role. By creating projects through the processes of weaving and sewing, the students have addressed one of the challenges they were given: *How might you make scientific data feel human?* Although hand-craft methods can help to facilitate an emotional, "human" connection to abstract data, they also fittingly represent the true nature of data as created by people (von Ompteda 2019a).

When data is physically manifested, all senses can be involved in information gathering (Jansen et al. 2015). The t-shirt invites the user to explore the data with their fingers; to run them upward and downward along the "water ripples" of sea level rising, experiencing anxiety in the process. The blanket also invites tactile engagement, warming the user's skin and making a direct reference to global temperature increase. Sensory engagement can further be used to convey a larger range of meanings (Jansen et al. 2015). For example, as noted, the prickling itchy blanket communicates climate change data through an uncomfortable experience, connecting physical discomfort with an emotional one.

Taking advantage of the natural affordances of objects to communicate meaning (Vande Moere 2008), the blanket and t-shirt bring data into direct physical contact with the user in an intimate way. The inviting nature of a cosy blanket or a simple t-shirt, create a benign and even pleasant invitation into ominous information. By mapping data to these deceptively innocuous objects, threatening information is transported into previously safe places; our home and our clothing. The immediate closeness and emotional connotations that we have already established with such objects (Vande Moere 2008) are now translated into our engagement with climate change data. These textiles are wrapped around our bodies, and as such, the data is wrapped around our bodies. Our relationship with climate change is now close and personal.

When the blanket and t-shirt physically surround us with climate change data, we are compelled to explore our relationship with this information. The uncomfortable blanket asks its user to consider their stance on the subject of global warming. The ten-dollar t-shirt asks its user to consider their agency (i.e., carbon footprint). Both objects urge their user to not be complacent. Ultimately, the choice to engage in these disquieting experiences is left to individuals, analogous to the choice we are offered every day to engage with this global issue or not. Whether climate change data is wrapped around your body, or folded into the drawer and ignored, these data manifestations nonetheless elucidate who you are in this period of human-driven climate change.

Acknowledgments

I gratefully acknowledge the ingenuity and passion of my students, with special thanks to Cecilia Salcedo-Guevara and Justin Yoon. Thank you to Leslie Atzmon for inviting me to participate in this project—which spurred me to initiate the climate change brief—and invaluable contributions to the manuscript. I am grateful to Elizabeth Harvey for insightful feedback and rich discussions regarding this work. Thanks to Nicole Torres who has photographed many of my students' projects with me. I thank Saadia Kardar for her careful formatting, proofreading, and photo editing. Lastly, my profound thanks to John Obercian for his support.

Notes

1. NASA's Global Climate Change website updates its data regularly, therefore the data that students have based their projects on will differ from what is currently published online.
2. I write this on a snowless four degrees Celsius (39 degrees Fahrenheit) winter day in Canada.
3. Justin Yoon is pictured here wearing his t-shirt.

References

Bauck, W. (2017), "The Fashion Industry Emits as Much Greenhouse Gas as All of Russia," *Fashionista*, https://fashionista.com/2017/09/fashion-industry-greenhouse-gas-climate-change-sustainability (accessed on January 22, 2019).

Brügger, A., S. Dessai, P. Devine-Wright, T. A. Morton, and N. F. Pidgeon (2015), "Psychological Responses to the Proximity of Climate Change," *Nature Climate Change* 5: 1031–7, https://doi.org/10.1038/nclimate2760 (accessed on June 6, 2019).

Design and Science (2019/2020) [Exhibition] Ypsilanti: University Gallery, Eastern Michigan University, September 11– October 17; Philadelphia: Esther Klein Gallery at the Science Center, February 13–March 28.

Gifford, R. (2011), "The Dragons of Inaction: Psychological Barriers that Limit Climate Change Mitigation and Adaptation," *American Psychologist* 66: 290–302.

Groeger, L. (2012), "Making Sense of the World, Several Senses at a Time," *Scientific American*, www.scientificamerican.com/article/making-sense-world-sveral-senses-at-time/ (accessed on April 17, 2021).

IPCC (2007), "Summary for Policymakers," In *Climate Change 2007: Impacts, Adaptation and Vulnerability. Contribution of Working Group II to the Fourth Assessment Report of the Intergovernmental Panel on Climate Change*, Cambridge: Cambridge University Press.

Jansen, Y., P. Dragicevic, P. Isenberg, J. Alexander, A. Karnik, S. Subramanian, J. Kildal, and K. Hornbæk (2015), "Opportunities and Challenges for Data Physicalization," 3227–36, *Conference on Human Factors in Computing Systems (CHI'15)*.

Lupton, E. (2018), *The Senses: Design Beyond Vision*. New York: Princeton Architectural Press.

Mollerup, P. (2015), *Data Design: Visualising Quantities, Locations, Connections*. London: Bloomsbury.

NASA (2018), "The Causes of Climate Change," *Global Climate Change*, https://climate.nasa.gov/causes/ (accessed on December 23, 2018).

NASA (2021a), "Global Temperature," *Global Climate Change*, https://climate.nasa.gov/vital-signs/global-temperature/ (accessed on April 2, 2021).

NASA (2021b), "Climate Change: How Do We Know?" *Global Climate Change*, https://climate.nasa.gov/evidence/ (accessed on April 2, 2021).

NASA (2021c), "The Effects of Climate Change," *Global Climate Change*, https://climate.nasa.gov/effects/ (accessed on April 2, 2021).

NASA (2021d), "What is The Greenhouse Effect?" *Global Climate Change*, https://climate.nasa.gov/faq/19/what-is-the-greenhouse-effect/ (accessed on April 2, 2021).

NASA (2021e), "Sea Level," *Global Climate Change*, https://climate.nasa.gov/vital-signs/sea-level/ (accessed on April 2, 2021).

Nature Climate Change (2018), *The Price of Fast Fashion*, www.nature.com/articles/s41558-017-0058-9 (accessed on January 22, 2019).

Norgaard, K. M. (2013), "Climate Denial: Emotion, Psychology, Culture, and Political Economy," in J. S. Dryzek, R. B. Norgaard, and D.Schlosberg, *Oxford Handbook of Climate Change and Society,* 399–413, Oxford: Oxford University Press.

Plait, P. (2016), "Grasping Climate Change," *Slate*, https://slate.com/technology/2016/08/climate-change-is-slow-making-it-hard-to-grasp.html (accessed on April 17, 2021).

Rogers, A. (2018), "The Climate Apocalypse Is Now, and It's Happening to You," *Wired*, www.wired.com/story/the-climate-apocalypse-is-now-and-its-happening-to-you/ (accessed on April 17, 2021).

Talbot, J. (2015), "The Threefold Intangible Challenge of Climate Change," *Global Climate Adaptation Partnership*, https://climateadaptation.cc/blog/entry/the-threefold-intangible-challenge-of-climate-change (accessed on April 17, 2021).

Turkle, S. (2007), *Evocative Objects: Things We Think With.* Cambridge: MIT Press.

Vande Moere, A. (2008), "Beyond the Tyranny of the Pixel: Exploring the Physicality of Information Visualization," *IEEE Conference on Information Visualisation (IV '08):* 469–74.

Vande Moere, A. and S. Patel (2010), "The Physical Visualization of Information: Designing Data Sculptures in an Educational Context," *Visual Information Communication*, pp. 1–23.

Viégas, F.B. and M. Wattenberg (2007), "Artistic Data Visualization: Beyond Visual Analytics," *Lecture Notes in Computer Science* 4564 (15): 182–91.

Von Ompteda, K. (2022), "Data and Emotion: The Climate Change Object," in I. Gwilt (ed.), *Making Data*, London: Bloomsbury.

Von Ompteda, K. (2019a), "Data Manifestation: A Case Study," in T. Triggs and L. Atzmon (eds.), *The Graphic Design Reader*, London: Bloomsbury.

Von Ompteda, K. (2019b), "Data Manifestation: Merging the Human World and Global Climate Change," *IEEE VIS Arts Program (VISAP'19)*: 1–8.

14 Visual Thinking and the Art of Medical Diagnosis

MICHAEL CHANDLER

Never make a big diagnosis on a little test.

Douglas Chandler, M.D.

Introduction

The practice of medicine is an art form that strives to weave incongruent threads into whole cloth. The threads—which include a narrative of the natural history of a patient's condition, a depiction of observable findings, and imagining various unseen factors—are woven together using a traditional methodology, begun by the Ancients, that applies observation, reasoning, knowledge, judgement, and experience. Current advances in digital technology, however, have encouraged contemporary practitioners to rely more on empiric testing (CAT scans, MRI scans, etc.). Empiric testing alone, taken at face value, often displaces the imaginative thinking that lends coherence and richness to this exact same empirical data. The traditional modes of medical reasoning that require the mastery of specialized thought processes and habits of mind—individual knowledge based on observations, experience, creative interpretation, and reasoning skill—are then supplanted by these technologies at the expense of creative reasoning. The result is the degradation of traditional skills of medical observation and thinking ability among medical students and trainees, which accelerates the trend of relying on data points over reasoning, and can lead to simplistic, and possibly incorrect, diagnoses.

I argue, in this essay, for the revival of traditional modes of imaginative medical reasoning using the methodologies of visualization and thought experiments. Thought experiments, which meld real-world observations with imagined or unseen elements in the mind's eye, were popularized by nineteenth-century German physicist Ernst Mach. When considering a problem that could not be resolved by actual physical experimentation, Mach constructed an imaginary visual apparatus in the mind's eye that both reflects the thesis of the problem and allows for the manipulation of circumstances that were impossible to create in the tangible world (Brown 2010: 1).

Medical students or practitioners may have access to numerical and visual empirical data, but I contend that they, like Mach, must apply it to the situation of an individual along with their specific observations and imaginative thinking. This dual process benefits patients by transcending a cookbook approach to medical diagnosis and increasing the likelihood that complex, unusual, or unexpected medical conditions will be noticed. It also elevates medical practice by strengthening it both as science and as an art.

Traditional imaginative medical reasoning employs Bayesian analysis. Eighteenth-century Scottish clergyman Thomas Bayes developed this mathematical model to assign probabilities and to predict outcomes. Medical reasoning applies a non-numeric intuitive adaptation of the strictly algebraic Bayesian statistical process. In applying the Bayesian probabilistic method, clinicians develop an initial probability (a prior) that a patient has a disorder. In the application of medical reasoning, this prior is then examined using visual, physical, and verbal information that is obtained from the history, physical examination, and often diagnostic testing, of a patient in order to arrive at a final probability estimate. A deeply discerning process, medical Bayesian analysis uses both observable findings and visual imagination to enhance diagnosis and management of medical conditions. This secondary layer of observation and imagination (in conjunction with empiric data) is key to medical reasoning.

In this essay, I consider how medical reasoning can be understood as a Bayesian probability assessment blended with a Bayesian-based thought experiment. The paradigm of the thought experiment can offer an instructive look at the value of traditional medical reasoning. Further, I consider how the history of the thought experiment reveals certain visually-based habits of mind that are commonly utilized by designers and certain scientists that could be invaluable to contemporary medical education and practice.

Medical Reasoning and Visual Thinking Strategies

Harvard Medical School researchers Sheila Naghshineh et al. (2008) have observed a decline in the use of the Bayesian method and physical exam skills to the degree that the resulting medical reasoning is impaired. They conducted a study in 2008 to determine whether Visual Thinking Strategies (VTS) could be used by medical students in ways that could enhance their appreciation for, and skill in, physical examination and the Bayesian method. In VTS training, a technique that was developed by educational researcher Abigail Housen, participants analyze and interpret imagery in order to stimulate creative thinking and hone observation and imagination skills. The Naghshineh study demonstrates that VTS enhance observational skills of medical students, validate the virtues of performing a skillful physical exam, and contribute to medical Bayesian analysis.

VTS are also commonly an innate part of the thought-experiment process. Nineteenth-century chemist J.H. van 't Hoff used thought experimentation to imagine the way that nature arranges carbon atoms. Nineteenth-century physician and pathologist Rudolph Virchow likewise used thought experiments to visualize his sense that cells are derived from other cells by a process of division (Wilson 1947; Ventura 2000). Equally important, Virchow's methods and ideas had a profound impact on early-twentieth-century Canadian physician Sir William Osler, who is often called the Father of Modern Medicine. Explaining that "variability is the law of life," Osler contended that "no two individuals react alike and behave alike under the abnormal conditions which we know as disease" (Stone 1995: 270). Osler, who has been described as one of the "greatest diagnosticians ever to wield a stethoscope," emphasized open-minded observation and creative imagination in medical diagnosis (Tuteur 2008).

Osler's medical diagnosis necessarily deals with material entities, as well as with un-seeable or intangible structures and processes. As in design, a key element in Osler's process was the conjuring of images of those entities that are un-seeable in the physical world as a pivot between physical description/observation and insightful understanding. I contend that in order to produce this kind of rich medical narrative, Osler utilized visual, physical, and numerical medical information as part of a thought experiment akin to those used by Mach, van 't Hoff, and Virchow.

Quantifiable or digital medical data (ironically, much of which happens to be visual) can reveal the unseeable. It can also facilitate the understanding of disease processes in large populations of individuals. Mach's thought-experiment process suggests how students or practitioners may apply data gathered from medical tests and empirical knowledge to individual medical diagnosis. Mach described a VTS-like process of "continuous change of visual imagination," a flow of imaginary, intangible mental visuals during thought experiments, which aggregate in the mind and promote creative thinking and fruitful ideas (Mach 1897: 456). Associated insights generated during thought experiments, according to Mach, can suggest how things function in the physical world, a notion that is a critical part of design, and, I contend, individual medical diagnosis and Bayesian analysis.

The Bayesian Process, Thought Experiments, and Visual Thinking Strategies

As noted above, in Bayesian analysis clinicians develop an initial probability (a prior) that a patient has a certain disorder. Clinicians use the prior to come to a final probability estimate using visual, physical, and verbal information from the medical history and examination of a patient.

When a patient presents with a particular symptom—shortness of breath, for example—a trained physician employing medical reasoning instigates a Bayesian maze-like process in the mind's eye. Using visual imagination and visual thinking, they traverse this thought-experiment maze during the elucidation of the specific cause of a symptom in order to arrive at an appropriate treatment. The starting point of the maze leads to an ever-narrowing set of pathways that are directed by the medical narrative, physical findings, and specific interpretations that fit the natural history of a range of possible conditions. The practitioner works through the mental maze in the mind's eye using imagination and experience, along with what she has observed in the individual patient, in order to evaluate the likelihood of each condition. Shortness of breath that is caused by asthma, heart disease, kidney disease, neurological conditions, pneumonia, or poor physical conditioning all offer specific clues for which path to take through the maze. When observation and reasoning don't lead to a clear conclusion—and the "exit" from the maze is elusive—then empiric testing may help to define the correct path.

Practitioners get feedback through the Bayesian thought-experiment process, and they have the opportunity to refine their future diagnoses with lessons learned through reason. Iterative processes in design thinking and making allow for conceptual exploration that typically leads to more inventive design, as designers are precluded from settling on final outcomes too quickly. Similarly, iterative medical exploration and learning add depth to medical practice and prevent physicians from locking in on diagnoses too early, which can cause problematic outcomes. The problematic alternative methodology is first to create a list of diagnostic options, next to perform empiric tests to rule in or rule out possibilities, and then to arrive at a diagnosis. Doing so adds little to refinement in thinking and methodology, utilizes limited iterative learning, locks in a diagnosis too quickly, and must often be repeated in the same way each time to address similar issues in patients.

The strict Bayesian process is mathematical and probabilistic. But when it is used in medical reasoning, as described in the shortness-of-breath example above, it becomes clear that creative thinking and mental imagery play a role in the development of the priors and the generation of iterative diagnostic processes, as well as in the selection and contextual interpretation of specific medical tests (the thought experiment apparatus). Physicians Herrle et al. in fact, explored the extent to which students, trainees, and practicing physicians value visual comprehension of physical examinations and observations when using the medical Bayesian approach:

> In this study, trainees and experienced physicians similarly underestimated the impact of examination findings when estimating condition probabilities and, as a consequence, often chose to order additional diagnostic testing to reduce diagnostic uncertainty. A better understanding of when and how physicians apply examination findings in their assessment of condition probability may provide the foundation for improving the way physicians use these observations in everyday clinical practice.
>
> 2011

These researchers concluded that to be successfully applied, physical examination in medicine requires examiners to appreciate the clinical signs they observe and to assign meaning to them in the context of the patient's circumstances. The practitioner in my shortness-of-breath example pays attention to specific qualities of the shortness of breath, and gathers information on other important symptoms. Their observations may suggest asthma, heart disease, kidney disease, neurological conditions, pneumonia, poor physical conditioning—or even an unknown to be further investigated. They may then order tests to confirm their diagnosis or to inquire further.

Like design thinking, this medical process requires imagination and visualization to synthesize numerical and visual data with examination findings into a comprehensive picture of the medical issue at hand. VTS may be key to this process.

Abigail Housen studied how VTS can enhance observational capability and critical thinking among the general population, and assessed the transfer of content and context into non-art domains. Using visual imagery as a prompt, Housen's VTS process utilizes three foundational questions: What is going on in the image? What do you see that makes you say that? What more can you find?

Housen wondered whether such observational and critical thinking skills could be transferred from one situation to another. She considers, in particular, the context and content of observational and critical understanding. Context transfer, according to Housen, involves critical and creative thinking strategies in a social setting that is different from the one in which such thinking was learned. Content transfer involves critical and creative thinking that is applied to new social contexts in a different subject domain. Housen found evidence of both types of transfer—context and content—after training in VTS techniques. She concludes that the most productive visual thinking processes are "question-based, give the learner repeated opportunity to construct meaning from different points of view, take place in an environment that supports looking in new and meaningful ways, and are inspired by rich, varied, and carefully chosen [visuals]" (Housen 2007: 16). These aspects of VTS augment viewers' ability to visualize and then aggregate both real and imaginary visual factors, a process that is common in design and also has much potential to enrich medical reasoning.

The evaluation of an individual patient is, in great part, a process that requires this sort of visualization and creative thinking on the part of the physician. Patients describe their situations in lay language, and physicians or students align the story with their knowledge and experience of medical science and natural history, and then examine the patient for signs or clues. The goal is to then use Bayesian and other reasoning skills such as VTS to transform this verbal, physical, and visual information into a thesis that will lead to a diagnostic and therapeutic plan.

Thought Experiments in Science and Medicine

Scientists have frequently employed thought experiments to understand and explain theoretical concepts that resonate with the instinctive knowledge and heuristics, or rules of thumb, that humans learn and apply to life experiences (Irvine 1991). We all know that pigs cannot fly, yet we can imagine what that might actually look like.

Philosopher of science James Robert Brown investigates the application of "merely imagined" thought experiments to scientific and philosophical matters. In particular, he considers an imagined experiment that could have been performed by Galileo, involving the rate at which a small musket ball and a cannonball fall from a tower, and physical experimentation involving a constructed apparatus (1991: 122–3). This hypothetical Galilean experiment could have been performed to demonstrate that all objects fall at the same rate regardless of weight. In reality, Galileo devised his own thought experiment—he challenged and disproved the Aristotelian notion that heavy objects fall faster than light ones by mentally tying the two objects together.[1] Like Galileo, Mach, and Einstein, physicist Erwin Schrödinger employed thought experiments to visualize improbable or invisible phenomena that could not be (re)created in the physical world.[2]

To understand the role that thought experiments play in scientific creativity, it is essential to explore the practical and philosophical underpinnings of thought experiments more closely. Brown, in fact, argues that all thought experiments are visual (1991: 122). He offers a taxonomy of thought experiments of two general types, destructive and constructive. The destructive type experiment is intended to disprove an idea, as Galileo did with his tethered balls. Constructive thought experiments are exemplified by Newton's explanation of the moon's orbit around the earth. Newton asks us to consider a cannonball fired from a mountain top. The distance at which it falls to earth depends on the speed of the projectile. If the projectile is moving fast enough, it could actually circle the earth repeatedly, just as the moon does.

How does engaging in new modes of thought and experience—such as the thought experiments that employ visual thinking and imagination described above—enhance specific skill development in medical trainees? The history of science offers some insights into this question. Physiologist Robert Root-Bernstein (1985) distinguishes between two types of scientists: those who focus on granularity and discrete fact gathering and those who focus on discerning patterns. The latter group of scientists imagine how facts could be transformed by creative imagination into conceptual and, often visual, illuminations of their thoughts—that is, thought experiments. Root-Bernstein observes that many of these creative thinkers exhibited talents in the artistic, literary, or musical arenas, and argues that their adeptness in non-linear pursuits was either essential or intrinsic to their scientific contributions.

Specifically, Root-Bernstein compares and contrasts the practical and philosophical constituents of two well-known nineteenth-century chemists, J.H. van 't Hoff, who I mentioned earlier in this essay, and Hermann Kolbe. In an inaugural lecture at the University of Amsterdam in 1878, van 't Hoff presented a speech devoted to imagination in science. Van 't Hoff defined imagination as "the ability to visualize mentally any object with all its properties so that one recognized it with the same great certainty as by simple observation" (Root-Bernstein 1985: 50). In other words, he argues that vivid mental images function exactly like real-world observations. Indeed, architect Kyna Leski points out in her book *The Storm of Creativity* that for both designers and scientists "Imagination is an instrument for seeing. It takes you on a ramble through choices and leads you to concrete attributes, from percept (the object of perception) to concept and then to something realized that you can perceive again" (2015: 116–17).

Van 't Hoff used visualization to imagine how nature might actually arrange carbon atoms. Kolbe, on the other hand, asserted that atoms could not be visualized, and it was therefore useless to imagine their appearance (1985: 50). Although thought experiments may sometimes be difficult to (re)produce in the real world, the *conclusions* drawn from these experiments can be tested in the physical world to explain and refine an underlying thesis to great effect. Van 't Hoff's theory of the atomic structure of carbon atoms was ultimately demonstrated empirically. His process suggests how visual imagination aids scientific inquiry.

Imagining the unknown or unseen permits scientists to understand, explain, and expand upon data rather than merely to observe and describe it. Van 't Hoff argued that facts may be understood as the foundation of scientific inquiry, but that "Imagination [is] the building material; the Hypothesis, the ground plan to be tested; Truth or Reality, the building" (van 't Hoff as quoted in Root-Bernstein 1985: 50). Imagination, particularly visual imagination, allows scientists and physicians, no less than designers, to communicate ideas that reach our consciousness by conveying meaning in a visual fashion.

Ernst Mach was similarly devoted to thought experimentation. He was particularly interested in using thought experiments as a pedagogical tool—in part, as a reaction to the prevailing educational philosophy of the rote learning of classical material. He observed about rote learning: "What they have acquired is a spider's web of thoughts too weak to furnish sure supports, but complicated enough to produce confusion" (Matthews 1988: 254). Mach established a journal of pedagogy for secondary schools that contained a thought experiment in each edition. The thought experiments were intended for classroom use to enhance the educational experience for students and teachers. He felt strongly that "Experimentation in thought is important not only for the professional inquirer but also for mental development as such" (Mach 1976: 143; Matthews 1988: 254).

Designers, artists, and other creators of novel objects fashion these products, in part, by utilizing thought experiments. A thought experiment may serve the purpose of establishing the validity of a premise or, equally important to both design and medical practice, to discredit a fallible one. In design, it is paramount that there be a safe and workable outcome for users; in medicine both establishing a correct diagnosis and avoiding an incorrect diagnosis are crucial for the patient.

Thought Experiments in the History of Medical Science

Nineteenth-century medical science offers insightful examples of such mental developments. Famed nineteenth-century German physician and pathologist Rudolph Virchow was a talented visionary who used visualization to stimulate novel thinking.[3] Through meticulous study and observation, he adopted the view that cells formed the basis of living organisms and that organs were collections of these cells: the "ultimate irreducible form of every living element" (Ventura 2000: 550). Virchow also postulated that all cells were derived from other cells by a process of division. Like van 't Hoff's ideas about the structure of carbon atoms, Virchow's concept of cellular division was novel, it was unwitnessed except in his mind's eye, and his ideas were ultimately proven to be accurate.

Virchow's work on cell theory was simultaneous with his introduction of the orderly process of postmortem examinations, including the microscopic examination of tissues, which formed the basis of the scientific study of human health and disease. Physician Hector Ventura suggests that Virchow used visual observations and other data from multiple dissections to derive a general theory of disease processes (2000: 550). These foundational achievements, which had their basis in thought experiments and visual imagination, profoundly influenced the training and outlook of early-twentieth-century physicians such as Sir William Osler, who adopted Virchow's clinico-pathological methodology and applied it to the clinical practice of medicine.

Sir William Osler instituted a rigorous scientifically based education program at Johns Hopkins School of Medicine that drew heavily on the so-called German model established by Virchow. Ventura writes that "The influence of his [Virchow's] work has been deep and far reaching, and in one way or another has been felt by each one of us [physicians]" (2000: 8b). The 1892 *Osler's Textbook of Medicine*, which went through sixteen editions, carefully articulates the necessity of detailed and accurate physical and visual observation and fidelity to scientific understanding.

At the time that his textbook was conceived, observation and a deep understanding of the natural history of disease processes and their physical manifestations demonstrable by examination formed the basis of medical practice. Post-mortem examination—enhanced by Virchow's ideas that were based in observation and conceptualization—provided definitive, and always visual, evidence of the clinical conclusions. The iterative process of obtaining a clinical history, performing a detailed examination, generating a differential diagnosis in the mind's eye, followed by a prognosis and treatment plan, became the standard model of medical practice.

Osler described how creative observation, interpretation, and conceptualization lead to proper diagnosis:

> It is important to recognize that there is nothing mysterious in the method of science, or apart from the ordinary routine of life. Science has been defined as the habit or faculty of observation. By such the child grows in knowledge, and in its daily exercise an adult lives and moves. Only a quantitative difference makes observation scientific—accuracy; in that way alone do we discover things as they really are.
>
> 1920: 47

Osler also addressed the issue of observation and interpretation in a commencement address given to a medical school class in 1885 to attach meaning and methodology to refined observation skills: "You will do well if you bring a knowledge which is practical, senses which have been trained to exact observation, habits painstaking and careful, and above all an appreciation of method in work" (1885 speech to students, Silverman et al.: 194).

Osler applied Virchow's visually imaginative and scientifically rigorous methodology to clinical observation and physical diagnosis. Osler described this skill in a physician:

> He does not see the pneumonia case in the amphitheater from the benches, but he follows it day by day, hour by hour; he has his time arranged so that he can follow it; he sees and studies similar cases and the disease itself becomes his chief teacher, and he knows its phases and variations as depicted in the living, he learns under skilled direction when to act and when to refrain, he learns insensibly principles of practice and he possibly escapes a "nickel-in-the-slot" attitude of mind which has been the curse of the physician in the treatment of disease.
>
> Osler as quoted in Stone 1995: 271

As in Robert Root-Bernstein's elaboration of J.H. van 't Hoff's approach, Osler encouraged imaginative visualization and discouraged a focus on mere fact gathering. Like Mach, Osler urged against rote learning. Using Osler's approach, medical observation, as in viewing an image, may occur at several levels. *That is a pretty picture of a garden;* It can also be viewed in the designer, artist, or patient's context with meaning imputed to elements of that which is observed employing specific vocabulary that are specific "terms of art." Coupling observations with attached meaning is a necessary precursor to describe, discuss, and analyze the conceptual meaning of the observed items and creates the substrate for the thought experiment.

Physician Marvin Stone sums up Osler's process with a lyric from composer Stephen Sondhelm: "The art of making art is putting it together" (1995: 274). Indeed, the creation of the thought experiment requires the mental retention of imagery and meaning to be melded with the thought experiment. VTS as applied to the preternatural thought experiment process in medicine may require some methodological instruction because medical observation and diagnosis, like visualization in design thinking, is unfamiliar to the uninitiated. As Osier wrote, "the old art cannot possibly be replaced by, but must be absorbed in, the new science" (Stone 1995: 274)

Medical Pedagogy, Visual Thinking, and Imagination

Contemporary medical education has diminished the promotion of critical and creative practices that previous generations of students acquired in their training. Medical students are also not typically encouraged to study the humanities and the arts. Are contemporary students channeled in their studies so early that there is less opportunity to explore intellectual and artistic pursuits, or has the process attracted participants that eschew these alternate intellectual styles as unnecessary pedagogical burdens that distract from their linear and goal-oriented path? Osler, in fact, warned against a narrow intellectual focus among physicians.

> The extraordinary development of modern science may be her undoing. Specialism, now a necessity, has fragmented the specialties themselves in a way that makes the outlook hazardous. The workers lose all sense of proportion in a maze of minutiae. Everywhere men are in small coteries intensely absorbed in subjects of deep interest, but of very limited scope.
>
> Osler speech, Silverman et al. 2007: 49

Osler specifically cautioned against overspecialization and, as noted above, granular study as detrimental to the artful application of science to medicine. Naghshineh's study offers empirical evidence that developing visual thinking skills in medical trainees may lead to enhanced performance of their particular professional tasks, including assessment of facts in various contexts. In the tradition of Root-Bernstein's creative thinkers, she suggests, it is possible to create a thought experiment to examine the facts and construct a model to assess the medical situation at hand. Naghshineh's team's ideas spotlight the roles that imagination and visual and non-linear thinking can play in elevating medical education, reasoning, and practice beyond the merely rote consideration of a series of facts—Osler's "nickel in the slot." This perspective echoes themes elucidated by her intellectual forebears, Virchow and Osler, who likewise evolved from totally rote traditions of scientific education that were lamented by Mach, and refuted by Van 't Hoff's creative approach to scientific inquiry.

Conclusion

In *The Storm of Creativity*, Leski likens the creative process to the manner in which a storm builds. She begins with a quote from Stoic Greek Philosopher Epictetus (ca 55–135): "It is impossible for a man to begin to learn that which he thinks he already knows" (2015: 167). That is, shedding preconception and embracing uncertainty is a prelude to creativity in both design and medicine. Elemental forces then gather and coalesce to create patterns and circumstances that literally blow in the wind with unseen forces of nature acting in repetitive and unexpected ways. Leski goes on to suggest that attentiveness is a key element in the creative process. Unlearning what you know to make space for new thoughts aids the process. Defining and framing a problem come next. A creative endeavor may then include gathering and tracking data, and with the application of intelligence and imagination, something previously unseen takes shape. What Leski is describing is what happens in thought experiments, and it is crucial to good medical diagnosis.

I would like to conclude with a brief anecdote from my own medical practice speaking of the value to patients of diagnostic thought experiments. Decades ago, while in training, I was evaluating a patient who described shortness of breath. I did a thorough evaluation and examination. Listening to his chest with a stethoscope, I heard sounds unlike anything I had experienced. The lung is configured like an upside-down tree in which the branches, or bronchial tubes, get smaller and smaller the farther you go out. Like a pipe organ, bronchial tubes of different sizes create different pitches. What I heard was all monochromatic pitches. In a kind of thought experiment, I attempted to imagine what kind of process could result in such monotony and I could not. I did know it required explanation. A chest X-ray revealed a lung tumor that was compressing the surrounding bronchial tubes to a uniform caliber. Thus, the monotonous sounds.

Twenty years passed, and I was examining another patient who was referred to me with a diagnosis of asthma. In the process of evaluation, I excluded asthma on the basis of the presentation and symptoms, but on examination, I heard the same monotonic breath sounds. I sent her for a chest X-ray that day and found a curable lung tumor. Conducting a thought experiment with patient one added to my diagnostic repertoire, which benefitted patient number two. As it should be. Like tectonic plates that are constantly reforming the earth's crust, flexibility, curiosity, and visual imagination drive creative openness in medical reasoning.

Ernst Mach devised vivid thought experiments to unleash creative thinking in nineteenth-century German secondary school students trapped in a universe of rote learning. Contemporary medical students cope with a similar burden of rote learning and linear thinking. Drawing on the argument for imaginative thinking outlined in this essay, I encourage medical students to dissect not just lifeless cadavers, but also the lively spirit of thought and meaning that can be discovered in visual thinking and imagination.

The Bayesian-based thought experiment, applied to medical reasoning, and conducted in a virtual visual mental universe—based on observation and coupled with knowledge and experience—leads to diagnosis, treatment, and the refinement of heurisitics. This kind of exposure could re-kindle a thoughtful and creative spirit in medical education. Adoption of creative methodology, whether on the clinical wards or in the designer's practice, creates an intellectual platform to propel the intuitive leaps that move us into a creative realm. Fortune favors the prepared mind.

Notes

1. According to Aristotle's view, the lighter ball would slow the heavier ball, the heavier ball would speed up the lighter ball, and the combined balls should fall faster together. This reveals internal contradictions of the Aristotelian theory that nullify its validity.
2. According to science writer Ali Sundermier,

> [t]his is something Einstein started thinking about when he was just sixteen years old. What would happen if you chased a beam of light as it moved through space? If you could somehow catch up to the light, Einstein reasoned, you would be able to observe the light frozen in space. But light can't be frozen in space, otherwise it would cease to be light. Eventually Einstein realized that light cannot be slowed down and must always be moving away from him at the speed of light. Therefore, something else had to change. Einstein eventually realized that time itself had to change, which laid the groundwork for his special theory of relativity.
>
> 2016

For more information see J. D. Norton, (2013), "Chasing a Beam of Light: Einstein's Most Famous Thought Experiment," www.pitt.edu/~jdnorton/Goodies/Chasing_the_light/ (accessed on July 6, 2021).

In his TED talk physicist Chad Orzel explains that:

> Austrian physicist Erwin Schrödinger is one of the founders of quantum mechanics, but he's most famous for something he never actually did: a thought experiment

involving a cat. He imagined taking a cat and placing it in a sealed box with a device that had a fifty percent chance of killing the cat in the next hour. At the end of that hour, he asked, "What is the state of the cat?" Common sense suggests that the cat is either alive or dead, but Schrödinger pointed out that according to quantum physics, at the instant before the box is opened, the cat is equal parts alive and dead, at the same time. It's only when the box is opened that we see a single definite state. Until then, the cat is a blur of probability, half one thing and half the other.

2014

For more information see P. Yam (2012), "Bringing Schrödinger's Cat to Life," *Scientific American*, www.scientificamerican.com/article/bringing-schrodingers-quantum-cat-to-life/ (accessed on June 25, 2021).

3. Virchow was, incidentally, also an anthropologist and archaeologist.

References

Brown, J. R. (1991b), "Thought Experiments: A Platonic Account," in Horowitz, T. and G. Massey (eds.), *Thought Experiments in Science and Philosophy*, 119–28, Savage, MD: Rowman and Littlefield Publishers.

Brown, J.R. (2010), *The Laboratory of the Mind: Thought Experiments in the Natural* Sciences. London: Routledge.

Herrle, S. R., E. C. Corbett, Jr., M. J. Fagan, C. G. Moore, D. M. Elnicki (2011), "Bayes' Theorem and the Physical Examination: Probability Assessment and Diagnostic Decision-Making," *Academic Medicine* 86 (5): 618–27, www.ncbi.nlm.nih.gov/pmc/articles/PMC3427763 (accessed on June 25, 2021).

Housen, A. (2007), "Art Viewing and Aesthetic Development: Designing for the Viewer," in *Visual Thinking Strategies,* https://vtshome.org/wp-content/uploads/2016/08/2Housen-Art-Viewing-.pdf (accessed on June 25, 2021).

Irvine, A. D. (1991), "On the Nature of Thought Experiments in Scientific Reasoning," in T. Horowitz and G. Massey, *Thought Experiments in Science and* Philosophy, 149–65, Savage, MD: Rowman and Littlefield Publishers.

Leski, K. (2015), *The Storm of Creativity* (*Simplicity: Design, Technology, Business, Life*). The MIT Press, Kindle Edition.

Mach, E. (1976), "On Thought Experiments" (1897), in *Knowledge and Error* (trans. T. J. McCormack and P. Foulkes), 134–47, Dordrecht, Holland: Reidel.

Matthews, M. (1988), "Ernst Mach and Thought Experiments in Science Education," *Research In Science Education* 18: 251–7.

Naghshineh, S., J. P. Hafler, A. R. Miller, M. A. Blanco, (2008), "Formal Art Observation Training Improves Medical Students' Visual Diagnostic Skills," *Journal of General Internal Medicine* 23 (7): 991–7, www.ncbi.nlm.nih.gov/pmc/articles/PMC2517949/ (accessed on March 25, 2021).

Norton, J.D. (2013), "Chasing the Light: Einstein's Most Famous Thought Experiment," in J. R. Brown, M. Frappier and L. Meynell (eds.), *Thought Experiments in Science, Philosophy, and the Arts*, 123–40, London: Routledge.

Orzel, C. (2014), "Schrödinger's Cat: A Thought Experiment in Quantum Mechanics," TED talk, ted.com/talks/chad_orzel_schrodinger_s_cat_a_thought_experiment_in_quantum_mechanics/transcript?language=en (accessed on June 25, 2021).
Osler, W. (1920), "The Old Humanities and the New Science," Presidential Address to The Classical Association, May 16, 1919, Boston and New York: Houghton Mifflin Company, https://medicalarchives.jhmi.edu:8443/osler/oldhum.htm (accessed on June 25, 2021).
Silverman, M. et al. (2007), *The Quotable Osler*. Philadelphia: American College of Physicians.
Root-Bernstein, R. S. (1985), "Visual Thinking: The Art of Imagining Reality," *Transactions of the American Philosophical Society* 75 (6): 50–67, The Visual Arts and Sciences: A Symposium Held at the American Philosophical Society.
Stone, Marvin J. (1995), "The Wisdom of Sir William Osler," *The American Journal of Cardiology* 75: 269–76.
Sundermier, A. "These 5 mind-melting thought experiments helped Albert Einstein come up with his most revolutionary scientific ideas," *Business Insider*, businessinsider.com/5-of-albert-einsteins-thought-experiments-that-revolutionized-science-2016-7#imagine-youre-standing-on-a-train-2 (accessed on June 25, 2021).
Tuteur, A. (2008), "Listen to your patient," The Skeptical OB, https://web.archive.org/web/20120319170242/http:/open.salon.com/blog/amytuteurmd/2008/11/19/listen_to_your_patient (accessed on March 26, 2021).
Ventura, H. O. (2000), "Rudolph Virchow and Cellular Pathology," *Clinical Cardiology* 23: 550–2.
Wilson, J. W. (1947), "Virchow's Contribution to the Cell Theory," *Journal of the History of Medicine and Allied Sciences* 2 (1): 163–78, https://doi.org/10.1093/jhmas/II.2.163 (accessed on June 25, 2021).
Yam, P. (2012), "Bringing Schrödinger's Cat to Life," *Scientific American*, www.scientificamerican.com/article/bringing-schrodingers-quantum-cat-to-life/ (accessed on June 25, 2021).

15 The Scientist's Social Network: Reimagining Crystallographic Diagrams Ahead of the 1951 Festival Pattern Group Collaboration

EMILY CANDELA

In 1951, just six years after the end of the Second World War, the national exhibition, the Festival of Britain, opened across the United Kingdom. Its epicenter was on London's South Bank, where futuristic constructions such as the flying-saucer-shaped Dome of Discovery accompanied exhibitions celebrating British achievements in industry, the arts, and science. At the time of its launch, the country was still recovering from World War II, and the Festival set out to raise the spirits of the British populace and jumpstart the postwar economy. Today, the Festival looms large in the historical memory of British design in the period as future-facing: embodying qualities of the "contemporary" or "Festival" style—which is associated with color and light contours and reflecting aspects of contemporary science. One project in particular has come to symbolize, and is perhaps partly responsible for, this last point: the Festival Pattern Group (FPG) project for the Festival. The FPG was a discipline-boundary-crossing venture that brought together collaborators from industrial design and the science of X-ray crystallography to produce pattern designs derived from the atomic and molecular structures of the physical world.

The FPG, which was organized by the quasi-governmental body the Council of Industrial Design (CoID), was an ambitious effort bringing together twenty-eight manufacturers from industries ranging from glass to textiles to wallpaper. They worked with a scientific consultant, the crystallographer Helen Megaw, who had originally proposed the idea to translate crystal structures into pattern design, and who provided crystallography diagrams to the FPG manufacturer's designers. The group produced prototypes for dozens of pattern designs derived from diagrams emanating from X-ray crystallography, including dress fabric printed with a pattern based on a crystallographer's diagram of the structure of the protein methaemoglobin, and a wallpaper pattern based on the structure of boric acid (Figs. 15.1 and 15.2). These crystallography-derived designs were on show throughout the 1951 Festival. They were incorporated, for instance, into the design of some science exhibits. Their true home, however, was the Regatta Restaurant on London's South Bank. There, the carpets, curtains, cutlery, and even waitresses' collars bore patterns based on

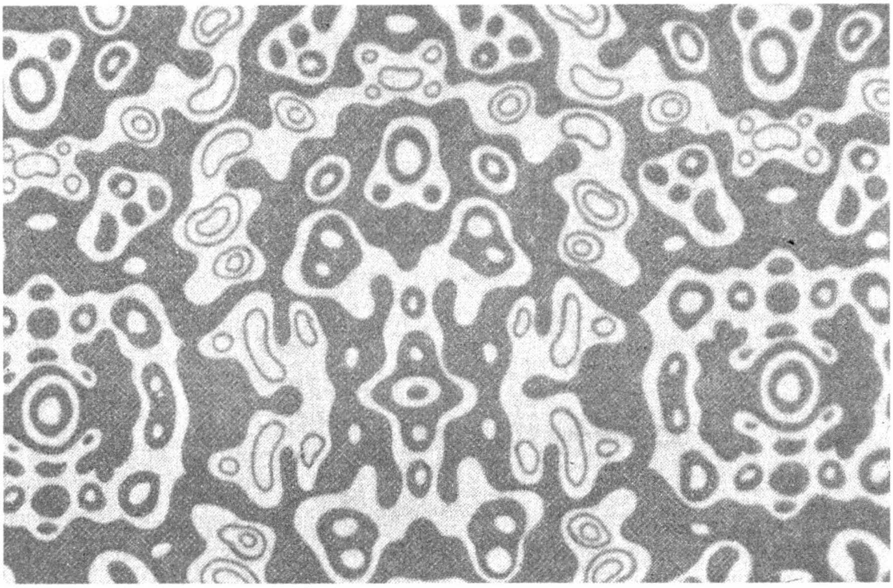

Fig. 15.1 Printed dress fabric with pattern based on an X-ray crystallography diagram of horse methaemoglobin, produced as part of the Festival Pattern Group project. Designed by S. M. Slade for British Celanese Ltd., 1949–1951. © The Design Council Slide Collection at Manchester Metropolitan University Special Collections.

the sub-microscopic structures of substancessfrom minerals to biological mattermstudied at the time by X-ray crystallographers (Fig. 15.3).

Despite its debut at such a high-profile event, however, the FPG was a short-lived endeavor. Most of the group's prototypes were never commercially produced, so they did not ultimately contribute to the Festival's economic goals, and they are unlikely to have had a wide-ranging influence on design at the time.[1] But the story of the FPG, particularly of how the project was catalyzed, illuminates a topic that has seen little in-depth research: how postwar British industrial design cultures engaged with science. In doing so, it presents an opportunity to hone approaches for generating richer understandings of interactions between design and science disciplines, both in the past and the present.

On the surface, the FPG can look like a neat interdisciplinary union, one that is in keeping with the historical memory of a generalized modernist interest in science in the design culture of postwar Britain. Several British designers in the period, such as textile designer Lucienne Day, deployed natural or quasi-scientific forms. And in the US, which might inflect the popular memory of British design, several now-iconic objects designed in the period reflect so-called atomic or molecular forms—including George Nelson's "Ball Clock" and Ray and Charles Eames's wire "Hang-it-all" wall-mounted hooks with spherical finials. But to explain postwar design through recourse to science as an element of 1950s "period style" can yield an incomplete or even misleading historical

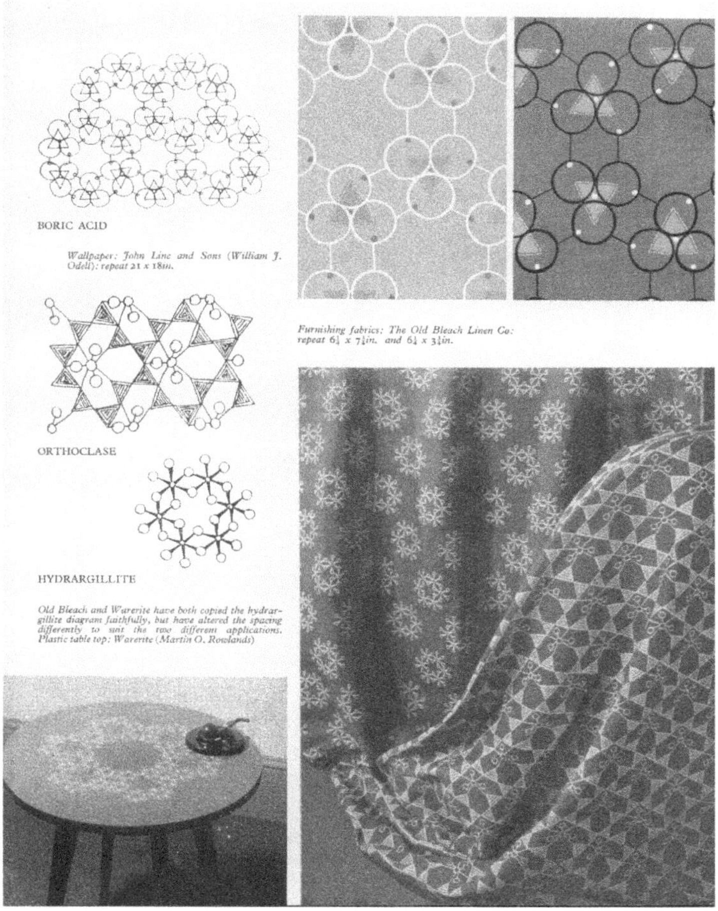

Fig. 15.2 Images of FPG textiles, wallpaper, and plastic laminate, alongside reproductions of crystal structure diagram drawings provided to the group's designers by Dr. Helen Megaw. Mark Hartland Thomas, *The Souvenir Book of Crystal Designs* (London: Council of Industrial Design, 1951), 9 © Design Council Archive by permission of the University of Brighton Design Archives.

narrative. Such an imagination of the past pictures the history of design as a progression of styles aligned with particular eras, while excluding consideration of numerous social, material, political, cultural, and other historical contingencies. This approach fails to illuminate questions about how the "science" got into "design," and why designers might have been interested in contemporaneous developments in scientific fields in the first place. But, the notion of period style sets the tone for many historical accounts of postwar British science-inflected industrial design, such as the so-called atomic furnishings bearing wire forms and ball-feet or finials produced in the period.

A more nuanced picture of exchange between design and science fields emerges upon close examination of the cross-disciplinary dialogues that took place in the time

ABOVE: *The interior of the Regatta Restaurant, South Bank. Here can be seen the general effect of some of the materials and designs used in combination*

BELOW: *Settees upholstered in PVC coated fabric. Imperial Chemical Industries Ltd, Leathercloth Division (Fabric designer: G. Garnier). At the Exhibition of Science, South Kensington*

MEMBERS OF
THE FESTIVAL PATTERN GROUP

Barlow & Jones Ltd
E. Brain & Co Ltd
British Celanese Ltd
Carter & Co Ltd
Chance Bros Ltd
Dobroyd Ltd
Dunlop Rubber Co Ltd
Elkington & Co Ltd
General Electric Co Ltd
A. C. Gill Ltd
Goodearl Bros Ltd
C. A. Harvey & Co (London) Ltd
Imperial Chemical Industries Ltd (Leathercloth Division)
Arnold Lever (London) Ltd
John Line & Sons Ltd
London Typographical Designers Ltd
Old Bleach Linen Co Ltd
R.H. & S.L. Plant Ltd
Spicers Ltd
Stevens & Williams Ltd
James Templeton & Co Ltd
Vanners & Fennell Bros Ltd
Vernons Industries Ltd
Warerite Ltd
Warner & Sons Ltd
Josiah Wedgwood & Sons Ltd
Wood Bros Glass Co Ltd
A linoleum manufacturing company

Addresses of members may be had upon application to Mr Hartland Thomas, at the Council of Industrial Design.

16

Fig. 15.3 Photographs of the interior of the Regatta restaurant and Exhibition of Science at the 1951 Festival of Britain featured in Mark Hartland Thomas, *The Souvenir Book of Crystal Designs* (London: Council of Industrial Design, 1951), 16 © Design Council Archive by permission of the University of Brighton Design Archives.

leading up to the FPG's formation. A detailed, critical view of interdisciplinary relationships can be generated through looking at networks of practitioners and the ideas, objects, and images that moved among them. Network models, which have their roots in the sociology of science, can aid research on cross-field collaboration because they are based on the idea that institutions and objects are constituted by and embedded in the circulation of ideas, people, practices, and things. This chapter looks at the cross-disciplinary exchange leading up to the FPG collaboration through the lens of networks (in this case, zeroing in on what were literally social and professional networks). Although this historical account does not adopt a sociology of science methodology, such as Actor-Network Theory, it draws upon an approach to descriptions of entities or events as, in sociologist John Law's words, "heterogeneous set of bits and pieces each with its own inclinations" (Law 1992: 386). As Glenn Adamson, Giorgio Riello, and Sarah Teasley write, from the perspective of design history, network models can illuminate "how knowledge (of any form, from a decorative pattern or method of weaving to an industrial technique or piece of proprietary software) is transmitted across cultures" (Teasley, Riello, and Adamson, 2011: 4). Where historical relationships between science and design are concerned, such an approach allows researchers to push beyond the identification of parallels between the production of two fields to more complex empirical understandings of the transmission and translations of knowledge between them. Relationships between fields are not only about finding common ground; they are also about networks of people, institutions, ideas, and objects, and how their differing backgrounds and inclinations shape their encounters.

This history of the FPG explores the project's germination within boundary-crossing constellations of figures from science, architecture, design, and art circles, and the varied interests and aesthetic ideologies that guided their cross-disciplinary exchange.[2] It traces the circulation of Helen Megaw's proposal for a crystallography-inspired pattern design project and accompanying diagrams (which eventually spurred the FPG's establishment) within postwar design networks in the late 1940s. It pushes past the notion of a generalized interest in science among postwar designers in the period (after all, not all designers were keen to engage with science). I focus on the conditions of the interactions between people and diagrams at the center of the story, and the aesthetic, ideological, and even institutional impetuses of those who gravitated to Megaw's diagrams and her proposal for a project uniting crystallography and design.

In this story, the crystallographic diagram acts as a "trading zone," that is, as a catalyst and meeting place for cross-cultural exchange (Long 2011: 8). It is a meeting place where we do not always see the clear connections or shared ideas that might be expected when it comes to the topic of relationships between science and another field. Practitioners from science and design involved in the exchanges leading up to the FPG's collaboration did not necessarily gravitate toward the prospect of crystallography diagrams as a basis for pattern designs for the same reasons. Each person in the network highlighted below, including Helen Megaw herself, saw the crystallographer's diagrams through his or her own frame of reference. The emphasis here is not on shared ideas and ambitions across disciplines, but on the different ideas—animated largely by the formal qualities of the crystallographer's diagram—that drove engagement with science by those in the design field. Therefore, rather than focusing on the FPG's prototypes themselves, which have been written about elsewhere (Jackson 2008; McGill 2007; Schoeser 2001), I explore the dynamics of both overlapping and divergent interests of people from different fields in the period preceding the FPG's interdisciplinary collaboration. In tracing Megaw's proposal's reception across design networks, it is necessary to touch on a number of events and areas of art and design practices in postwar Britain, several of which are subject areas that historians have studied in their own right. Rather than going into depth on the histories of each area, I create a slice across them, indicating ways in which they are interwoven. In the above ways, this account of the FPG's history reframes the story, centering on what historians, designers, and scientists who study or participate in relationships between design and science can learn from the FPG when it comes to reflecting critically on cross-field exchange.

Visualizing the Invisible

In the postwar period, Britain was a hotbed of cutting-edge research in X-ray crystallography, a technique which had been developed in the early 1910s. Although the study of crystals and their internal structures had existed long before the twentieth century, the application of X-rays to crystallography opened up dramatic new possibilities for deep investigations of matter at the smallest scale. Scientists working in the field in the postwar era elucidated the underlying structures of materials, including naturally occurring crystalline substances such as minerals, as well as synthetic polymers, and laboratory-grown crystals of organic materials, such as proteins like DNA. X-ray crystallography was used in physics, chemistry, and biology.

A crystal is made up of a regular arrangement of atoms that repeats in three dimensions. X-ray crystallography involves a set of specific techniques centered on directing X-rays through crystals, most commonly to generate data about the structures of their atoms. In the early- to mid-twentieth century, data about this scale of matter that could not be observed visually was generated by sending X-rays through a crystal, which diffract off the atoms inside, leaving a trace on a photographic plate. The resulting "diffraction photograph," usually a complex arrangement of dots and dashes, required much interpretation in order to visualize the atomic structure that had produced it. And because X-ray crystallographers were envisioning information about matter at scales that had not been visualized before, and certainly could not be *seen*, they developed a visual language to interpret and represent this new vision of nature, drawing in part upon existing conventions for representing chemical and crystal structures through diagrams and three-dimensional models. The diagrams at the heart of the FPG collaboration are artifacts of this visual language.

Born in Dublin in 1907, Helen Megaw practiced within a culture of X-ray crystallography that was defined, in part, by the techniques described above, in which hand-drawn diagrams and three-dimensional modeling were key parts of many research processes. She specialized in mineral structures, conducting research on ice early in her career (which resulted in an Antarctic Island, Megaw Island, being named after her). She went on to undertake innovative research on minerals with ferroelectric properties, which have applications in electronics, and on the structure of feldspar minerals (Glazer 2009). Megaw was one of several female X-ray crystallographers in the period—such as scientists Dorothy Hodgkin, Kathleen Lonsdale, and Rosalind Franklin—who were producing leading research that had resonance beyond the field. Although the number of women in crystallography was not high in absolute terms (Julian 1990), early- to mid-twentieth-century crystallography is remembered in part for its relative openness to women compared with other physical sciences at the time. This is often attributed to the progressive attitudes of gatekeepers, such as crystallographer and leftist activist J.D. Bernal, along with a culture of progressive politics in the field. Megaw had done her PhD at Cambridge with Bernal, and later joined his laboratory at Birkbeck College in London before moving to the Cavendish Laboratory in Cambridge in the late 1940s as their first female scientist (Crowther 1974). She has been less visible in histories of X-ray crystallography than her colleagues Hodgkin and Lonsdale, however, who have been the subjects of biographies. There is consequently room for historical research into Megaw's role in twentieth-century crystallography, which this account does little to fill, although, as I show below, aspects of her scientific practice play a role in the history of the FPG.

The Crystal-gazers of Postwar Art and Design

As Megaw's diagrams circulated among postwar design communities, they sparked cross-disciplinary dialogues. In the process, the meaning of Megaw's crystallographic diagrams shifted as they were inserted into new contexts, encountering actors in fields outside of her own. They began their journey through the design world in 1946, when Megaw contacted the London design consultancy, the Design Research Unit (DRU), with a proposal: "I should like to ask designers of wallpapers and fabrics to look at the patterns made available by X-ray crystallography," she wrote, enclosing several crystallographic diagrams and recommending that they be used as a basis for pattern design (Megaw 1946a). Although it is not clear exactly which diagrams Megaw sent to the DRU, her letter mentions that they included some structures she researched. Fig. 15.4, which shows the atomic structure of the mineral afwillite drawn later by Megaw for the FPG, suggests what the diagrams received by the DRU may have looked like.

Megaw's diagrams and accompanying message reached members of a specific social network that connected scientists, design figures, and artists who shared a set of overlapping political and aesthetic interests, and in which members of the DRU were deeply embedded. Although they ultimately did not pursue her idea to design patterns based on the representations generated by X-ray crystallographers, the DRU's circle was sympathetic to the overture from an X-ray crystallographer. Their responses to Megaw's proposal and diagrams indicate what the members of this design consultancy might have seen in the diagrams emanating from X-ray crystallographic research that arrived at their studio in 1946. Many designers outside the network weren't interested in science, let alone in the inscrutable (to the non-expert) diagrams of the underlying crystal structures of materials. The reception of these diagrams, therefore, illuminates the aesthetic and political frameworks that shaped approaches to science by this particular constellation of figures from postwar design and fine art cultures.

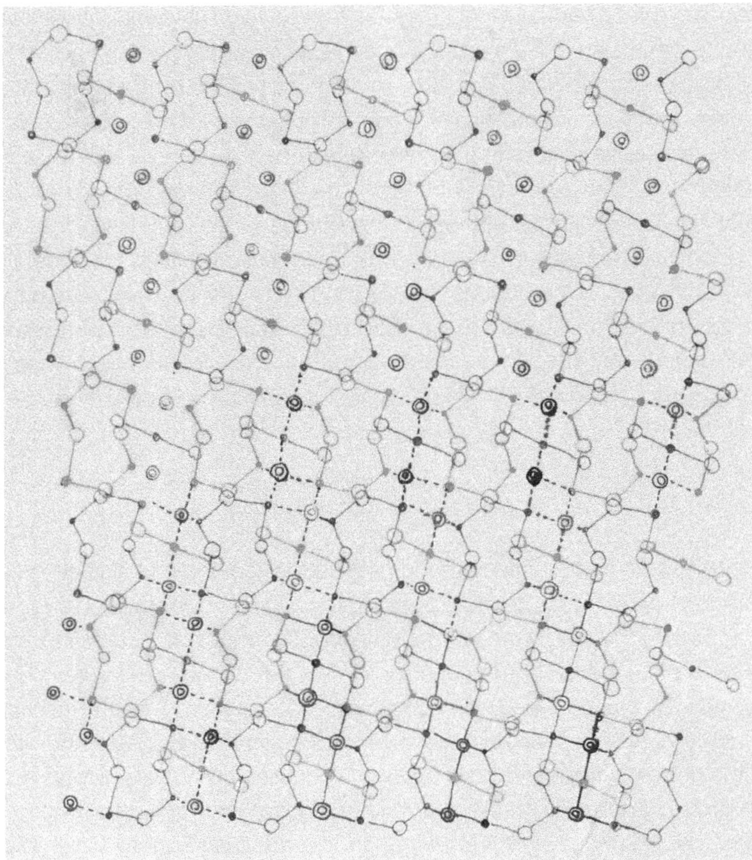

Fig. 15.4 Diagram of the mineral afwillite by Helen Megaw for the Festival Pattern Group, AAD 1977/3/429 © Colin Wilson, Estate of Helen Megaw / Victoria and Albert Museum, London.

Upon receiving Megaw's message and enclosed diagrams, the DRU's Marcus Brumwell was intrigued. "Your ideas about the beauty of shapes in nature is of course an absolutely first class one," he wrote, and set about making plans for her to discuss the proposal with the director of the DRU, Herbert Read (Brumwell 1946). When Megaw contacted the DRU, she had an inkling that they were open to dialogue with scientists. One of the DRU's founding aims was "to bring artists and designers into productive relation with scientists and technologists" (Cotton 2011: 29) in pursuit of their aims to reform and rebuild British industrial design after the war. And they had, in fact, originally contacted Megaw about working with them as a scientific consultant (at their friend Bernal's recommendation) before she sent them her proposal. Designer Misha Black, who would later play a role in the development of several FPG pattern designs as a designer of the Festival's Regatta restaurant, was a founding member of the DRU, along with the designer Milner Gray, and Brumwell, the head of an advertising agency. Its director, Read, was an art and design critic, poet, co-founder of the Institute of Contemporary Arts, and champion of British modernism.

The DRU's inclination toward current science reflects their sympathy with the sentiments of their friends from constructivist fine art practice at the time. The DRU members were bound through friendships and professional ties to a cross-disciplinary network of scientists, design figures, and artists that included Bernal and the "St Ives" circle of constructivist and abstract artists (so named because they had decamped to the Cornish coast from London during the war), which included Barbara Hepworth, Ben Nicholson, and Naum Gabo. Their associates in the St Ives circle positioned themselves within a British legacy of drawing upon nature in their work. But they also kept a keen eye on contemporary developments in science, harboring a fascination with new scientific practices of envisioning and understanding forms and patterns in nature beyond the visible. Writing in the 1944 book *This Changing World*, edited by Brumwell, Herbert Read proclaimed that "Science has taught us that underneath the shifting appearance of nature [. . .] there is a system of law" (Read 1945: 8). Scientific ideas animated the artists' own explorations of underlying "laws" of form. Many in the St Ives group were among the British avant-garde's enthusiastic readers of the biologist D'Arcy Wentworth Thompson's *On Growth and Form* (1917), a mathematical exploration of recurring morphologies across animate and inanimate nature.

For many in the circle linking the DRU and St Ives artists, interest in contemporary science went beyond the visual. It also extended to political concerns about the social role of science. They looked to contemporary scientific advances as a potential peaceful force, one that could help usher in an imagined social utopian future. As editor of *This Changing World*, Brumwell, a friend and supporter of the work of many St Ives constructivists, brought together contributors from a range of topics, including the sciences, to imagine their future shape in postwar Britain. He included an essay by Bernal (a member of the extended network described here who was a close friend of Brumwell's and Hepworth's), laying out the potential of science to contribute to the distribution of resources within a future socialist system: "the ends for which people are striving—food, work, security, and freedom—are gifts which science has put within reach of all" (Bernal 1945: 16). Members of this network, such as Hepworth, also expressed concerns about science's social role and the potential dangerous ramifications of the rapid development of scientific research that was estranged from other cultural realms. This issue hastened their intent to engage with science and scientists (Barlow 1996). As Barbara Hepworth wrote to her friend Margaret Gardiner, "the speed is out of proportion in the world of invention to the detriment of poetry and aesthetic vision ... I cannot see any hope of stopping this suicidal impulse unless Art & Science stand firm together" (Gardiner 1982: 28). A sense of the unbridled but "parallel" progress across science, the arts, and other areas underpins Brumwell's book as well (Brumwell 1945: 1). Within this circle, Megaw's proposal was appealing, because it forged an intersection and engagement between the work of contemporary scientists and the arts.

Visual qualities of crystallography diagrams also linked, on a symbolic level, to the socialist utopian political convictions of many associated with the St Ives circle. It is difficult to know exactly how deeply those in art and design fields who were active in this network engaged with scientific knowledge—and they certainly did not engage with crystallography as scientists would—but it is clear that Hepworth and Gabo, in particular, harbored an enthusiasm for crystal structure that was fed through their relationship with Bernal (Burstow 2014; Barlow 1996; Hammer and Lodder 1996). Art historian Robert Burstow writes about the reasons for their enthrallment to crystallography, arguing that geometric form, including that found in crystals, symbolized "order, precision, predictability, universality" for the constructivists, and thus the social utopian potential they saw in science (2014: 60).[3] Indeed, when Brumwell forwarded Megaw's diagrams and proposal to his friend Barbara Hepworth, she wholeheartedly encouraged what she called Megaw's "marvelous" idea ("why hasn't anybody thought of it before?," wrote the artist). Hepworth advised, "The main point seems to me to produce them [. . .] exactly as they really are. To me they are more beautiful than any man-made pattern" (Hepworth 1946). For Hepworth, the geometric structures found in Megaw's diagrams and proposal would have cut right to the heart of her explorations of form, but also to her political convictions and imagination for the future.

Whether it was crystallographic symbolism, the promise of uniting science and "art," or both, Brumwell perceived a resonance with the concerns of the constructivists in the patterns Megaw had drawn and sent to the DRU. He responded to her proposal by lending her his beloved copy of the 1937 book *Circle: International Survey of Constructive Art*, edited by Gabo, Nicholson, and the architect J.L. Martin. *Circle* articulates science's place in the ethos and ambitions of the constructivists, expressing both their enthusiasms and anxieties about contemporary trajectories of scientific research. Brumwell's interpretation of Megaw's proposal in light of the constructivist outlook on science is also revealed by his statement, in response to Megaw's proposal that, "The general idea is one which interests Herbert Read profoundly" (Brumwell 1946). Read shared with the circle of St Ives artists a preoccupation with natural forms—and with crystal structure and its associated symbolism for the St Ives constructivists. Read's novel *The Green Child*, for instance, has the protagonist travel to a subterranean utopian society in which crystals and "the science we call crystallography – the study of the forms, properties, and structure of crystals – was the most esteemed of all sciences," comprising the foundation for the civilization's very ideas of beauty and truth (Read 1935: 173). Megaw's proposal to import crystallographic forms into pattern design was aligned with Read's outlook. For him, the aesthetic frameworks of abstract art, associated with artists such as Hepworth and Gabo, extended to his vision for the future shape of industrial design—the area that Megaw's proposal for crystallographic wallpaper and textile patterns directly impinged upon. In his book *Art and Industry*, Read called for "new aesthetic standards for new methods of production" in industrial design derived from abstract artists' investigations of form (Read 1956: 9). Brumwell expressed his excitement to share

Megaw's ideas with Read, "the man to help us to fructify them" (Brumwell 1946), and Megaw was soon invited to write an essay for a DRU monograph edited by him.

Megaw's impetus for proposing a cross-disciplinary pattern design project was in many ways distinct from that of her interlocutors in the DRU who were so enthusiastic about the proposal. Although she was closely associated with Bernal, it is not clear whether she shared the socialist utopian views of the milieu described above, or the modernist concerns about form and abstraction that would have driven the DRU's interest in her diagrams. Megaw's writing on the subject suggests that she was driven by an abiding interest in the aesthetic qualities of crystallography diagrams, which was embedded in both her scientific practice, and, as I explain below, her personal engagement with decorative art and amateur craft. The visualization of crystalline structures through diagrams was a vital element of her scientific work, aiding her research on the structures of minerals. In the process of drawing diagrams for the purposes of her scientific research, Megaw was attuned to the aesthetic qualities of the patterns that emerged. In her initial letter to the DRU, Megaw wrote, "I am constantly being impressed by the beauty of the designs which crop up in the course of the [scientific] work without any attempt of the worker to secure anything more than clarity and accuracy" (Megaw 1946a). One of Megaw's fundamental aspirations in proposing the use of crystallographic diagrams in pattern design was to communicate publicly about this aspect of X-ray crystallography practice. As she later wrote in an essay, "Pattern in Crystallography," for the (ultimately abandoned) book project by the DRU, "It is hoped that [the structures elucidated by crystallographers] may suggest to designers ways in which to broadcast to a wider public some of the aesthetic pleasure found in the subject by crystallographers themselves" (Megaw 1946b).

Megaw's proposal to the DRU was not the first time she had pursued the decorative applications of crystallography diagrams. Her ongoing engagement with pattern design through amateur craft frequently drew upon her scientific practices of representation. As Lesley Jackson writes, Megaw had once given her friend, the X-ray crystallographer Dorothy Hodgkin, a linen cushion embroidered with the structure of aluminum hydroxide, and she had made Christmas cards with the same crystal structure (Jackson 2008). During retirement, when Megaw discovered the plant *Perovskia*, its leaves also found their way onto her Christmas cards (Glazer 2002). Megaw exhibited a wider knowledge of pattern design as well. In her letter to the DRU, she invoked the nineteenth-century arts and crafts designer William Morris, who devised wallpaper patterns based on repeated motifs drawn from nature. She suggested that textiles with patterns derived from crystallography diagrams should be named after the substance represented, "just as the William Morris patterns were called after their constituent flowers" (Megaw 1946). Her allusion to Morris points to the Victorian slant of Megaw's reference points for design, in contrast to the modernist outlook of the DRU.

In circulating her diagrams to the DRU, however, she tapped into an existing network bound by an already-porous boundary between crystallography and design, and where crystalline structure—of the kind she presented in her diagrams—was a fulcrum for cross-field relationships and exchange. But as Megaw's exchange with the DRU shows, this kind of exchange does not necessarily mean that scientific knowledge, in the form of the crystallographer's diagram, operated in design or art circles in the same—or even similar form—to the way it operated in scientific communities. What begins to emerge from Megaw's dialogue with the DRU's circle is the multivalent character the crystallographic diagram took on as it traveled through design networks before the establishment of the FPG. Between Megaw's dispatch of the diagram as a visual messenger of the "aesthetic pleasure" underpinning her scientific practice of visualizing structures of matter, and the DRU's studio, Megaw's proposal and diagrams become overlaid by the modernist aesthetic frameworks that shaped the DRU's outlook on science. The DRU's particular modernist frameworks do not, however, describe the outlook of postwar British design culture as a whole when it comes to engagement with contemporary science, and it was not the only one that played a role in the FPG story, as I explain below.

Mark Hartland Thomas's Crystalline Aesthetic

Megaw's diagrams resurfaced in the design world in 1949 when her colleague Kathleen Lonsdale presented images of them, along with Megaw's proposal that they be applied to pattern design, at a talk organized by the Society of Industrial Artists, a professional association for industrial and graphic designers. These diagrams caught the eye of Mark Hartland Thomas, the CoID's Chief Industrial Officer, a member of a Festival of Britain planning committee, and the person responsible for industrial design exhibits at the Festival. He wrote to Megaw after Lonsdale's lecture, eagerly asking if he could pursue their transformation into pattern design with manufacturers. Megaw agreed, and Hartland Thomas began assembling a group of manufacturers up to the task of producing patterns based on crystallographic diagrams, with the ambition of launching them at the Festival. Megaw became the group's official "Adviser on Crystal Structure Diagrams." Working closely with Megaw, Hartland Thomas was an enthusiastic steward of the project right up until its Festival debut.

Hartland Thomas plays an important role in the networks crisscrossing science and design fields in the FPG's pre-history, but his own reasons for engaging with scientific subject matter are not immediately obvious. The CoID represented a different environment for the negotiation of scientific subject matter than that of the DRU. Bringing science and design together was not a particular aim of CoID-promoted design. And although the Festival celebrated British achievement in science, industry, and the arts, the merging of scientific form with industrial design *in individual objects* was not an overarching goal for the CoID for its industrial design exhibits. The idea to pilot such a collaboration between industrial design and science was actually so removed from the exhibition's aims that Hartland Thomas was initially worried that the FPG's use of scientific diagrams in pattern design might overstep his remit. Thomas, in fact, was concerned not to "give offence to my scientific colleagues" planning the Festival's science exhibits, as he later wrote, so he sought the approval of Ian Cox, Director of Science and Technology at the Festival, before embarking on the project (Hartland Thomas 1950). Given that such a close collaboration between science and design for the Festival was not an objective in its planning, and even seemed out of the ordinary to the Festival planner who organized it, the question of why Hartland Thomas so enthusiastically took on coordination of an elaborate cross-disciplinary collaboration, and how he reconciled the project's emphasis on scientific form with the institutional dictates of the CoID, is crucial to understanding the roots of the FPG collaboration.

Hartland Thomas had strong aesthetic convictions underpinning his response to crystallographers' representations of the underlying structures of matter, and his background as an architect committed to tenets of the modern movement is key to understanding his aesthetic outlook.[4] After the Second World War, Hartland Thomas belonged to a cadre of architects calling for the adoption of classical ideals of geometric proportion. This was a response to the postwar industrialization of architecture and the state planning of housing, which had sparked anxieties among architects about their authorial role, and the identity of the profession as an aesthetic endeavor (Neumann 1996). A rallying call to take up the commitment to geometric proportion came from Hartland Thomas two years after the end of the war, when he called for a new aesthetic based on ancient ideals of "Scale, Modulus, Proportion [. . .] Symmetry, and Balance" (Hartland Thomas 1947a: 37). This revived interest in the deployment of classical geometrical order in architectural design methods had antecedents in theories of continental modernist architecture, most closely identified with Le Corbusier. This interest in geometric order also reflects the legacy of the neoclassical Beaux-Arts tradition in architectural education in Britain that was still alive in the period when many postwar architects, including Hartland Thomas, were trained.

The crystallographic diagram aligned with many of the geometric ideals with which Hartland Thomas was preoccupied, such as symmetry and the repetition of modular elements. He highlighted these qualities in *The Souvenir Book of Crystal Designs*, an illustrated guide to the FPG sold at the Festival, lauding the diagrams Megaw provided for the project for having "the discipline of exact repetitive symmetry" (Hartland Thomas 1951: 2). Even before setting eyes on Megaw's diagrams, however, Hartland Thomas had identified a parallel to the tenets of a fundamental pillar of his aesthetic system, "modulus," in the structures of crystals. Just as a repetition of a crystal's single unit cell produces a larger crystalline structure in nature, modulus involved the election of a single "dimensional unit" as the basis for proportions throughout an individual building (Hartland Thomas 1947b: 79–80). In his 1947 book *Building Is Your Business*, Hartland Thomas identified the antecedents of his architectural ideal in nature, "at a much smaller scale in crystalline structures, and at a vastly larger scale in celestial geometry" (Hartland Thomas 1947b: 74).

In addition to their appeal on the level of his aesthetic ideology, Hartland Thomas was also driven by an institutional interest in his enthusiasm for crystallographic diagrams and subsequent intent to organize the FPG collaboration. He saw in the crystallographic diagrams' aesthetic a practical purpose one linked with the institutional objectives of the CoID. The conventions of crystallographic diagrams offer a visual coherence across the representations of numerous structures. He admired this "remarkable family likeness" of crystallography diagrams, and throughout the FPG's working period, emphasized that this formal coherence should be maintained in the final designs (Hartland Thomas 1951: 5). In the FPG project, he deployed crystal structure diagrams in this way to create a visual identity across a collection of products from different industries. The FPG thus reflected the CoID's objective of forging stronger links with industry, intended to ensure widespread uptake of the tenets of "good design" that the Council promoted among manufacturers. In fact, the CoID's Director, Gordon Russell, saw the FPG's value primarily in its encouragement of stronger relationships across industries, rather than in the decorative use of crystallographic diagrams. He underplayed what he called "the decorative possibilities of the patterns themselves," while praising the project's potential for relationships between the FPG's industrial constituents: "It is this aspect of the matter making the project a sort of Design Centre that I think most important," he wrote (Russell 1950). The latter spoke more directly to the CoID's core aims than did the notion of adapting scientific forms to pattern design. Hartland Thomas, on the other hand, combined the two; he sought to unify the products of different industries using the crystallographic diagram as an aesthetic tool. This too mirrored aspects of Hartland Thomas' practice in architecture; two years after the Festival, he worked to standardize components across the building industry through an organization called the Modular Society, just as he sought to unify the products of different kinds of manufacturers in his administration of the FPG using the crystallographic diagram as an aesthetic instrument of standardization.

As a Festival planner, Hartland Thomas also responded to the FPG's potential to contribute to the Festival's drive to showcase British accomplishments. The Festival commemorated the 1851 Great Exhibition of a century earlier. Whereas the 1851 event was an explicit celebration of imperial power exhibiting objects from the British Empire, the 1951 Festival, set at the beginning of the empire's dissolution, was comparatively national-focused.[5] Since X-ray crystallography was, as Hartland Thomas wrote in the *Souvenir Book*, "particularly highly developed in Britain," that meant it contributed to the Festival's national narrative, and had in fact already been slated to be featured in the exhibition's science exhibits as a result (Hartland Thomas 1951: 6). As Jo Littler has pointed out, "the perpetuation of imperial narratives of discovery and the heroic adventurer discovering new lands" was palpable within the Festival's exhibition narrative despite its focus on Britain rather than the empire, "only now it resided in 'science'" (Littler 2006: 26).

As in the case of the DRU, Megaw's diagrams prompted Hartland Thomas to read into the scientific image his own codes, meaning, and potential uses; they appealed to his aesthetic ideology as a modernist architect, and were deployed as tools to pursue the institutional goals he worked toward as a CoID officer and Festival planner.

He is thus a part of an extended network that cuts across design cultures of postwar Britain, through which the crystallographic diagram circulated, shifting in its significance and perceived potential—aesthetically, conceptually, politically—as it moved through environments associated with varied modernisms. Tracing the circulation of Megaw's diagrams among design networks in the late 1940s, it is clear that there is no notion of a generalized preoccupation with science among design cultures of the period. There is no single way that scientific knowledge operated in postwar British industrial design circles. Artifacts of the crystallographer's visual language appealed to several contemporary cultures of practice (some of which did overlap). Their varied convictions included hopes for a social utopian future, institutional imperatives for the future of British industrial design, and classical ideals of proportion emanating from postwar architectural debates.

"Undisciplined" Dialogues

Within the networks of practitioners from different fields and disciplinary cultures who play roles in the FPG's backstory, the crystallographic diagram acted as a mobile "trading zone." The concept of the "trading zone"—which has been employed in anthropology and subsequently the history of science—refers to a site that allows for the exchange of knowledge between different cultures. These might be physical sites—coffee shops, for instance—or symbolic ones. Historian Pamela O. Long, for instance, evokes the concept of the "trading zone" in research on how knowledge exchange between artisans and "learned men" affected the development of the Scientific Revolution in early modern Europe (as discussed in Lee Chichester's essay for this volume). Long describes the Vitruvian tradition in architecture as a symbolic "trading zone" between these groups (Long 2011). The story of the FPG's pre-history reveals the crystallographic diagram as a symbolic trading zone, as a site for communication and exchange between practitioners in different fields and between visual languages and the differing aesthetic, political, professional, and bureaucratic aims that drove them.

The FPG represents an opportunity to explore how scientific representations functioned outside the laboratory—"in public"—in cultures of practice associated with design production and policy, including that of the members of the DRU and of Hartland Thomas (as the modernist architect *and* as the CoID officer). Designers and artists represent particular *publics* for science. As such, this history of their involvement in cross-field dialogues preceding the establishment of the FPG undermines outdated notions of the passive reception of scientific knowledge by publics outside of scientific practice. A fuller picture of the reception and negotiation of scientific knowledge in public can only be achieved though understanding the audiences for scientific knowledge as "active consumer[s]" (O'Connor 2009: 335). An active public for science may be motivated by interests other than or in addition to gaining accurate understandings—in a scientist's view—of a scientific subject. As historian of science Katherine Pandora writes, "Encounters with science in the everyday world can be multifarious, miscellaneous, overlapping, partial, and contradictory—in fact, *un*disciplined" (Pandora 2009: 347). The "trading zone" is an important site for such "undisciplined" encounters—one in which complex dynamics of agency between the approaches of different fields or cultures play out. This reality bubbles beneath Megaw's conversations with figures outside her immediate field. Although she initiated the dialogues recounted here and provided the diagrams and proposal at their center, her interlocutors negotiated crystallographic knowledge through the lens of their own interests and world views. This is an important point when it comes to how scientific knowledge, in the form of a diagram for instance, is received outside the walls of the laboratory in other fields. The knowledge signified by a representation in scientific practice can dissolve and be replaced by new sets of meanings in a non-scientific practice.

The story of the FPG offers an opportunity to develop more distinct understandings of the role and significance of science in practices and in institutional and intellectual cultures associated with postwar British design. These understandings move past the surface understandings of "style" to the aesthetic outlooks and ideologies, institutional objectives, and political frameworks that underpinned them. As this snapshot of the process demonstrates, there is much more to learn about the complex and varied ways in which scientific knowledge was negotiated within British industrial design of the period. Although this essay has focused on a cross-disciplinary exchange that took place in the past, the FPG story's potential resonance extends to the present. Today, notions of interdisciplinarity shape many efforts in arts and humanities research and the cultural sector that create or study links between science and design.[6] The current enthusiasm for interdisciplinarity has much positive potential for research that crosses discipline boundaries in productive ways. The impulse to celebrate interdisciplinary connections, collaboration, and exchange must not obscure the complex aspects of encounters between disciplines, however, such as those at the root of the FPG collaboration. Interrogating both moments of resonance *and* disjuncture can aid understandings of how disciplinary cultures shape cross-field interactions, as well as the values that underpin impulses towards cross-field exchange itself.

Notes

1. Paul Reilly, the CoID's information officer in the 1950s, remarked in reflections on the Festival twenty-five years later that the FPG can "hardly be said to have laid the foundations for a new school of design—indeed they barely survived the Festival year" (Reilly 1976: 61). As Lesley Jackson points out in her detailed history of the FPG, several factors may have contributed to the fact that many FPG prototypes were not commercially produced, including continued wartime materials shortages, and restrictions on the domestic furniture market (Jackson 2008).
2. Where other texts on the FPG have touched on the cross-disciplinary exchange before the FPG's formation as background to the topic, the focus has been weighted toward

description, leaving much room for interpretation on the topic. Further scholarship on the FPG includes Jackson (2008), McGill (2007), Schoeser (2001), and Forgan (1998).

3. Burstow acknowledges however that its precise imprint is difficult to pinpoint in their sculpture, especially as the artists studied mathematical models, the geometric forms of which are clearly evident in their work and resemble the geometric forms of crystals (Burstow 2014: 60).
4. Hartland Thomas was an active member of the MARS group (Modern Architectural Research), the British arm of the modernist architecture body, Congrès internationaux d'architecture moderne (CIAM).
5. As Jo Littler has observed, however, a subtler version of Imperial Britain was on display, for instance, in the imagery of Britannia that was integral to the Festival's logo (Littler 2006).
6. This is evident in recent research interest in relationships between histories of science and art (which accounts for a larger body of scholarship than histories on science and design). In 2002, historian Ludmilla Jordanova identified the research attempting to bridge histories of art and science being published at the time as part of a broader cultural engagement with "art and science" (Jordanova, 2002: 341). Examples include work by art historians Arthur I. Miller (2014) and Martin Kemp (2006). It is embodied also by the work of institutions, such as the Wellcome Collection, a museum and gallery that opened in London in 2007, which is concerned with the "connections between science, medicine, life, and art" (The Wellcome Collection 2018). And in 2008, the Museum of Modern Art in New York staged an exhibition, "Design and the Elastic Mind," which included several design projects sitting on the border with science (Antonelli and Aldersey-Williams, 2008).

References

Antonelli, P. and H. Aldersey-Williams (2008), *Design and the Elastic Mind*. New York: Museum of Modern Art.

Barlow, A.J. (1996), "Barbara Hepworth and Science," in D. Thistlewood (ed.), *Barbara Hepworth Reconsidered*, 95–107, Liverpool: Liverpool University Press.

Bernal, J.D. (1945), "Transformation in Science," in M. Brumwell (ed.), *This Changing World*, 15–26, London: George Routledge & Sons.

Brumwell, M. (1945), "Forward," in M. Brumwell (ed.), *This Changing World*, 1–2, London: George Routledge & Sons.

Brumwell, M. (1946), Letter to Helen Megaw, March 1, Archives of Art and Design, Victoria and Albert Museum, London, UK, 1977/3/13.

Burstow, R. (2014), "Geometries of Hope and Fear: The Iconography of Atomic Science and Nuclear Anxiety in the Sculpture of World War and Cold War Britain," in C. Jolivette (ed.), *British Art in the Nuclear Age*, 51–80, Surrey: Ashgate.

Cotton, M. (2011), *Design Research Unit 1942–72*, Cologne: Buchhandlung Walther König.

Crowther, J. G. (1974), *The Cavendish Laboratory, 1874–1974*. New York: Science History Publications.

Forgan, S. (1998), "Festivals of Science and the Two Cultures: Science, Design, and Display in the Festival of Britain, 1951," *The British Journal for the History of Science* 31 (2): 217–40.
Gabo, N., B. Nicholson and J.L. Martin, eds (1937), *Circle: International Survey of Constructive Art*. London: Faber and Faber.
Gardiner, M. (1982), *Barbara Hepworth: A Memoir*. Edinburgh: Salamander Press.
Glazer, M. (2002), "Helen Megaw" (obituary), *Independent*, March 28, www.independent.co.uk/news/obituaries/helen-megaw-9178295.html (accessed on March 30, 2018).
Glazer, M. (2009), "Megaw, Helen Dick (1907-2002)," in L. Goldman (ed.), *Oxford Dictionary of National Biography*, 712–14, Oxford: Oxford University Press.
Hartland Thomas, M. (1947a), "Aesthetics the Vanguard Now," *Architectural Design*, February: 36–7.
Hartland Thomas, M. (1947b), *Building Is Your Business*. London: Allan Wingate.
Hartland Thomas, M. (1950), Letter to Brian Peake, June 2, Design Council Archives, Brighton, UK, 5384.
Hartland Thomas, M. (1951), *The Souvenir Book of Crystal Designs*. London: Council of Industrial Design.
Hammer, M. and C. Lodder (1996), "Hepworth and Gabo: A Constructivist Dialogue," in D. Thistlewood (ed.), *Barbara Hepworth Reconsidered*, 109–133, Liverpool: Liverpool University Press.
Hepworth, B. (1946), Copy of letter to Marcus Brumwell, enclosed in letter from Brumwell to Helen Megaw, March 15, Archives of Art and Design, Victoria and Albert Museum, London, UK, 1977/3/15.
Jackson, L. (2008), *From Atoms to Patterns: Crystal Structure Designs from the 1951 Festival of Britain*. Somerset: Richard Dennis.
Jordanova, L. (2002), "And?," *British Journal for the History of Science* 35 (126): 341–5.
Julian, M. M. (1990), "Women in Crystallography," in G. Kass-Simon and P. Farnes (eds.), *Women of Science: Righting the Record*, 335–83, Bloomington: Indiana University Press.
Kemp, M. (2006), *Seen/Unseen: Art, Science, and Intuition from Leonardo to the Hubble Telescope*. Oxford: Oxford University Press.
Law, J. (1992), "Notes on the Theory of the Actor-Network: Ordering, Strategy and Heterogeneity," *Systems Practice* 5: 379–93.
Littler, J. (2006), "'Festering Britain': The 1951 Festival of Britain, Decolonisation and the Representation of the Commonwealth," in S. Faulkner and A. Ramamurthy (eds.), *Visual Culture and Decolonisation in Britain*, 22–42, Hampshire: Ashgate.
Long, P.O. (2011), *Artisan Practitioners and the Rise of the New Sciences, 1400–1600*. Corvallis: Oregon State University.
McGill, T. (2007), "Design Under the Microscope: The Festival Pattern Group 1951: The Council of Industrial Design and the Mechanics of Industrial Liaison," *The Decorative Arts Society Journal* 31: 92–115.
Megaw, H. (1946a), Letter to Marcus Brumwell, February 20, Archives of Art and Design, Victoria and Albert Museum, London, UK, 1977/3/12.
Megaw, H. (1946b), "Pattern in Crystallography," November 1946, Design Council Archives, Brighton, UK, 1466.
Miller, A. I. (2014), *Colliding Worlds: How Cutting-Edge Science is Redefining Contemporary Art*. New York: W. W. Norton & Company.

Neumann, E. (1996), "Architectural Proportion in Britain 1945–1957," *Architectural History* 39: 197–221.

O'Connor, R. (2009), "Reflections on Popular Science in Britain: Genres, Categories, and Historians," *Isis* 100: 333–45.

Pandora, K. (2009), "Popular Science in National and Transnational Perspective: Suggestions from the American Context," *Isis* 100: 346–58.

Read, H. (1935), *The Green* Child. New York: New Directions.

Read, H. (1956 [1934]), *Art and Industry*. London: Faber and Faber.

Read, H. (1945), "Threshold of a New Age," in M. Brumwell (ed.), *This Changing World*, 7–14, London: George Routledge & Sons.

Reilly, P. (1976), "The Role of the Design Council," in M. Banham and B. Hillier (eds.), *A Tonic to the Nation: The Festival of Britain 1951*, 58–61, London: Thames and Hudson.

Russell, G. (1950), Letter to Dr. R.S. Edwards, June 26, Design Council Archives, Brighton, UK, 5384.

Schoeser, M. (2001), "The Appliance of Science," in E. Harwood and A. Powers (eds.), *Twentieth Century Architecture 5: Festival of Britain*, 118–26, London: Twentieth Century Society.

Teasley, S., G. Riello, and G. Adamson (2011), "Introduction: Towards Global Design History," *Global Design History*, New York: Routledge.

Thompson, D. W. (1961 [1917]), *On Growth and Form*. Cambridge: Cambridge University Press.

"About Wellcome Collection," *The Wellcome Collection*, https://wellcomecollection.org/about-us (accessed on April 6, 2018).

16 Cinematic Data Visualization

CATHERINE GRIFFITHS

Introduction

Data visualization, in which data are translated into visual form, is an essential instrument for communicating information. Data visualization typically uses stationary abstract forms to simplify data and make it more accessible, and to amplify comprehension of the information. These are accomplished by transforming numerical information into a form that uses color, shape, scale, and spatial organization to reveal trends, relationships, and orders of magnitude in the information. Data visualization is a tool for the representation of the known, but it is also a means of revealing new insights through the design process.

Over the past few years, data sets have become bigger, and information has grown more complex and nuanced. Data has also become more widely available, which points to the fact that context is critical to fully understanding data—especially for big, nuanced, and complex data sets that are accessible to a broad audience. These observations suggest that the streamlined, static, and often context-limited graphic languages that are used for data visualization need to be reconsidered. Despite the fact that graphic language can sometimes render data more legible, data visualizations of complex information are frequently too visually or statistically abstract to comprehend—especially for the public. Scientific data from environmental monitoring, and models developed from scientific studies, for example, are consequential to the public, but they are frequently impervious to laypeople.

The tradition in data visualization has been to render statistical meaning and scientific insight more accessible, however, visualizations may be too abstract because the data is decontextualized. When data is collected, it has typically first been extracted and then disconnected from its source so that it can be analyzed objectively. Being disconnected from its origins separates the social, environmental, and human contexts of data from data visualizations. In this essay, I will explore how context is critical to a thorough understanding of scientific data.

Although some abstraction may be necessary for analyzing and configuring mathematical and scientific data, we designers need to experiment with multi-sensory modalities that communicate clearly, while also taking context into consideration when representing data. Design research and practice, in fact, support a reflexive process of examining design questions critically and contextually—such as how to best represent complex data—while also using abductive reasoning in which there is clear premise, but a to-be-determined conclusion. These aspects of design research and practice make it an ideal way to explore avenues for communicating data clearly and effectively.

Cinematic data visualization, an effective approach that came out of design research, uses cinematographic aesthetics to recontextualize how designers can render data visualizations, and how audiences understand them. Combining techniques from legacy photochemical film processes and contemporary virtual visualization capabilities, cinematic media offers designers a rich aesthetic and atmospheric palette of communication methods. A visual, time- and motion-based medium, cinema can be deconstructed into elements of narrative, place, photography, motion, framing, depth of field, and computer-generated effects. In cinematic data visualization, designers can utilize these elements as visual tactics to present data as a sensual narrative that is situated in a relevant contextual environment.

In this essay, I first discuss how cinematic data visualization contextualizes data. I present two of my projects—*Alluvium* and *LA River Nutrient Visualization*—as case studies that elaborate these ideas using representational modalities. Each of these projects utilizes a range of design and data visualization tactics to construct situated visualizations, which I explore here using the language of cinema.

Cinematic Data Visualization

Cinematic data visualization is part of the recent turn in image culture from photography to data—from lens-based, photographic images that function as representations through mimesis (images that resemble physical phenomena), to computationally generated digital images that exist as data structures. These data structures are formed by pixel arrays and, in addition to their representational possibilities, serve as computational mechanisms for restructuring visual information. Pixel-array images, such as digital photographs, can be filtered and manipulated by applying algorithmic methods to the pixel data. This means that they can be treated like other data sets, which is not the case for traditional photographic imagery.

In *Technologies of Vision: The War Between Data and Images*, digital media scholar Steve Anderson considers the oppositional relationship between data and fixed images, which, he argues, sets up an unproductive competition between computability and mimesis. Anderson warns that "data and images," in reality, "are complementary—and at times functionally congruent—existing in a dynamic interplay that drives technological development and media making" (2017: 19). He suggests the term *parallax view* as a "conceptual toolkit that enables us to make sense of the evolution of imaging technologies" (2017: 42). Anderson also conceptualizes what he calls the *negotiated mode*, which is to be used as a strategy for "negotiating the unstable relationship between data and images," one that allows for the perception of sameness, difference, and points of convergence, without privileging either (2017: 26). In this essay, I discuss two of my projects that use cinematic data visualization to elude this binary between the computational and the representational, the virtual and the real. I do this by compositing images and data, mixing photographic cameras (physical lens-based) and virtual cameras (computed in a software to frame and record a particular perspective), simulating lens-like effects and perspectives onto data, and using data both to represent and construct a cogent cinematic data visualization environment.

In her concept of the *cinematic humanities*, cinematic arts researcher Holly Willis discusses using cinematic modes as a form of critical design inquiry. "Humanistic enquiry" is "enhanced through the practices and modes of cinema," Willis contends, "even as cinema continues to expand into what has been dubbed 'the post-cinematic.' The cinematic humanities include examples of critical visual work that integrate space, time, and the methods of design to produce new ways of knowing" (Willis 2015: 63). Post-cinema reveals how cinema—the twentieth-century's primary mechanism of storytelling—is incorporating new technologies and alternative spatial configurations in the twenty-first century that redefine media expression. Foregrounding data and computational methods in a cinematic practice as I do here, for instance, is one way that cinematic data visualization fits within post-cinema and the critical humanities.

Ideas from situated practices also contribute to cinematic data visualization as a communication method. Situated practice foregrounds connections to space, place, and context. It is an approach that favors using knowledge that is derived from original contexts, rather than being extracted, abstracted, and transferred to another context in which its site-specific meanings are lost. As a situated practice, cinematic data visualization grapples with ways to foreground relationships between the analysis and representation of data and its environmental and local situations. Digital culture scholar Jill Walker Rettberg argues that *situated data analysis* can be a method for critically engaging with data and its representation. "Situated data analysis is a new method for analyzing these architectures," according to Rettberg, "that emphasizes how data is always situated, both in how it is constructed and how it is presented in different contexts" (Rettberg 2020: 1). Situated data analysis preserves the relationship between data and where it comes from—that is, it retains information on its local context—and eschews the problematic practice in which data is considered agnostic and objective.

This situated approach to data analysis also draws attention to the constructed nature of data representations. The notion of data objectivity implies a universalizing perspective. Contextualizing data to a specific site, on the other hand, reveals the subjective construction of perspective, and likewise, of its representation. Contextualization also points to a shift in the operational use of data. In other words, instead of focusing on the aesthetics of data visualization and the practice of making data look more captivating, contextualization centers on how visualizations function and what purpose they serve.

Experiments on *situated visualization* practices also suggest how users of decision support systems—in which data visualization emphasizes seeing data in situ—can effectively communicate the relationship between data and its contextual referents. These experiments suggest that including spatial and semantic information in data visualizations helps the users of data visualizations to make more effective decisions and to communicate their decision-making process (Marques et al. 2019). *Situated visualization* can likewise represent context through the figure-ground relationship.[1] Figure-ground relationship, which comes from gestalt design theory, is applied to data visualization in mobile and augmented devices. Using figure-ground this way reveals semantic, spatial, and temporal relationships between data (the figure) and its environment (the ground). This technique thus allows data analysis to take place in situ (White, 2008). In the next sections, I discuss how I use the cinematic data visualization concepts and techniques discussed above in my projects *Alluvium* and *LA River Nutrient Visualization*.

Case Study 1: *Alluvium*, A Particle Simulation

In my cinematic flow animation *Alluvium*, I present the results of a geological study on the impact of diverted flood waters on a sediment channel in Death Valley, California. The results of the study—which have significant implications for global warming—are rendered more accessible, and their meanings made more palpable, through a sequence of hybrid compositions of landscape cinematography and computationally produced simulations.

Alluvium presents geoscientists Noah Snyder and Lisa Kammer's findings on the consequences of a 1941 forced diversion of a desert wash at Gower Gulch in Death Valley. This diversion, which was intended to protect downstream infrastructure, blasted a 5-foot-deep (1.5 meter) channel upstream from Zabriskie Point in Death

Fig. 16.1 ***Alluvium*****. Visualizing data in situ by compositing and camera-matching a particle simulation to landscape cinematography. Image courtesy of author.**

Valley (2008). The sudden, massive volume of water and debris that passed through this narrow channel, instead caused massive erosion in Gower Gulch. Snyder and Kammer found that—along with flash floods and the sensitive conditions of the terrain—this forced diversion brought about geological change over several decades that would otherwise have taken thousands of years. Their study reveals "a situation analogous to a step-function change in climate" (2008) and effects that likely mimic the impact of climate change on river flooding and discharge.

Situated in the Mojave Desert in California, Death Valley National Park is a place of extreme conditions. It has recorded the highest ambient temperature in the world at 134° Fahrenheit (57° Centigrade) at the lowest altitude in North America, 282 feet (86 meters) below sea level, and it receives average rainfall of less than 1.5 inches (4 centimeters) annually. It also has an intrinsic and unusual relationship with water because of annual flash floods. My cinematic data visualization begins by addressing the context of this sensual environmental conflict: feverous desiccation that coincides with extreme flooding. In my video, the filmed scenes show the arid quality of this geography, and the simulations depict the flooding.

Alluvium is an experiment in cinematic data visualization that connects Snyder and Kammer's study about geological change at Gower Gulch to principles of design. My *Alluvium* visualization combines two assessment scales to best depict their findings: wide-angle landscape photography that provides a geographic context of the vast transformations that have taken place (see Fig. 16.2), and close-up macro-photography that shows the materiality of the terrain and its morphological possibility.

Fig. 16.2 ***Alluvium*****. A situating shot of Gower Gulch to capture both scales of assessment: wide-angle photography shows the geomorphological consequences of flood water on the landscape, while macro photography details the granular role of sedimentation.**

Design Tactics

Alluvium employs a range of cinematographic design tactics. For *Alluvium*'s scenes, I used particle simulations that are point-cloud data structures, which are sets of data points in space that represent motion, in this case, the flow of water. Particle simulations are used to simulate movement and to capture "fuzzy" phenomena, for example, atmospheric conditions such as weather or air flow, or water conditions such as ocean/river flows. The *Alluvium* point cloud is driven by a fluid-dynamic physics simulation that creates a time-based flow of information—in this case, information about water flow. I employed camera-matching techniques, which is a method that adjusts photographic and virtual cameras in order to produce a visualization that synchronizes their information and imagery. A virtual camera is an animation software that behaves like a camera or digital camera would in the real world.

The final video simulates these series of particle flows layered over filmed sequences of the eroded landscape of Gower Gulch to convey a notion of data in situ. I designed these particle flow/landscape sequences using compositing, camera-matching—placing 3D objects as if they are in a background image—along with the fluid dynamic simulations. I also used rotoscoping—animation done by tracing over live-action footage frame by frame—and immersive cameras. I used these techniques together to situate the data back onto Gower Gulch, the place from which it was originally collected.

At the center of my visualization is a careful calibration between the properties of the physical camera: the lens, frame, tracking, and recorded image, with the properties of its digital counterpart: a virtual camera and a 3D model of the site. I generated this 3D model of the site from topographic data; this model is actually invisible in the final video. The site model contains the particle simulation mentioned above, which uses millions of particles to evoke water flowing through the site, and in which these particles interact and collide with the detailed shape of the site. I then composited and spatially embedded photographic and digital scenes so that the particle simulation appears to flow through the real landscape.

I utilized rotoscoping, an animation technique that is used to create realistic live-action effects, to create a sense of spatial depth and detail in how the particle simulation moves. Rotoscoping is *situated visualization* that can be used to set up a figure-ground frame. I used it to create a semantic and spatial dynamic between the particle simulation and its environment, the invisible 3D model of the site that is part of the simulation workflow. In *Alluvium*, rotoscoping made it possible, for example, for the digital particles to flow below and around the rock that is in the foreground of the video (see Fig. 16.3). The particles thus appear to be in the physical environment.

Scale and Frame of Reference

I also positioned the cameras so that they produced, for the audience, a sense of immersion into the simulation and landscape. To do this, I set up the cameras to

Fig. 16.3 ***Alluvium*****. A situated visualization. The rock in the foreground, photographed by a physical camera, is rotoscoped, so that the particle simulation in the virtual camera appears to flow around it. This creates the effects that the particles are embedded in the natural environment.**

record from the viewer's frame of reference instead of placing them overhead or at a distance. Whereas Snyder and Kammer's 2008 study primarily focused on changes to the channel's width as recorded from aerial data scans, my visualization takes a perspective from inside the channel, and shows material details of terraces and river beds. This visual information provides more granular detail and human-scale perspective than the study did.

Coloration and Parallax View

To help viewers visualize the force, directionality, and turbulence that are characteristic of fluid dynamics, I mapped the velocity of the particles to coloration in a pink to purple spectrum. The coloration reveals how the invisible flow of water interacts with the sediment and eroded morphology of the terrain at a granular level. As a time-based simulation, the colorized particle visualization captures *before* and *after* sediment erosion information in the same image. The colorized virtual particle simulation also appears to carve into the physical terrain (see Fig. 16.4), creating the sense of a feedback loop between the simulation and the visual representation of the environment.

The fuchsia tones in the simulation also stand in stark contrast to the natural tones of the desert environment in the filmed sequences. This contrast points to a *parallax view,* a dynamic interplay between data and images, mimesis and computability, virtual and the real elements. In *Alluvium*, the simulation mode depicts the history of

Fig. 16.4 ***Alluvium*****. The velocity of the particles is mapped to their coloration, visualizing the force, directionality, and turbulence characteristic of water. The simulation is matched to a particular site of undercut erosion, so that the particles appear to carve the physical terrain.**

the water forces that sculpted the sediment, and the cinematographic mode reveals their contemporary consequences on the landscape. This kind of binary design workflow is often used to produce photorealism. I used this design strategy to negotiate the aesthetic differences between the two modes—data and mimesis—without privileging either.

Cinematic Aesthetics

Alluvium has a cinematic aesthetic because I used slow tracking shots of the landscape that I filmed with a shallow depth of field to create a classically cinematic pace, atmosphere, and sense of narrative time. The point cloud simulation of water is calibrated to this same tempo and focal depth (see Fig. 16.5), which grounds the scientific study in a sensual cinematic narrative of how the landscape was transformed over time in this place. To further embed the data into the site imagery, I mapped a shallow depth of field from the photographic camera to the particle simulation. Doing so added to the cinematic aesthetic; it also focuses the viewers' attention onto specific interactions between the particles and the terrain and creates a grounded human-scale perspective. The human-scale is also visualized in human-based time.

My first intention, when designing *Alluvium*, was to make scientific insights accessible that might otherwise be obscured and intangible because they are too abstract. Visualizing the phenomena of water and its geomorphological consequences at a human scale and over cinematic time, means that the study's findings are understood in relation to their original context. Retaining the context in *Alluvium*

Fig. 16.5 *Alluvium*. A detail shot of the flows and collisions between the particles and the landscape, which is actually a 3D topographic model of the site, made invisible. The camera's motion and the simulation are carefully paced between scenes, and the shallow depth of field from the photographic camera is also replicated in the virtual camera to create a cinematic tempo, atmosphere, and sense of narrative.

reveals meanings to viewers that are generated through place, origin, and the particulars of situation. The torrential water flow in Gower Gulch can otherwise only be perceived in retrospect, through the evidence of erosion and deposition of sediment. In this sense, *Alluvium* compresses time, providing viewers a sense of a geological scale of time.

Cinematic aesthetics convey a sensual and humanistic quality, which might at first seem to be an unusual choice for communicating scientific analysis of flood data. In *Alluvium*, though, I purposefully move beyond statistical abstraction in order to speak to a broad public. As the twentieth-century's primary mechanism for subject identification and narrative structure, cinema has used framing, focus, and mood to help the audience to better relate to the subject matter, and this strategy is applied here to analysis of scientific material.

Case Study 2: *LA River Nutrient Visualization*

LA River Nutrient Visualization presents and analyzes nutrient levels in the water of the Los Angeles River. The project charts publicly available data that show changes in the river's health based on water quality standards. My video provides an overview of the scope of available annual geographic, seasonal, and public water-quality data, as well as the limitations of this data. *LA River Nutrient Visualization* also makes a case for greater community engagement with water testing: this data visualization, in fact, revealed that community-organized water testing was both more comprehensive and effective than government testing, a realization that positions community groups to be the leaders in LA River restoration efforts.

To create *LA River Nutrient Visualization*, I used existing remote-sensed data and water-quality data from the Los Angeles watershed. Remote-sensed data is collected by special cameras and sensing technologies, which are typically mounted on satellites or aircraft, to capture information about land, the atmosphere, oceans, and other planets. For *LA River Nutrient Visualization*, I began by visualizing, and then combining, the remote-sensed data and water-quality data with filmed sequences of several water monitoring sites along the river. I spatialized the water data geographically, then gave it 3D form using data-driven geometric parameters that I animated over time. Next, I post-processed the video with cinematic scenes of urban-scale models that I created using a human perspective. In *LA River Nutrient Visualization*, I abstracted numerical data, which I then presented in the urban-scale context of the river.

Community-led Testing

For this project, I generated an accurate 3D terrain model using a digital elevation model of the LA River watershed (see Fig. 16.6). Working with publicly available data sets from the Water Quality Portal, Friends of the Los Angeles River, and the organization Science Land, I also discovered that there are thirty-nine monitoring stations along the LA River that have water-quality data from 1966 to 2014.

The most recent comprehensive studies of the LA River water chemistry, though, were organized by two community groups who employed professional water-quality

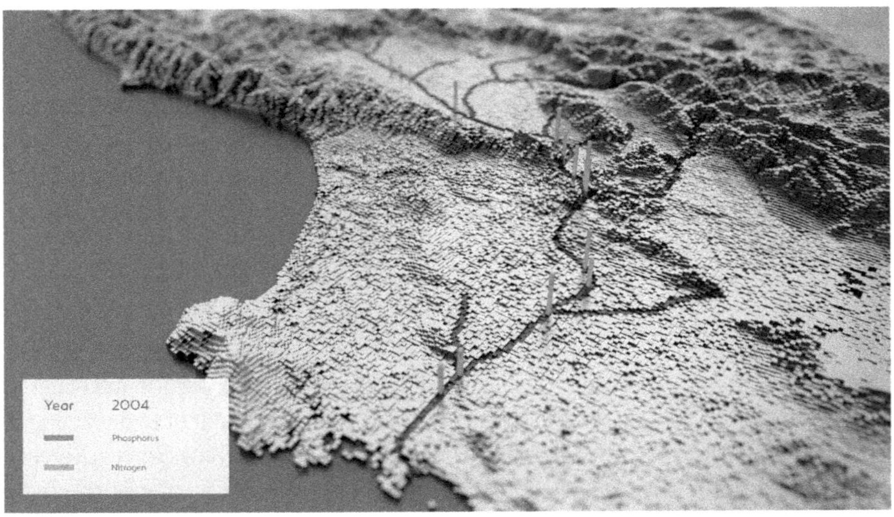

Fig. 16.6 ***LA River Nutrient Visualization*****. Reconstruction of the site of study, the Los Angeles River watershed from digital elevation data, combined with nutrient data from river monitoring sites.**

laboratories to test the water. *LA River Nutrient Visualization* articulates how this community-led testing augments government testing. Indeed, this kind of community-led water testing of the LA River empowers people with a stake in the watershed to take charge of their local environments, understand when consequential changes occur, and take action, if need be.

Pollution of the LA River is primarily caused by non-point source pollution, in which pollutants are passed into the water through precipitation and land run-off, rather than coming from a singular discrete source. Non-point source pollution can, in part, be remediated by the actions of local communities through local hazardous waste management. Citizen scientists who employ both raw scientific data, along with the possibility of using situated data visualizations such as *LA River Nutrient Visualization*, have had good success connecting communities with their environments.

Design Tactics

Certain information visualization design approaches can also both increase awareness and spur people to alter community and governmental behaviors, promote more knowledgeable practices, and campaign for environmental change. To create a user-friendly video, in *LA River Nutrient Visualization,* I first reconstructed the local geography from a digital elevation dataset into a 3D heightfield. In a 3D heightfield, terrain is rendered in 3D form by transcoding (converting) grayscale pixel data into an elevation image. The grayscale value of each pixel is a measure of the elevation of the landscape. I transcode the pixel value into a measure of height to create the

Fig. 16.7 ***LA River Nutrient Visualization*****. An animated visualization of nutrient data at a single water monitoring site over time. The visualization evokes a data urbanism in which the virtual camera is placed within the data to convey a sense of physicality and space.**

landscape. This is a common technique in geographic surveying to create accurate terrain models. I then visualized both the water-nutrient data and landscape using a cuboid structure in order to establish a spatial dialog between the two data types. I next animated the water-data readings over time at different sampling sites in each moving camera sequence. This process established a familiar cinematic tempo and mood for the content presentation. Using a cinematic tempo and mood also connects the animation of the data, the camera's smooth motion, and the post-processing effects, which include filters and effects that enhance a video. Animating the water data this way created a narrative-like and atmospheric quality to which a broad audience can relate. Along with these design tactics, I also attended to a public audience by constructing a sense of scale and place using a low third-person point of view, which situates viewers in the visualizations of the river-water data and urban context. As in *Alluvium*, I post-processed cinematic affects onto the data visualizations, simulating a shallow depth of field, ambient "dusk-like" lighting, and shadows.

I also juxtaposed shots from virtual cameras and photographic cameras, which made it possible for me to marry data that I had gathered from the virtual cameras with drone-captured video scenes of water testing sites along the river and its tributaries. I presented these composites as in-situ temporal vignettes, which overlaid local geographic reference points from the urban datascape onto real-world local geographic reference points. As in *Alluvium*, blending the perspectives of the data analysis visualization and the local context—in *LA River Nutrient Visualization* these are photographic aerial views—gives the data a sense of scale. Toggling between the two perspectives invites viewers to consider each data collection site as it is visualized in the datascape (see Fig. 16.7) in relationship to its real-world origin (see Fig. 16.8).

Fig. 16.8 ***LA River Nutrient Visualization*. A drone-captured aerial view of a water monitoring site to create an in-situ vignette of the site's local context in juxtaposition with the data visualization.**

Fig. 16.9 ***LA River Nutrient Visualization*****. Virtually, cameras are post-processed to add a shallow depth of field and atmospheric lighting and shadows. A low third-person perspective is used to position the data better in its urban context.**

The *LA River Nutrient Visualization* renders scientific data as cinematic scenes that convey local, urban, and community contexts in a critical manner. Nuances in the environmental site data, such as the particular state of the river at that location, the amount of vegetation in the river versus empty canal, and whether the river environs are within a community, or are industrial or rural, are essential content for viewers. But this project also presents this information in an accessible and humanistic form. By doing so, this project makes water quality and data collection accessible to a public audience.

Engaging the public in this way promotes the practice of citizen science, in which everyday people participate in scientific research. The shift in aesthetics, between photographic and virtual, also supports the shift in perspective between community and institutional roles by incorporating relatable cinematographic vignettes into the data visualization. The abstract scientific data is synchronized with its original locality (see Fig. 16.10) rather than being extracted from it. Local and situational information also relate well to local people and their community perspective. As noted above, in spite of the availability of water data from government funded research, the water-quality data that was collected by community organizations and citizen-science groups was the most consistent and effective for my analysis.

Conclusion

Cinematic data visualization incorporates ideas from design research and practice and the cinematic humanities to critical questions about how people visualize and

Fig. 16.10 ***LA River Nutrient Visualization*****. Video of each data site puts the data back into communication with its local neighborhood, which also speaks to the visualization's findings that community organization and citizen science is an effective means of data collection and should be recognized in the future redevelopment of the LA River.**

apprehend scientific data. In *Alluvium* and *LA River Nutrient Visualization* I offer two different approaches to data visualization as a situated practice. *Alluvium* uses particle simulations and camera-matching to embed the findings of a scientific study of flood waters and sediment erosion into its real-world environment. It preserves the relationship between data and its real-word environmental origins. *LA River Nutrient Visualization* uses remote-sensed data and water quality data to structure an urban datascape. This datascape is combined with cinematographic vignettes of the local environmental context, a process that retains the entangled meanings that come from the mix of origin, place, and community. Cinematic tactics augment and humanize viewers' experience of data. Enhanced meanings that come from contextual information are critical to thorough communication, but they also call into question the presumed objectivity of scientific data by showing how context can change information. In *Alluvium* and *LA River Nutrient Visualization*, I use cinematic modes to think critically and creatively about negotiating data, modes of representation, and information structures to create compelling cinematic information visualizations of scientific data.

Note

1. Figure-ground uncouples foreground from background, for example, for black type on white paper, the black type is the "figure," and the white of the paper is the "ground."

References

Anderson, S. (2017), *Technologies of Vision: The War Between Data and Images*. Cambridge: MIT Press.

Marques, B., B. S. Santos, T. Araújo, N. C. Martins, J. B. Alves, and P. Dias (2019), "Situated Visualization in The Decision Process Through Augmented Reality," *23rd International Conference Information Visualization (IV)*, 13-18, doi: 10.1109/IV.2019.00012 (accessed on June 25, 2021).

Rettberg, J. W. (2020), "Situated data analysis: a new method for analyzing encoded power relationships in social media platforms and apps," *Humanities and Social Sciences Communications* 7 (5): 1-13, doi: 10.1057/s41599-020-0495-3 (accessed on June 25, 2021).

Snyder, N.P., and L.L. Kammer (2008), "Dynamic adjustments in channel width in response to a forced diversion: Gower Gulch, Death Valley National Park, California," *Geology* 36: 187–90.

White, S., S. Feiner (2009), "SiteLens: Situated Visualization Techniques for Urban Site Visits," *Proceedings of ACM CHI 2009*, 1117–20, Boston, MA.

Willis, H. (2015), "Writing Images and the Cinematic Humanities," *Visible Language* 49 (3): 63–77.

Index

Entries in italics refer to Figs.

www.ingramcontent.com/pod-product-compliance
Ingram Content Group UK Ltd.
Pitfield, Milton Keynes, MK11 3LW, UK
UKHW020905190926
475560UK00010B/291

9 781350 061927